D1426443

Innovations in concrete

David Bennett

Published by Thomas Telford Publishing, Thomas Telford Ltd, 1 Heron Quay, London E14 4JD. URL: http://www.thomastelford.com

Distributors for Thomas Telford books are
USA: ASCE Press, 1801 Alexander Bell Drive, Reston, VA 20191-4400, USA
Japan: Maruzen Co. Ltd, Book Department, 310 Nihonbashi 2-chome, Chuo-ku, Tokyo 103
Australia: DA Books and Journals, 648 Whitehorse Road, Mitcham 3132, Victoria

First published 2002

Also available from Thomas Telford Books
Manual of numerical methods in concrete.
M.Y.H. Bangash. ISBN 07277 2946 2
Historic concrete: background to appraisal. Edited by James Sutherland, Dawn Humm and Mike Chrimes. ISBN 07277 2875 X

A catalogue record for this book is available from the British Library

ISBN: 0 7277 2005 8

Typeset by Keyword Publishing Services, Barking, Essex
Printed and bound in Great Britain by MPG Books Ltd, Bodmin, Cornwall

Preface

This book is a state-of-the art summary on recent and past innovations in reinforced concrete technology, many of which have not been fully exploited or widely appreciated by the construction industry. It is based on the author's researches and his association with the promotion and marketing of structural and architectural concrete whilst working for the British Cement Association (BCA) and his own consultancy. The book describes in readable essay form a number of case studies and practical examples where speed of assembly and lean construction have been engineered into a project. It highlights the outcome of some original and pioneering researches, to showcase the advances concrete has made in the past two decades.

Topics such as high strength concrete are taken from a 130 MPa concrete for columns in skyscrapers to a 500 MPa concrete that has been developed to replace plastic and ceramic products in the future. A new super-strength concrete, described in the book, is currently being used to construct the sensational Bridge of Peace in Seoul for the opening of the 2002 World Cup. Reinforcing bar and post-tensioning strand using glass-reinforced plastic and carbon fibre alternatives are reviewed, as well as conventional reinforcing systems. On fast-build there is a study of the four-day floor construction cycle and how to assess propping loads to determine the early strength and integrity of suspended floor slabs. Lightweight aggregate concrete, shearhead technology, reinforcement rationalisation, composite precast construction, architectural concrete, tilt-up construction and the new concept in air rights construction are just some of the many subjects that are covered.

There is no mention of translucent concrete, which has been reported in the architectural press recently. Concrete by definition must use cement and aggregates, both of which are opaque. To become translucent the cement will, one presumes, have to be replaced with a clear resin-based binder.

Innovations in Concrete is essentially an amalgam of good ideas in concrete technology that have inspired and influenced the author over the years. They have been proven either in research, in practice or as design prototypes. Not all of these concepts are ready to be adopted into every construction, and they may require further development to unlock their potential. By using the background information and references in the book, the enquiring reader can pursue a particular subject in greater detail. It is very likely that designers and construction professionals may not be familiar with all these concepts, nor have an appreciation of their relevance or potential in everyday construction usage. It is hoped that, by using the book, design and construction professionals will increase their awareness of new developments in concrete construction technology and be in a better position to design and build more competitively in the future.

For terminology and units I have used MPa (N/mm^2) and GPa (kN/mm^2) for concrete strength and concrete modulus, and conventional SI units for all other material strengths and loadings.

In compiling and drafting the material for this book the author is indebted to a number of people and organisations for granting permission to publish text and images. They are: Martin Clark, formerly of the BCA, now Chief Executive of the BPCF; Martin Southcott of the Reinforced Concrete Council; Bendt Aarup of CRC Technology; Lafarge UK; TermoDeck; Bekaert Steel Fibres; Liapor Aggregates; Richard Rogers Partnership; John Clarke of the Concrete Society; Peter Sheard of Eurocrete; Denise Chevin of *Building Magazine*; *New Civil Engineer*; Connell Wagner Mott McDonald; Creteco; Rapidobat; Tarmac Precast; Taylor Woodrow; Wacker; and Skilling Ward Magnusson Barkshire.

Two people deserve special mention. The first is Ian Guy, who worked painstakingly hard and meticulously in copy-editing, collating figures and tables and proof-checking my draft. The second is the late Derek Birchall, whose brief but illuminating paper on high strength cement and work on MDF cement has been the most thought-provoking piece of concrete research I have come across in the past thirty years. It is a shame we never met. This book is dedicated to his memory and to the members of the special project team at ICI in Runcorn in the 1970s — Kevin Kendall and Anthony Howard — who still possess the only MDF guitars in the world!

David Bennett
May 2002

Contents

1 Concrete: the material

Material specification

The concrete specification — what's it all about?

The constituent materials of a modern concrete can be quite diverse. There is a choice of cement types, any one of which can be combined with pozzolana, PFA and slag. Aggregates include quarried materials that can be crushed such as limestone, granite and waste concrete, or dredged from the sea or won from land-based gravel pits. Some aggregates, such as Lytag and Leca, are made by heating and expanding the products of industrial waste. Fine materials can occur naturally like sands, or be refined from the by-products of industrial waste like china clay and furnace bottom ash. These materials have been developed over past decades by ready-mixed concrete companies and aggregate suppliers in order to remain competitive in a market that relies on high volume turnover and small profit margins. The increasing complexity of concrete material specification, typified by the wide range of permissible cement types, admixtures and aggregates, reflects the advances in concrete material technology and the commercial pressures of a competitive construction industry.

No other industry in the construction world has customised its products quite as comprehensively as the ready-mixed concrete industry. From a single batching plant a ready-mixed concrete supplier can service customers within a 25-mile radius, offering them a doorstep delivering service the day the order is placed. The product can be a low grade blinding concrete, a high performance structural concrete, a lightweight aggregate concrete or a flowing concrete for an industrial floor, or any one of 50 different mixes. These products must comply with a plethora of British Standards. For aggregates the standard is BS 882, for cement and cement replacements it is BS 12, BS 146, BS 4027, BS 3892, BS 6699, BS 6588, and so on.

The pressure to broaden the limits of the standard, to permit a greater variety of material to be used in concrete, is driven by the commercial needs of aggregate suppliers and the ready-mixed industry to seek out alternative raw material sources as established quarries become exhausted. The combination of these products must satisfy minimum performance specifications, such as compressive strength, durability and workability for example, which are required for structural concrete. Concrete specifications are framed around BS 8110 standard clauses, with the addition of particular clauses based on practice and experience, or on the solutions to problems with certain materials in concrete which have established a precedent. Typically, there may be a problem with sea-dredged aggregates containing sodium chloride, or with high cement content mixes that contain fine aggregates with a high silica content, or one of exclusion of concrete admixtures such as superplasticisers because, in the opinion of the specifier, they are not considered essential in concrete.

A concrete specification document is, in essence, largely a cobbled collection of clauses some of which refer to BS 8110, some to recent research findings, while some are based on the specifier's own experience. It may seem unnecessary to have every facet of the concrete work included in a contract document, as many of these clauses will not be appropriate. From the engineer's point of view, however, this is a pragmatic approach, where nothing is left unsaid or open to third-party interpretation, just like the conditions of sale you see in small print on the back of a hire purchase agreement. And, like the hire purchase agreement whose clauses are written in the interests of the loan company, specification clauses are written to protect the engineer's interests in the concrete. There is nothing wrong with that since the engineer's interest, it can be argued, are in the best interests of the project, but there is one proviso: what is stated must be understood, measured and achieved by a contractor employed to do the work on site. That is the key point of a good specification. In my experience many specification clauses that are drafted are vague, lack clarity and often convey misleading information.

Moreover, an understanding of the overall economic benefits of material choice and concrete practice in a specification must go beyond comparisons of material cost differences and the constraints of outmoded working practice. The specification should be regarded as the accelerator pedal of economic construction, and how it is drafted will either encourage the contractor to innovate or severely curtail the opportunity to make any efficiency gains in the construction.

Review of the key specification clauses

The purpose of this review is to highlight some of the benefits to be gained, in time, cost and quality of workmanship, through adopting key clauses in the specification. Emphasis of the clauses has been directed to multi-storey in situ construction, because there is a greater need for improvement in this sector.

Cement and cement replacements

Cements are commonly the most expensive constituents of a concrete mix. Ready-mixed suppliers will look for the most economical way of meeting the minimum strength criteria laid down in the specification. Combinations of cement and cement replacement will be assessed to find the most economic mix proportions. Often ordinary Portland (OP) cement will be substituted by a slower setting and cheaper cement such as Portland blastfurnace (PBF) or ground granulated blastfurnace slag (GGBS) cement, or used in combination with, say, pulverised fuel ash (PFA) to reduce the OP content which is more expensive. Provided the concrete meets the minimum 28-day strength requirement and satisfies any durability standards, for example by maintaining a low water/cement ratio, then the concrete supplied will be to the specification. The client and the project will benefit from the competitive price of concrete specified in this way.

As the properties of composite cements differ in both fresh and hardened state from those in concrete made solely from OP, their use may not be an economic advantage to the contractor in terms of quick release of formwork. The slower setting composite cements, such as GGBS and PFA cements, require more time to achieve the minimum striking strength in the early days than an OP mix and are likely to delay formwork striking by at least a day or two as compared to an OP mix. See Figs 1.1(a) and 1.1(b) for a typical comparison of OP cements with 0%, 40% and 70% replacement with PFA and GGBS.

If the contractor wishes to opt for the more expensive OP concrete to speed up the forming cycle, then the specification must encourage this option. If fast build and quick turn-around of formwork is to be encouraged, then the specification must permit early striking based on the

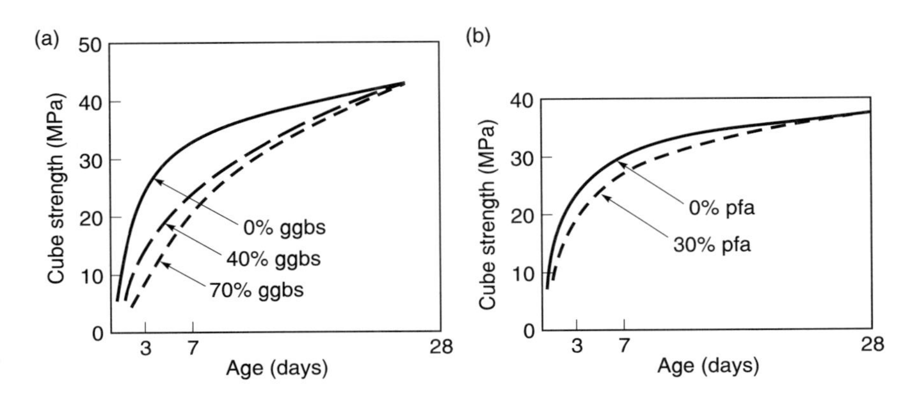

Fig. 1.1. Typical strength gain curves of concrete made with ordinary Portland cement partially replaced with (a) pulverised fuel ash, (b) ground granulated blastfurnace slag

strength of the concrete, and large area pours. There must be the opportunity for the contractor to justify the extra cost of using OPC, by the real saving in time and overheads to be made in adopting faster construction methods.

Formwork is probably the single largest cost component in assembling a concrete frame. Priority given to formwork economics and its efficient use will encourage the contractor to build efficiently and reduce the total cost of construction. Savings in construction time benefit the client from earlier rental income. Limiting the pour size and imposing restrictions on formwork striking time, on the basis of conservative tables in BS 8110, is analogous to designing the ultimate fast car and specifying square wheels.

Formwork striking

Formwork striking tables that appear in BS 8110 are by their very nature conservative and should not be used where speed in construction is critical. The basic charts for determining the maturity of concrete in such tables have been derived from cements that were made in the 1960s, giving maturity curves that are closer to those of the slower hardening composite cements of today. Moreover, when a particular grade of concrete is plotted, say 30 N, the curve is plotted to give 30 N strength at 28 days. In practice the concrete strength for 30 N grade will have been set much higher, with a margin for quality control, resulting in a target strength closer to 35 N. Modern OP cements are more finely ground and have a higher C_3A content than in the past, resulting in higher early strength gains. In fact modern OP cement behaves more like the rapid-hardening cements of the past. These important factors could be overlooked if the specification limits striking times on the basis of tables in BS 8110 or similar.

If striking times of 48 h and 76 h are being planned, then it is critical to monitor the actual strength gain of the concrete (see Fig. 1.2). In situ pull-out tests like the Lok test, or temperature-matched cubes, maturity meters or basic cubes cured with the actual slab pour, are the options available. Temperature-matched cubes, cubes that are cured alongside the slab and pull-out tests have been proven to be reliable methods of monitoring the in situ strength of concrete. Direct reading pull-out tests would appear to be the more versatile method, as one does not have to rely on laboratory cube crushing tests before the slab strength is known. The direct reading pull-out meter is calibrated to give strength of concrete based on the resistance of pulling out a cone inserted into the top 50 mm of the concrete slab — the weakest zone — giving some assurance that the bulk of the slab must be more mature. The pull-out meter is calibrated against a concrete cube of known strength.

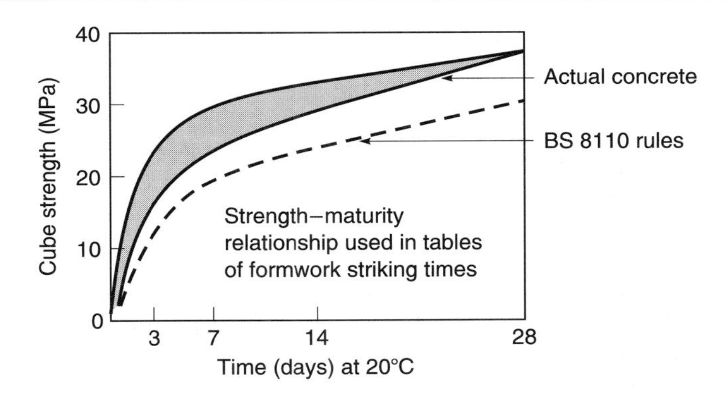

Fig. 1.2. Strength development envelope for ordinary Portland cement concrete

For very quick formwork cycling, where large panels of formwork are removed and repositioned to give a fast floor cycle, a full engineering assessment of the loading history of the immature slab is required. Refer to Chapter 2 for further guidance.

Draft specification

Formwork to suspended slabs can be removed only after the concrete has achieved a minimum compressive strength of ... (60%) of structural strength specified and in such a manner that it will not damage the concrete.

Striking time of soffit forms shall be based on measurements taken of the actual strength of the concrete, as determined by compressive strength testing of cubes of equal maturity to the slab concrete or by direct pull-out tests or other in situ tests to the approval of the Engineer.

The contractor shall submit to the Engineer a typical formwork layout, showing the position of support legs, the contact area of sole plates and the load to be transferred to the supporting floor. The contractor shall advise the Engineer in writing what precautions are to be taken to ensure that the supporting concrete floor can safely resist the punching shear and forces from the construction load plus the wet concrete, without damaging or cracking the concrete.

If in situ test measurements are not taken or fail to comply, then striking time of formwork shall be based on CIRIA Report 67: 'Tables of minimum striking times for soffit and vertical formwork'.

Large area pours

To crack or not to crack. There has been a great deal talked about the cracking of concrete, particularly flat slabs. What the engineer wants to see, if the eye can get close enough, is fine hair cracks of not more than 0.2 mm width. The architect and client do not want to see any cracks at all, while the services engineer is fairly relaxed about cracks so long as the slab has not deflected much and is quite level across its surface. Reinforced concrete under flexure has been designed to crack, but the crack width is controlled by distribution reinforcement placed near the face of the concrete.

Concrete will also creep and shrink with time. Creep is related to the coarse aggregate and cement bond in the mix and contributes to the deflected profile of the concrete, whereas drying shrinkage can cause long-term shrinkage cracks to appear where the slab is restrained from movement. Both these phenomena are unrelated to the size of the concrete pour. Creep deflection may increase if the slab is loaded at an early age. Drying shrinkage is related to the mix constituents, water/cement ratio and cement content.

Restraints. Risk of cracking of concrete at a early age can be due to thermal contraction and to plastic cracking. Plastic cracking is caused by rapid drying of the wet concrete surface during placing and is dependent on the prevailing weather, characteristics of the mix, and the rate of evaporation of water from the surface of the placed concrete. It has nothing to do with pour size. Risk of early-age thermal cracking, caused by the temperature rise of hydrating cement, is more prevalent in deep foundation pours and the like and is exacerbated by restraint to movement as the concrete cools. For relatively thin suspended slabs this risk is not great, as the temperature rise of the concrete above ambient temperature is generally not large. However, where there is a potential risk of crack formation, the structure must be designed to minimise the chance of large indiscriminate cracks occurring.

Thermal cracking will arise if the tensile strain capacity of concrete is less than the tensile strain imparted to the slab because of restraint R preventing the slab from contracting as it cools. Restraint to movement can be attributed to rigid structural elements of the frame such as vertical shear walls, core walls, massive columns and the hardened edges of previously cast slabs. Generally, the formwork panels and support columns are comparatively flexible elements which allow the cooling concrete to move without much restraint.

Whatever the pour size specified, the coefficient of thermal expansion, the tensile strain capacity and the drop in temperature will be the same, because these are characteristics of the concrete mix and the air temperature. The only thing that is influenced by pour size is the restraint factor. Rigid restraints are assigned values of $R = 0.5$, whereas areas under formwork and columns are assigned a partial restraint factor of $R = 0.2$. In practice, complete freedom from restraint is unlikely.

A fully rigid restraint value of $R = 1.0$ might seem appropriate for rigid elements, but only if the onset of tensile strain coincides with the peak temperature rise of the concrete at the moment when the concrete goes from a plastic to a hardened state. As this does not happen in practice, a reduction factor is applied (a) to allow for the concrete still being in a plastic state when it begins to cool, (b) for heat transfer to the adjacent slab joint, reducing the peak temperature rise at the restraint, and (c) to acknowledge that the peak temperature is not uniform through the depth of the slab. In concrete sections thicker than 1 m, changes in temperature between the core and the surface can become significant, giving rise to internal restraint. Forecasts of temperature differences between inner and outer surface should be checked in addition to the overall thermal contraction of the pour area.

The direction of the induced strain imparted to the concrete is parallel to the restraint face or slab edge. Adjacent to an edge restraint a free edge will tend to warp, causing a diagonal restraint at the corner, just like the effects seen on retaining walls. The restraints for many slabs were modelled by the National Physical Laboratory (NPL). After a study of these results, simplified working rules were adopted on the basis of a triangular distribution of restraint patterns, proposed by the author.

Assessing the distribution of restraint calls for engineering judgement, because nothing is neat and exact in this complex topic. Rather than spend time on complex and wasteful analysis, it is reasonable to assume a 45° distribution angle from the edge of the restraint.

For practical purposes, suggested values of restraint factors for thin suspended slabs are given in Table 1.1.

Table 1.1. Restraint factors for thin suspended slabs

Restraint type	Direction	R	Distribution
Horizontal slab edge, length of wall	Parallel to edge	0.5	45° triangular
Between rigid ends	Between restraint	0.5	Area of slab confined
From formwork and slender columns	Whole area	0.2	

Plots of restraint areas for a series of small pours for a flat slab with a central rigid core can be seen in Fig. 1.3(a). The first pour area has low restraint as the concrete is free to contract on all sides. Successive pours increase the restraint area of each pour due to the restraint from the previously cast slab edge or the core wall. If the slab is poured in three larger pour areas, as shown in Fig. 1.3(b), the area of slab restraint for each pour becomes less as a proportion of the pour area, and the restraint reduces. Ultimately, if the slab is cast as one pour, Fig. 1.3(c), the area of restraint as a proportion of the whole slab is a minimum.

Clearly, it can be seen that small area pours increase the proportion of restraint to free contraction in suspended slabs, whereas large area pours reduce the restraint area. One point to check when planning a very large area pour is the amount of contraction at the slab edge and how this might affect the warp at the top of the columns. There is a potential risk of cracks developing in the column head just below the slab. Also, if a large slab pour is restrained at opposite ends by stiff shear walls, it may be best to pour the slab in two areas to avoid the risk of an indiscriminate crack appearing in the middle of the slab, unless adequate distribution reinforcement is provided to control the size of the cracks.

Estimating temperature drop

In the absence of site monitoring or qualitative data, the temperature drop for the slab can be estimated from tables in CIRIA Report 91. The early temperature rise of a section of concrete is due to the internal heat generated by the cement during hydration, and the concrete is cooled by external heat loss to the air. The temperature drop depends on a number of factors, namely the placing temperature of the concrete, the average air temperature (day and night), the wind chill or solar gain on the surface of the slab, the type and volume of cement, the insulating or conducting value of the formwork and any surface covering or protection to the concrete. In thin suspended slabs, the temperature drop will tend not to be large or critical for the slab, except perhaps in winter when direct heat and surface insulation are applied to keep the fresh concrete temperature well above freezing when the air temperature is below 0°C.

BS 8110 Part 2 gives a maximum temperature drop to avoid early thermal cracks forming. According to the British Cement Association (BCA), for gravel aggregates, there has been no reported evidence of early thermal cracking of fully restrained (thin) slabs when the temperature drop has been less than 20°C. The empirical 20°C rule has often been used in practice, which suggests that nothing need be done if the temperature drop is less than 20°C.

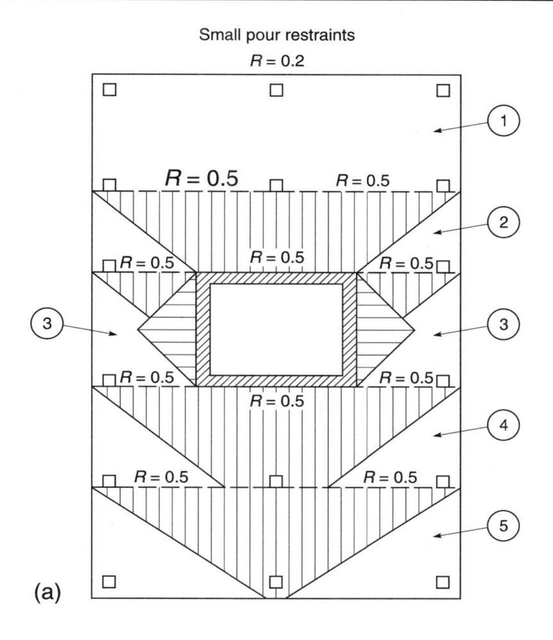

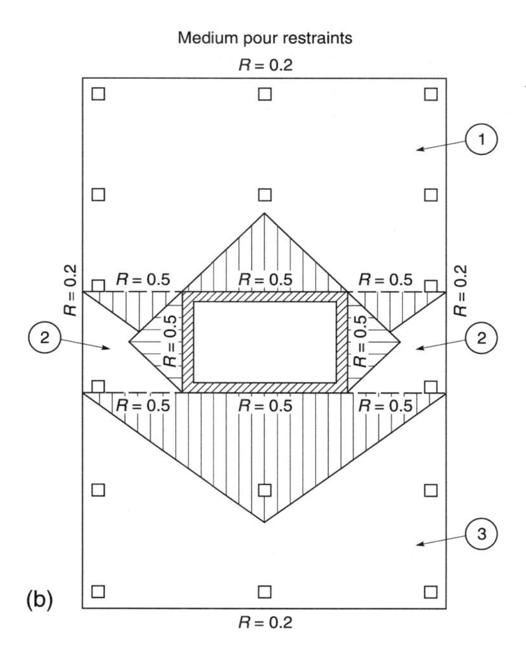

Fig. 1.3. Restraints for (a) small and (b) medium pours

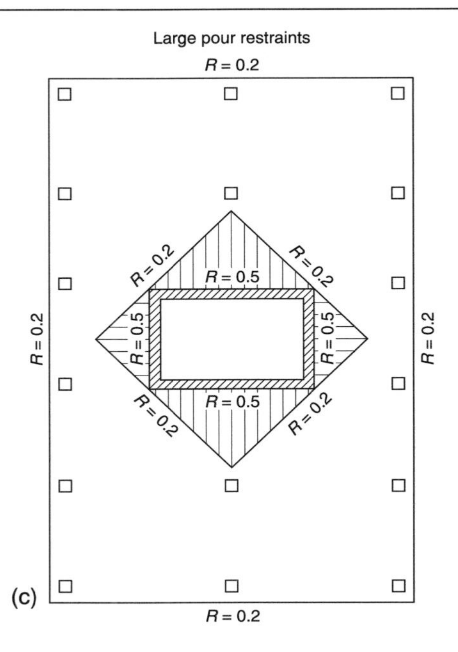

Fig. 1.3 (continued). (c) Restraints for large pours

The coefficient of expansion and tensile strain capacity of concrete made with different aggregates are shown in Table 1.2 and can be used to estimate the maximum temperature drop and crack risk of a particular concrete.

Suggested design approach

There is no need to restrict areas of suspended slab pours, as this is likely to increase the risk of early-age thermal cracking. The following approach has been recommended by the BCA:

Table 1.2. Typical values of tensile strain capacity and thermal expansion coefficient for concrete made with different aggregates

Aggregate	Tensile strain capacity	Thermal expansion coefficient α
Gravel (flint)	70×10^{-6}	$12 \times 10^{-6}/°C$
Limestone	90×10^{-6}	$8 \times 10^{-6}/°C$
Lytag	110×10^{-6}	$7 \times 10^{-6}/°C$

- ☐ Identify the pour configuration for economic construction, based solely on the contractor's ability to place, finish and protect the proposed area of concrete in a working day.
- ☐ Check the crack potential based on the likely temperature drop of the slab, using the coefficient of thermal expansion and tensile strain capacity of the concrete and assuming a restraint value of 0.5. Refer to CIRIA Report 91.
- ☐ If there is a risk of cracking, then identify the slab area at risk and calculate the minimum reinforcement required to control cracking using the following rule:

$$p_{\text{crit}} = p_{\text{ct}}/p_y$$

where p_{crit} is the critical minimum steel area ratio, p_{ct} the tensile strength of concrete at early age, and p_{y} the characteristic strength of reinforcement, i.e. 460 N/mm^2 for high-yield steel and 250 N/mm^2 for mild steel.

Check whether sufficient reinforcement has been provided in the right direction to control crack formation. If not, add reinforcement to the high restraint zones of the slab. For guidance on maximum permissible crack widths refer to BS 8110 Part 2 and to BS 8007 for water-retaining structures.

> *Draft specification clauses*
>
> There shall be no restriction on pour size.
>
> The contractor shall submit to the Engineer for approval a layout showing the position of proposed construction joints, the sequence of construction and a method statement on the arrangements to supply, place, finish and protect the intended pour area within a working day.
>
> The number of construction joints shall be kept to the minimum necessary for the execution and progress of the work.

Aggregates and admixtures

Aggregates. Natural aggregates, whether land-based or marine-dredged, are the most commonly available aggregates. Where possible, specifications should encourage the use of locally available materials to ensure the price of concrete is competitive. Restrictions on certain types of aggregates, particularly marine-dredged material, may encourage the supply of imported aggregates which will attract high haulage charges, thus pushing up the price of the concrete.

The water absorption of an aggregate does not necessarily give a guide to its suitability when used to produce a durable concrete. Naturally dense aggregates which have high water absorption may be weak and would probably fail the strength requirements in BS 882. For this reason a limit on water absorption in the specification is considered unnecessary. On the other hand, specifying lightweight aggregate to BS 3797 for concrete will increase the cost to about three times as much as when natural aggregates are used. Supplying, say, a C40 grade lightweight aggregate concrete to a site is likely to cost between £15 and £20 per cubic metre more than a C40 normal weight concrete, and yet there could be saving in the total construction cost. The reduction in concrete density from 2400 kg/m^3 to 2000 kg/m^3 reduces dead load and, in combination with post-tensioned floors, can reduce the slab depth in the order of 20% over normally reinforced concrete. On a low-rise building of six or less floors there is no advantage in the reduction in floor depth, but once the building reaches ten floors and more in tight planning zones, with height restrictions, then savings in dead load, cladding and service cost become significant.

Draft specification for aggregates

The aggregates used shall comply with one of the following British Standards:

Normal weight concrete

BS 882 Aggregates from natural sources for concrete
BS 1047 Air cooled blast furnace slag coarse aggregate for concrete

Lightweight concrete

BS 3797 Lightweight aggregates for masonry units and structural concrete

Admixtures. There are a number of reasons why admixtures are of benefit to concrete. Apart from reducing the free water content of the mix and increasing workability, they can reduce the cost of concrete by reducing the cement content necessary to achieve a given strength and allow the use of more marginal aggregates to make satisfactory concrete.

Because of the influence of pulverised fuel ash, ground granulated blastfurnace slag and certain cement types on the performance of some admixtures it is recommended that trial mixes be carried out to ensure the desired properties are achieved.

In most west European countries engineers have to draft special clauses in the specification if they want to preclude the use of admixtures in concrete. It is accepted practice in these countries for the ready-mixed suppliers to automatically dispense admixtures into every load of concrete supplied.

Draft clause on admixtures

The admixture to be used shall comply with the following British standards, as applicable:

BS 5075 Concrete admixtures

Part 1: Specification for accelerating admixtures, retarding admixtures and water reducing admixtures
Part 2: Specification for air-entraining admixtures
Part 3: Specification for superplasticizers

The chloride content of any admixture shall not exceed 2% of chloride ions by mass of the admixture. The alkali level of the admixture shall be stated. The value shall be added to the free lime equivalent of the cement where the Engineer has confirmed that alkali–silica reaction is a potential problem.

The Engineer shall be provided with information about the admixture regarding the performance of the concrete relating to

(a) the effects of under and over dosage,
(b) the chemical name(s) of the main active ingredients,
(c) whether the use of the admixture leads to entrainment of air, if it is not an air-entraining type,
(d) any known side effects.

High strength, high performance concrete

Introduction

We are not talking about wonder cements, but about the development of really high performance concretes with a practical pay-off, in construction, of high early strength gain and high long-term strength. I introduced the concept of high strength concrete into the UK following an intense but highly valuable study tour to Chicago during one hot week in July 1986. Fired up by the ease with which Material Services

Corporation, the largest ready-mixed company in Chicago, and the Portland Cement Association (PCA) at Skokie were talking about 150 MPa (N/mm^2) to 250 MPa concrete, I initiated a literature search and the first practical high strength concrete trials at the BCA, upon my return (Fig. 1.4). The test trials were sponsored by Boral Lytag, Elkem Chemicals, Pioneer Concrete, Square Grip and PSC Freyssinet with not a consultant in sight, although the usual names were approached. It is thanks to the shared belief of this small group that the future for high performance concrete is assured today, judging by the spate of research projects on high strength concrete in the UK.

At the time high strength was thought to be useful in reducing the size of columns in office buildings, providing more lettable space, making fast runway repairs, or reducing the depth of prestressed concrete bridge beams. Recent test work by Taywood Engineering and Arup suggests that high strength could also be practical for suspended floors by taking advantage of the enhanced concrete modulus and higher tensile stress. Now that is real progress. We also consider developments in lightweight aggregate concrete, self-compacting concrete and macro defect-free concrete; the latter is still in its development stage, but has produced some astonishing results.

A literature review of high strength

The definition of high strength, assumed in a report drafted in 1988 by Dr Les Parrott, was for a workable concrete with a compressive strength equal to or greater than 70 MPa (Fig. 1.4(c)); it could be supplied in a ready-mixed truck, and uses normal concrete materials. The age at which high strength concrete (HSC) is measured can be greater than 28 days and is tested using either concrete cubes or cylinders. An edited and shortened version of Dr Parrott's report follows.

Introduction

The notion of high strength concrete is not very new; concrete of high strength and density had been used in the nuclear industry two decades ago and in 1967 Nasser compiled a bibliography, mainly of US work on the subject. Serious interest in HSC first started in North America when the economic advantages of reducing column size and the cost of reinforcement in high rise columns was recognised. From 1980 onwards many other countries, such as Japan, Scandinavia, Russia and the UK, were beginning to show an interest in the development of HSC.

High strength concrete has now been used in high rise buildings, bridges, piles and heavily loaded columns because of its structural

(a)

(b)

Fig. 1.4. High strength concrete trials at the BCA, Wexham Springs: (a) test mix; (b) test columns

Material		Quantity kg/m^3
OPC		540
Crushed rock	20mm	865
	10mm	375
Marine sand (7% m/c)		525
Admixture Fosroc Conplast 430		9.0 litres
Free water		155 litres

(c)

(d)

Fig. 1.4 (continued). (c) Mix design for C80 concrete; (d) comparison of column sizes for 40 MPa and 70 MPa concrete

advantages. Other usage has been in the application of wear-resistant concrete and the production of moulds for metal castings. In 1988 the highest design strength specified for ready-mixed concrete appears to be 131 MPa, although laboratory controlled mixes have gone higher than 200 MPa. This concrete was used for intermediate columns on the upper floors of the Two Union Square Building in Chicago. It should be noted that a specified design strength of 70 MPa produces a concrete that has a target strength of about 90 MPa for compliance.

What makes high strength concrete? In order to achieve high strength in concrete the water/cement ratio must be kept very low to minimise voids and to improve the strength of the cement bond, always assuming that the cement in the mix will fully hydrate at such low water/cement ratios. In order to make HSC workable at low water/cement ratios superplasticisers have to be used. It is generally agreed that OPCs with high C_3S (tricalcium silicate) yield higher strength concrete and that full cement hydration benefits from the inclusion of mineral admixtures such as PFA, blastfurnace slag and silica fume. Careful selection of aggregates is vitally important as this can affect the ultimate compressive strength.

Mix constituents

The essential mix constituents of HSC are are follows.

Cement. Cement particle grading affects reactivity and water demand. Particles of intermediate size are particularly beneficial to the development of high strength concrete.

Admixtures. The dominant admixture used in HSC production is a superplasticiser. Its primary function is to make concrete with a low water/cement ratio workable. It is possible to reduce the water content by 20% and to raise the 28-day strength by 20% or more. A number of researchers suggest that sulphonated napthalene formaldehydes are more effective plasticisers than those with a melamine base. However, it is usual to select the most effective combination of superplasticiser and cement type on the basis of testing. A rapid and convenient method of selection is by the plastic consistency of cement paste. Pentella has suggested that the Vicat needle test can also be used for this type of assessment. It has been observed that cement replacements containing blastfurnace slag seem less sensitive to the choice of superplasticiser than those containing silica fume and PFA. Retarders are often used in combination with superplasticisers to control and delay the early hydration of concrete.

The time when a superplasticiser is added to a concrete mix can greatly influence its effectiveness. It is general practice to dose the concrete with superplasticiser shortly before placing, because of the loss of workability with time once the admixture is introduced. It has a pot life of about one hour, after which time the mix becomes harsh and unworkable. Current practice in the UK is that half the dosage is introduced into the mix as the truck mixer leaves the batching plant and the remaining dosage is added when the mixer arrives on site.

A superplasticiser does not markedly change the surface tension of water in the mix, nor does it affect the setting of the concrete except when used with cements having a very low C_3A (tricalcium aluminate) content. It does not affect shrinkage, creep, modulus of elasticity, freeze/thaw resistance or durability of concrete. The admixture is thought to be adsorbed on the surface of the cement and of other fine particles, causing them to mutually repel each other. In this temporary state the fine particles slide over one another, rather than floc and stick together, to give the concrete its workability.

Research by Watanabe has shown that, for a given slump and workability, superplasticised HSC mixes require prolonged periods of vibration, as they can be sticky in character. Superplasticised mixes probably require two or three times longer to vibrate than normal concrete mixes, unless special high frequency poker vibrators are used.

Cement replacements. PFA can slightly reduce the water demand for HSC and can contribute significantly to strength development after 28 days. Furthermore, the use of PFA will usually moderate heat evolution and help to reduce the early-age temperature rise of cement-rich mixes. This is important in the control of early-age thermal cracking. As PFA is invariably cheaper than OP cement, its inclusion in HSC is encouraged where it is feasible.

Ground blastfurnace slag has heat evolution advantages and long-term strength gains similar to those of pulverised fuel ash and can be used for HSC production in much the same way. Nakamura suggested that the effectiveness of a slag in HSC was dependent upon its particle size, and that blended slag cements with particles finer than 10 μm gave higher concrete strengths than Portland cement blended with slag cements.

It is widely accepted that the use of silica fume in HSC reduces workability and therefore requires additional dosage of superplasticiser. Some researchers have shown that the 28-day strength of superplasticised concrete containing silica fume with a given cement content is increased by as much as 10 MPa. Some of the highest strengths of HSC have been reported with concrete containing silica fume. Other investigators have concluded that the effects of silica fume are small, the results being more dependent on the cement binder content and the superplasticiser being used.

Aggregates. It is frequently stated that the rate of strength development can be influenced by the choice of aggregates and that the best aggregates for HSC are obtained using crushed rock. Using the best aggregates and most favourable cements, the rate of strength development can be almost that of the pure cement paste. Limitation on the long-term strength development of limestone aggregate concrete is only obvious beyond the age of three days with HSC. In normal strength concrete there is no detrimental effect upon the rate of strength gain when limestone aggregate is used. There are reports that gravels and rounded aggregate have produced concrete with a

(a)

(b)

Fig. 1.5. (a) Close-up and (b) distant views of the Brunswick Building, Chicago (1965). High strength concrete was used in the transfer beam and ground-floor columns

Fig. 1.6. First National Bank Building, Chicago (1969): high performance concrete evolving the character of high rise architecture

strength of 120–140 MPa. Selected synthetic aggregates such as expanded shales and sintered PFA aggregates can yield a high strength concrete of 80 MPa provided that the maximum particle size is limited to 10–15 mm.

A mechanically strong coarse aggregate will not produce high concrete strength if the aggregate-bonding matrix is relatively weak. It has been shown that the maximum size of coarse aggregate should be limited to 10 mm to obtain the highest strength. The choice of fine aggregates does not seem so critical, although a coarse sand is preferred to reduce stickiness and water demand of the fresh concrete. Examples of early use of HSC in Chicago are shown in Figs. 1.5 and 1.6.

Mix proportions

The proportion of water to cement binder in the mix is the dominant factor in the design of high strength concrete. A wide range of 28-day strengths can be achieved with different cement types for a given water/cement ratio. It is not usual to test high strength concrete at 28 days, since strength gains may continue beyond 56 and 90 days with the inclusion of certain admixtures. Long-term strength gain associated with the addition of PFA and finely ground blastfurnace slag suggests that strength testing at 56 and 90 days would be more appropriate for such mix designs.

Increase in strength with the lowering of the water/cement ratio below 0.2 is unlikely to happen since the mix becomes difficult to consolidate and compact fully. Although the compressive strength of HSC is dominated by the water/cement ratio, the workability of the fresh concrete is also important since it governs the lowest water/cement ratio that can be placed and compacted. Without a superplasticiser it is unlikely that cement binder contents above 550 kg/m^3 will give any further increase in strength. A superplasticiser is critical in HSC not only to reduce the water/cement ratio and maintain a workable concrete, but to give an increase in strength for the same amount of cement binder. For example, a mix containing 550 kg of cement and a water content of 175 kg could be dosed with a superplasticiser to lower the water/cement ratio from 0.32 to 0.26 without a slump loss. This is likely to give a rise in 28-day strength from 80 MPa to 105 MPa.

Adams suggested that substitution of PFA for Portland cement did not significantly alter the slump of mixes with 550 kg/m^3 of cement and 165 kg/m^3 of water. Nakamura suggested the same for normal and finely ground blastfurnace slag, although substitution of silica fume caused a reduction in workability.

Watanabe examined the relationship of various measures to test the workability of HSC and concluded that workability was not adequately characterised by the slump test. For a given slump an increase in cement produced a low flow and a higher degree of cohesion. Kakazaki reported that the relationship between slump and concrete flow was also affected by the use of mineral admixtures. The required slumps of superplasticised concrete containing either silica fume or PFA to give the same flow as a Portland cement concrete with a slump of 180 mm were 210 mm and 60 mm respectively. On the other hand, HSC concrete containing silica fume exhibited less segregation of aggregates than concrete containing PFA.

The optimum proportion of fine to total aggregate was reported by Watanabe to be in the range of 30–44%.

Properties of HSC

Tensile strength. Comprehensive testing of various HSC mixes, some with superplasticisers, some with PFA, silica fume and slag additions, have shown that results for tensile strength can be determined using the normal weight equation for tensile strength:

$$\text{Splitting strength} = 0.226f^{0.705} + 1.0\,\text{MPa}$$

where f is the compressive strength of the concrete in MPa. However, the flexural tensile strength of lightweight aggregate concrete was lower than that of normal weight concrete with the same compressive strength, and was severely reduced by dry curing.

Elasticity. The elastic modulus has been correlated with compressive strength in numerous studies. Elasticity of HSC is dependent on the relative volume and elasticity of aggregate and cementitious matrix. Exceptionally high values of elastic modulus have been obtained using bauxite or silicon carbide as the aggregate. The low elastic modulus of lightweight aggregate generally leads to a relatively low elastic modulus of lightweight concrete, even at high strength levels.

Shrinkage and creep. A considerable number of drying shrinkage tests have shown that the shrinkage of HSC is not significantly greater than that of lower strength concrete. There is evidence to suggest that drying shrinkage is linearly related to the free water content of the concrete. Use of superplasticiser was found to have little influence on drying shrinkage, although Alexander reported a slight increase at early days, but after prolonged drying the shrinkage was small. Silica fume was found to have little influence on shrinkage, but PFA gave a slight increase. Shrinkage of high strength lightweight aggregate concrete was found to be 15% more than that of normal weight HSC; but it was still within the range expected from normal weight concrete.

Results from a number of studies suggest that the creep factor reduces with an increase in compressive strength. However, the total strains within HSC at a particular stress/strength ratio may be similar to those encountered with normal strength concrete, since the lower creep strains are counterbalanced by the higher elastic strains. The choice of aggregate can significantly affect creep, but this can be accounted for by treating creep as a proportion of the elastic strain. Swamy reported that creep strains of high strength expanded shale aggregate concrete were comparable to those measured with dense aggregate concrete.

Thermal properties. High strength concrete is synonymous with high cement content and this leads to high heat of hydration and temperature rise. The use of silica fume, PFA and slag can help to reduce the peak hydration temperature. The thermal expansion of HSC was reported as 12% higher than that of normal strength concrete made from the same material.

Durability. The low porosity of HSC generally imparts good durability under many exposure conditions. Several studies indicate that chloride ion ingress could be substantially reduced with the use of HSC and that HSC was much more resistant to attack from ammonium nitrate solution when silica fume was added. Nakamura found that finely ground blastfurnace slag was the most effective mineral additive in HSC for reducing the permeability of concrete in water. Holland reported that erosion and abrasion resistance of concrete improved with increased compressive strength, but aggregate type was also important.

Although high strength concrete is quite durable, a large number of investigations indicate that air entrainment is necessary to ensure good freeze/thaw resistance. According to Robson, the freeze/thaw performance of HSC was impaired by the use of superplasticiser. The impairment was associated with the size of the air bubbles that were entrained.

In certain highway and marine applications, freezing and thawing is accompanied by soaking with chloride solutions and this can exacerbate deterioration in concrete. Whiting found that scaling was worse with higher strength concrete under these conditions. Only an air-entrained 40 MPa concrete gave reasonable performance when a 4% calcium chloride solution was used as a de-icing salt.

Hobbs reported that alkali–silica reaction caused cracking in some high strength concrete cubes stored in moist or saturated conditions. It is difficult to limit the reactive alkali content of HSC, with its high cement content, since the latter is essential to HSC. The most reliable way to prevent cracking due to alkali–silica reaction is to specify a non-reactive aggregate. Many aggregates that are suited to the production of HSC, e.g. crushed igneous rock and metarmorphic limestones, are not thought to be deleteriously reactive. It also seems prudent to avoid the use of siliceous fine aggregates, unless they are known to have satisfactory performance in high alkali levels.

Main conclusions

The main conclusions from the literature survey are as follows:

1. Much of current interest in HSC is concentrated in North America, Japan and Scandinavia. The most active application for HSC is in high rise buildings in the US and Canada. Concrete with a compressive strength of 130 MPa has been used in columns of high rise buildings. HSC offers advantages in certain forms of bridge construction, particularly arched and segmental construction, and where high abrasion resistance is required.

2. Typical mix designs for HSC will have a binder content of not less than 450 kg/m^3 and a water/cement ratio of less than 0.3.
3. Extended periods should be allowed, i.e. 56 and 90 days, for high strength concrete to reach its specified strength. This encourages efficient use of mineral admixtures such as fly ash, slag and silica fume.
4. The tensile splitting strength of HSC correlates closely with compressive strength.
5. The modulus of elasticity does not increase in proportion to concrete strength and is substantially affected by the stiffness and percentage of aggregate in the mix.
6. Drying shrinkage of HSC is within the range of shrinkage movement experienced with normal strength concrete.
7. Elastic creep is related to elastic strain in the same way as in normal concrete. However, drying creep of HSC, which is proportional to drying shrinkage, diminishes with increasing compressive strength and becomes negligible above 100 MPa.
8. The durability of HSC is greater than that of normal concrete, except for freezing and thawing when accompanied by soaking with chloride solutions.
9. The production and handling of HSC requires greater than average control, but there are numerous examples of its successful use. Superplasticisers should be added to the concrete mix shortly before placing to maximise their effectiveness.
10. HSC is now used regularly and with confidence in several countries, with standards and national codes on best practice.

High strength concrete in Chicago

Chicago, the birthplace of the skyscraper, home of Al Capone, the Blues, Babe Ruth, the Sears Tower, the inspirational architects Louis Sullivan and Frank Lloyd Wright and the pioneering high rise engineer Fazlur Khan, is also the city of high strength concrete. One cannot begin a discussion on high strength concrete without making reference to Material Services Corporation, the largest ready-mixed company in Chicago, and to a presentation given by John Albinger on 23 May 1989 to an enthralled audience at the Regent Crest Hotel, London. Here is an extract of that presentation.

In order to talk about high strength concrete, it is important to understand that its success is not only a result of technical advancement and fulfilling a need, but the result of a conscious marketing plan and company business philosophy.

The directors of Material Services Corporation realised quite early on in the 1950s that the future market for ready-mixed concrete was going to be the development of the inner city of Chicago. They commissioned a large concrete batching plant and built it in the city centre, adjacent to the Chicago river, where cement and aggregate could be inexpensively delivered by barge. For such a market they felt they needed a sophisticated product and a commitment to quality to become the market leaders. They formed a quality control department and hired structural engineers to work for them, who could communicate and network with design engineers in Chicago to find out what was wanted.

Close communication between design engineers and the ready-mixed industry has been an important factor in the successful development of high strength concrete. So it was in 1961 when William Schmidt, structural engineer and pioneer of HSC, approached Material Services to develop a concrete with a design strength of 42 MPa, which at the time was nearly 10 MPa higher than anything we had supplied. Schmidt wanted higher strength concrete to reduce column sizes in a high rise residential block, in order to increase the lettable floor space. The cement content was increased and fly ash and chemical admixtures incorporated. Four years later we did the same for the columns of Lake Point Tower (Fig. 1.7), supplying a concrete with a design strength of 52 MPa.

In 1974, 62 MPa concrete was supplied for Water Tower Place (Fig. 1.8), which was the tallest concrete building in the world at the time, and in 1976 76 MPa was provided for two columns in River City Plaza. These columns were fully instrumented, as were the columns in the Chicago Mercantile Exchange in 1982, which had 97 MPa concrete (Fig. 1.9). Over this period the design community of Chicago became more knowledgeable and confident in the use of high strength concrete, with 62 MPa as common in specifications as 28 MPa concrete.

Using much of the test information and production techniques that Material Services adopted for high strength concrete manufacture, ACI Committee 363 published 'State of the art in high strength concrete' in 1984. Today Material Services is developing concretes of 150 MPa and higher, and considering their applications for bridge beams and transfer structures in buildings.

This is what we found when making successful high strength concrete.

Cements

We have found that OP (ordinary Portland) cements had to be graded and blended in order to achieve the optimum strength per pound of cement. We took our lead following the researches of Blick, who tested

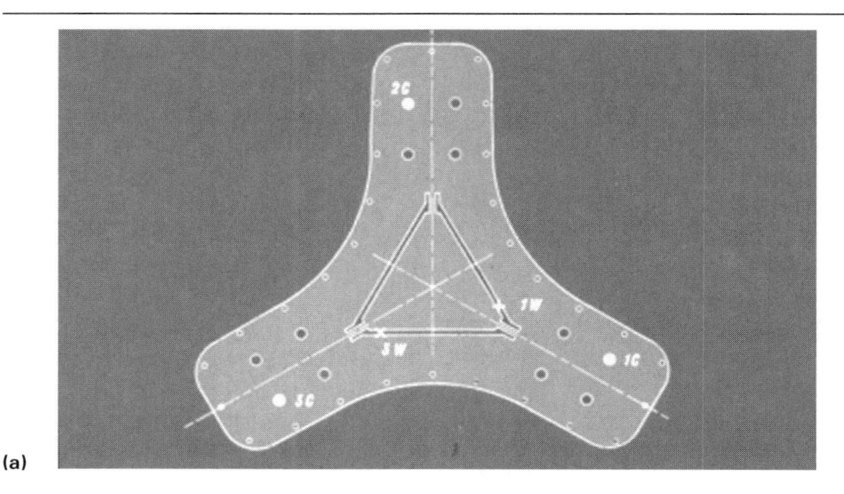

(a)

(b)

Fig. 1.7. Lake Point Tower, Chicago (1968): (a) floor plan; (b) the finished building

five various type 1 branded cements. He found that the optimum cement type, based on cement mortar tests, performed quite differently when used in a standard concrete mix. We replicated these test results in our test lab. We also found that when 20% of the cement by mass was replaced with fly ash (PFA), the mortar test results were quite different. Ironically, the mortar cube made with the highest strength cement

Fig. 1.8. Water Tower Place, Chicago (1974)

blend showed a loss of strength when mixed with fly ash, whereas the mortar cube made with the lowest strength cement showed the highest strength gain.

The conclusions we made were that the cement mortar test did not give an accurate picture of concrete strength gain, but that it can be useful in controlling cement quality once the blend of cement types has been selected. The optimum cement can only be determined through a series of laboratory trials. At least three cement contents for each cement type should be tested, ideally ranging from 136 to 409 kg/m^3 to give the strength efficiency per kilogram of cement.

If the cement content of a mix is below the optimum, then a higher strength concrete may be achieved by using a larger size aggregate. Conversely, if the cement content of the mix is above the optimum, higher strength may be achieved by using smaller aggregates. Once the optimum cement content is reached in a mix for a given water/cement ratio, the maximum strength will not be increased by additional cement. Moreover, too much cement will also cause the concrete to become too sticky and unworkable.

Fly ash

The benefit derived from fly ash varies with its chemical properties, particle size and compatibility with the cement used. Generally, class C fly ash with a high calcium oxide content results in higher strength concrete. The proportion of fly ash replacement is usually between 10 and 15% by mass of cement, but replacement as high as 25% has been successful with class C fly ash. All pozzolan should be considered a part of the cement mass when computing the water/cement ratio.

Micro silica

Micro silica, or silica fume as it is sometimes called, has been used to create high strength concrete. At Material Services we have preferred to develop HSC with fly ash rather than micro silica because of the cost benefits. The fineness of micro silica and its pozzolanic reactivity when combined with cement offers strengths in excess of 100 MPa. However, because it is such a fine material, a superplasticiser must be used to overcome the stickiness of the fine particles, which has a negative effect on the slump of the concrete. Micro silica has been used with fly ash but we can see no benefit from combining them, as the benefit of the fly ash is much reduced.

Chemical admixtures

The use of water reducers, retarding water reducers or a combination of both becomes necessary when mixing HSC concrete and trying to maintain the lowest practical free water content. High dosages of superplasticiser, above the recommended limit advised by the admixture manufacturer, have actually increased the ultimate strength without detrimental effect. Also, the use of ASTM type D retarding water reducer gave strength gains beyond 56 days, as this helped to delay the rapid early setting of concrete having such a high cement content. During hot weather such retarders are imperative as the hydration generates quickly and, if left unchecked, will cause premature early strength and serious curtailment of ultimate strength, as well as increasing the risk of thermal cracking.

Water reduction of around 30% is possible whilst maintaining a workable concrete. This is how Material Services was able to increase concrete strength from 62 MPa to 76 MPa.

Factors to be considered when planning high strength concrete production are: cement and ash compatibility, water reduction, setting time of the concrete, workability required, effective life of superplasticisers, time of addition of admixture and dosage level. Superplasticisers should be used in conjunction with retarders, and can be added to the concrete at the batching plant or at the job site.

Fig. 1.9. Chicago Mercantile Exchange (1982)

Air entrainers are not generally used because of the accompanying strength loss.

Aggregates

Careful consideration should be given to the shape, grading, surface texture and mineralogy of aggregates. Cubic-shaped crushed stone with a rough surface texture appears to produce the highest strength concrete. Smooth-faced, uncrushed gravel may be used to produce strengths of up to 70 MPa, but does not have the mechanical bond to go much higher.

Each cement content level will have an optimum aggregate size to achieve the highest compressive strength. In addition to carefully grading the coarse aggregate size for optimum surface area, further benefit can be derived by adjusting the proportion of coarse to fine aggregates for a particular mix. In our experience we found the recommended proportions of fine and coarse aggregates in ACI 211-84 do not work for high strength concrete.

Fine aggregates with a fineness modulus of 2.5 and under will produce concrete with a sticky consistency, less workability and lower compressive strength. Well-graded fine aggregates that provide good finishing characteristics in normal strength concrete are unnecessary as they may require more water than can be tolerated, to achieve workability. Fineness modulus of between 2.75 and 3.20 appears to be best for high strength concrete, but very angular sands should not be used since they will increase water demand for workability.

Material control

Once the materials for a high strength mix have been chosen, the consistency of the materials becomes critical. If variations become excessive the standard deviation could make the target strength too high to attain with the chosen mix. For example if 9000 psi (62 MPa) concrete is specified and the standard deviation is 697 psi (4.8 MPa), this gives a target strength of 10 200 psi (70.4 MPa). On the other hand, if the standard deviation were 1414 psi (9.8 MPa), this would require the target strength to be 12 000 psi (82.8 MPa).

A materials testing programme should be set up so that variation in material quality can be checked and dealt with before it starts to affect the concrete strength. All material suppliers should be made aware of the importance of supplying consistent material. They should also know that their material will be checked on a regular basis and the results forwarded to them. The handling and storage of materials need not be different from procedures used for conventional aggregates. Proper stockpiling of aggregates, control of moisture content and adequate segregation of cementitious materials are essential.

Mixing and transportation

High strength concrete is best produced in a semi-automatic or fully automated batching plant. It is very evident from our experience that fully automated plants always improve consistency of the mix. Normal batching procedures need not be changed but admixtures have been found to be most effective if they are introduced after the cement and water have been mixed. Truck mixers that are not efficient as concrete mixers should not be used for truck-mixed high strength concrete. It is far better to pan mix the materials before discharge into the truck mixer. The truck mixer should keep the mix agitated on route to the site.

As a rule, truck-mixed material should be mixed at maximum drum speed, at the rate of one minute for each cubic metre in the load.

When mixing high strength concrete at high ambient temperature, there is generally a higher water demand. If the water demand becomes excessive or slump loss is high, increased dosage of superplasticiser and retarder must be considered. If problems persist, then introducing ice blocks or chilled aggregates may be necessary to lower the concrete temperature.

High strength concrete must be delivered to the site with the minimum of waiting time (Fig. 1.10). Delays in transit or on site will lead to loss of slump and sometimes total loss of workability as the effect of the superplasticiser wears off. Water must not be added to the mix on site as it will be detrimental to the strength gain of the concrete. Any adjustment to the slump and workability must be made with a superplasticiser, but check whether the free water in the admixture significantly affects the water/cement ratio of the mix. In our experience the time allowed from loading the truck mixer to discharge should not be more than 90 min. If the concrete stays longer in the mixer but is still placeable it should be used, otherwise it has to be re-mixed with additional superplasticiser or discarded.

When ambient temperatures are low, superplasticisers and retarders may cause excessive retardation of the concrete. For mass pours and thick column sections this may not be a problem as the concrete will set adequately. However, test cubes may remain 'green' for some time and therefore should be cured alongside the structure or in a controlled environment until they can be transported to the test centre. Factors such as rate of slump/workability loss, initial setting time and ease of handling should be considered when adjusting the dosage and combination of admixture. One point to make is that if silica fume is used, it

Fig. 1.10. Truck mixer delivering high strength concrete to site

will produce a concrete that is much darker than a normal strength concrete and this may be a factor if the concrete is exposed.

Quality control

The concrete producer should be able to provide documentation as to the quality of the materials being used and verification on the accuracy of the weighbatching plant. It is important to check the capacity of the cube/cylinder test machine, because it may not have the strength to crush 100 MPa concrete. If the compressive strength is based on 56-day or 90-day criteria, then a relationship between the 7-day strength and the long-term acceptance test should be established, for early warning of problems or conformity of the concrete.

It is useful to call a site meeting to discuss all aspects of the work before concrete deliveries commence. What should be discussed and agreed is who accepts and rejects the concrete, what test regime is followed, the frequency of testing, and resolution for non-compliance. All concerned should be in attendance: the project engineer, main contractor, concrete frame contractor, concrete producer and testing agency. Cooperation both before and during construction is imperative to the overall success of the project.

High strength concrete is not without its problems and there are a few. Its use and delivery into the market cannot be based on the consequences of failure but rather on the confidence of success and teamwork.

Conclusions

We often ask ourselves why we produce higher and higher strength concrete when it represents no more than a fraction of our business. Promotion of this type of concrete requires sales professionals who are knowledgeable on structural design, capable of answering questions on high strength concrete and reassuring clients. The product is not cheap to produce, and must offer a cost-benefit to the building owner.

The indirect benefit of high strength concrete to the concrete producer is that it improves the quality of lower grade mixes, reducing the standard deviation and thus the amount of cement in a standard mix to achieve a particular grade of concrete. High strength undoubtedly improves the technical competence and image of a ready-mixed company. That is how we are able to justify our continuing investment and research into high strength concrete.

A high strength concrete flat slab

Large-scale testing of the strongest suspended slab ever produced in a UK test centre has confirmed that high strength concrete has benefits also in suspended floor applications. The slab that was tested measured 15 m × 15 m in plan, and was supported on four columns spaced 9 m apart. The columns were 300 mm square, with no drop heads.

The test formed part of a research programme to develop design rules for detailing reinforcement in high strength concrete structures, and in particular suspended slabs. The preliminary part of the research was a review of the provision for HSC in design codes across Europe and North America: BS 8110 (UK), NS 3473 (Norway), N 6720 (Holland), ACI 318 (USA), CAN 3A23.3 (Canada) and ENV 1992 (Eurocode 2). Some surprising differences were found, of which probably the two most interesting features were the differences in the stress blocks and the different partial safety factors that had been assumed.

Concrete test slab

The test slab (Fig. 1.11) was designed by Ove Arup and Partners and constructed at Taylor Woodrow's research laboratories by Taylor Woodrow Construction with the 110 MPa concrete supplied by Tarmac Top Mix. The test slab incorporated the results of extensive

Fig. 1.11. High strength test slab after removal of the formwork at 24 h (courtesy: Taylor Woodrow)

research into new design rules for reinforcement in high strength concrete structures.

Grade 110 concrete was supplied from a batching plant seven miles away. To maximise workability retention on site, only part of the melamine-based superplasticiser was added at the batching plant. The remaining plasticiser was added to the truck mixer once it arrived on site, using a metered pump. The concrete was thoroughly mixed to bring the workability to a flow of 60 mm on the flow table. The concrete was pumped into place and then compacted using an air-driven poker vibrator. The concrete pumped easily and pumping pressures were not significantly higher than those expected for conventional strength concrete. It was noticeable that, despite its high flow table spread, the concrete was viscous and needed considerable vibration to expel all the entrapped air. The pokers were inserted at closer than normal centres to ensure proper compaction. The surface of the slab was covered with polythene once the surface had been levelled with a screeding bar. It was important to do this without much delay to prevent plastic cracking and moisture loss, which can be a problem with high strength concrete.

The concrete rapidly gained strength and achieved 100 MPa in less than 24 h, reaching a temperature of 31°C. The rapid strength gain enabled formwork and falsework to be removed 24 h after casting.

The test slab was loaded to a uniform 1.5 kN/m^2 at three days to simulate typical construction loads. No back propping was used. The slab was loaded by hydraulically pulling the slab downwards, using a series of prestressing rods anchored to the strong floor. The measured deflection at the centre of the slab was only 2.5 mm. The load was maintained for 10 days but very little creep was observed. At 28 days after the slab was cast, the full working load of 13.125 kN/m^2 was applied to assess the slab performance at the serviceability limit state in accordance with BS 8110. The measured deflections and crack width were in good agreement with predicted design values.

Design criteria

Concrete characteristics. The characteristics of HSC used in design were:

- ☐ higher E modulus (Fig. 1.12(a));
- ☐ reduced creep coefficient (Fig. 1.12(b));
- ☐ increase in tensile strength (Fig. 1.12(c));
- ☐ higher bond strength for reinforcement;
- ☐ higher punching shear stress.

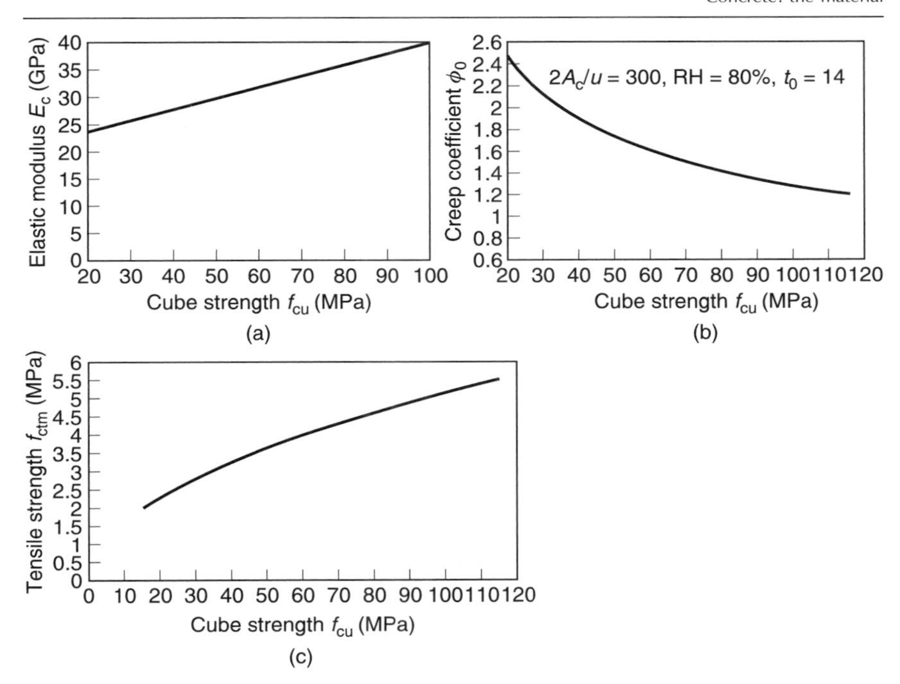

Fig. 1.12. Properties of high strength concrete: (a) elastic modulus; (b) creep coefficient; (c) tensile strength (Whittle, 1997)

The choice of concrete strength for the design was 100 MPa.The 28-day strength was 118.5 MPa and the 56-day strength 123 MPa. The mix constituents are given in Table 1.3.

Fire resistance. No changes in the current cover requirements, given in BS 8110, were considered necessary. However, it should be noted that there can be a risk of spalling — possibly explosive spalling — in a fire, if the moisture content of the concrete by volume is higher than 3%. As HSC is less permeable to vapour, vapour and pore pressure due to heat near the surface builds up quickly and spalling occurs once the pore pressure exceeds the tensile capacity of the concrete.

It has been found that the risk of spalling can be reduced by the addition of polypropylene fibres in the concrete. The mechanism by which this helps to prevent spalling is not well understood but it is thought that as the fibre melts it forms voids which relieve the pressure of the expanding vapour. It has also been found that the fire resistance of high strength concrete can be influenced by the choice of aggregates. For this test slab a nominal cover of 20 mm was provided to achieve a fire rating of $1\frac{1}{2}$ h, assuming the concrete had dried out sufficiently.

Table 1.3. Specimen mix design for 100 MPa concrete (Whittle, 1997)

Ingredients	Grade 100
Binder	
(a) OPC	295 kg/m^3
(b) GGBS	200 kg/m^3
(c) Microsilica (Emsac 500S) (MS/water ratio = 50/50)	80 kg/m^3
Superplasticiser (Rheobuild 716 SPA)	11.7 kg/m^3
Fine aggregate (natural flint river sand, zone 2/3)	525 kg/m^3
Coarse aggregate (crushed limestone)	
(a) 10 mm single-size	375 kg/m^3
(b) 20 mm single-size	865 kg/m^3
Water	96 litres
Slump	145 mm (after 30 min)
Flow table	420–480 mm
Total free water (including 0.65×SPA, 0.5×Emsac 500S)	143 litres
Freewater/binder ratio	0.27
28-day strength (100 mm cubes)	118.5 MPa
56-day strength (100 mm cubes)	123 MPa
Plastic density	2440 kg/m^3

Punching shear. BS 8110 limits the shear capacity for $f_{cu} = 40$ MPa concrete, as regards both maximum shear ($0.8 f_{cu}$) and tension shear capacity ν_c. Technical Report 49 of the Concrete Society recommends that there should be no limit to f_{cu} for maximum shear and that the limit for ν_c should be raised to correspond with an f_{cu} of 70 MPa.

Assuming the density of concrete to be 2500 kg/m^3 or 25 kN/m^3 and with a live load of 7 kN/m^2 applied on the test slab, the maximum shear stress at the column face is 4.8 N/mm^2. Based on BS 8110 rules, three rings of shear reinforcement are required because $f_{cu\ max}$ is 40 MPa.

For the test slab no special provisions are required for punching shear.

Flexure. The compressive stress/strain relationship alters as the strength of concrete increases above 60 MPa, as shown in Fig. 1.13.

The modified stress block can be used for flexure calculations exactly in the same way as with the normal method of BS 8110 except that the maximum strain e_u reduces with strength when $f_{cu} > 60$ MPa:

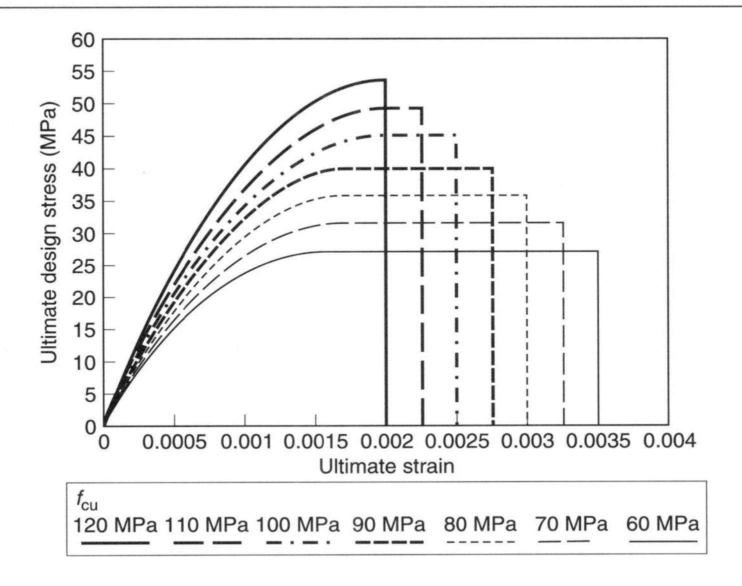

Fig. 1.13. Stress/strain relationship for high strength concrete (Whittle, 1997)

$$e_u = 0.0035 - \frac{f_{cu} - 60}{5000}$$

The existing rules for the spacing of reinforcement that are applicable to normal strength concrete are appropriate for HSC.

Depth of slab. Deflection is controlled by the span/effective depth limits given in BS 8110. The limit given in Table 3.9 of BS 8110 for continuous members is 26, based on normal strength concrete. For HSC this can be limiting, as the enhanced E value and tensile strain capacity of HSC confirms that there is a linear increase in the span/effective depth ratio with an increase in concrete strength.

For $f_{cu} = 100$ MPa, the limiting value of the span/effective depth ratio is 41. The reduction factor of 0.9 for flat slabs given in BS 8110 has been assumed for HSC. The modification factor for HSC with tension reinforcement is also applicable. But as the values given in BS 8110 are conservative for concrete above 50 MPa, the combination of the modification and reduction factors chosen for HSC balances out. Thus 41 is the design value for span/effective depth, and for a span of 9 m this gives an effective depth of 220 mm or greater.

Taking into consideration the cover of 20 mm, this slab has been designed with a depth of 250 mm. For a normal strength concrete the required depth would be 375 mm. This is a considerable saving in slab depth and dead load.

Vibration. Floor vibration is unlikely to be a problem for flat slabs if the slab depth meets the following requirements:

☐ the floor has at least four panels and is at least 250 mm thick;
☐ the floor has at least eight panels and is at least 200 mm thick.

Minimum tension reinforcement. The value of 0.13% given in BS 8110 for tension reinforcement is based on the tensile strength of concrete. Hence, as the strength increases, so should the minimum tension reinforcement..

Minimum shear reinforcement. Shear reinforcement should also increase with increase in concrete strength. The nominal spacing of links in columns measuring 300 × 300 mm should be reduced from 300 mm to 200 mm.

Double strength composite columns

This section draws on an informative and entertaining article written by Denise Chevin in *Building*, July 1990.

US structural engineers Skilling Ward Magnussen Barkshire (SWMB) have scooped most of the top prizes in building engineering in the past four years. They are arguably the most innovative high rise structural engineers in the world, and have pioneered a method of composite construction using steel floors with massive high strength soldier columns that can save millions of dollars on high rise building costs.

Jon Magnussen, the spokesman for the practice and a graduate of Berkeley University, is a real egghead and an expert on the design of high rise structures. He believes that if you have to build in steel, it is cheaper to build in a composite form for anything higher than five storeys. "What we are trying to do is to take the advantage of steel and bring in the economies of high strength concrete." A number of detailed studies were carried out by the practice to compare the cost of tall buildings constructed in recent years around the US with what they might have cost if they had been built in composite form the Magnussen way.

As a building gets taller it needs more steel to carry the gravity loads, of course, but it also needs more steel to control and limit the sway of the structure in gusting winds. Stability of SWMB's composite structures derives from four massive soldier columns positioned in the core of the building, which are tied together and braced by steel trusses. The bulk of the building's gravity load transfers to these massive soldier columns with steel floor beams spanning between the central core and the small perimeter columns which carry just gravity loads. The central columns are steel tubes filled with high strength concrete of strength ranging from 12 000 psi (83 MPa) to 25 000 psi (173 MPa).

"We were able to save the client \$14 million on the Columbia Seafirst Center [Figs. 1.14 and 1.15], completed in 1985, using this type of structure against an all-steel frame" says Magnussen. On the 56-storey Two Union Square building, finished in 1991, they clipped \$10.5 million off the construction cost of this \$98 million project, using 19 000 psi (131 MPa) concrete to fill the four steel tube columns. They made a saving of $\$2/\text{ft}^2$ on a modest 11-storey building in Oakland, California, and have no doubt that if SWMB had been given the job of designing the twin towers of the World Trade Center they could have

Fig. 1.14. Columbia Seafirst Center (now known as Bank of America Tower), Seattle (1985)

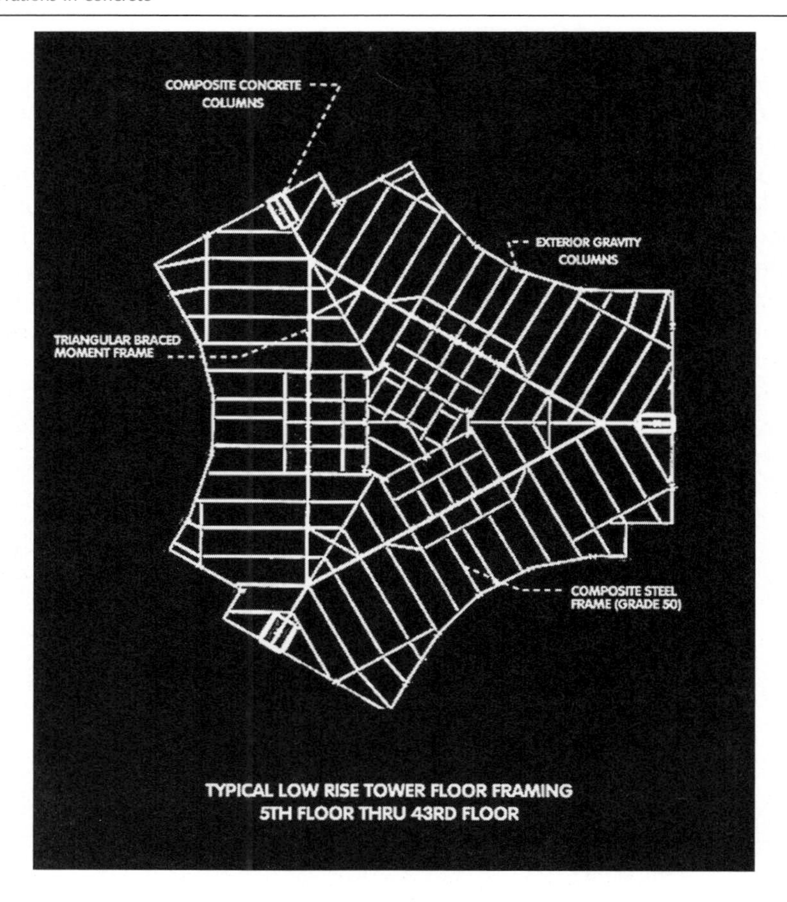

Fig. 1.15. Cross-section of Columbia Seafirst Center

made substantial cost savings on the steelwork. According to senior partner John Skirling, "The World Trade Center uses 180 kg/m^2 of steel in its original construction. If SWMB were to design it today, it could be built with half that amount of steel and would even have better structural performance. But you have to remember that when the World Trade Center was designed in the early 1970s, its structural design was a leap forward compared with what anyone else had done, with the exception of Fazlur Khan's Sears and John Hancock Towers."

Magnussen knows that technical innovation is driven by looking for ways to cut building costs—winning awards is a by-product. The pioneering approach of the firm is having confidence in high strength concrete and motivating suppliers to have it available in the city where they are designing a building. "If you can achieve 83 MPa concrete in Bangkok, as we have done, you can achieve it anywhere" says Magnussen. "In every city we have gone into and done a building with this structural system, we have found that the project ended up setting a new record for concrete strength. In San Francisco we got the strength up to 9500 psi (66 MPa), in San Diego up to 10 000 psi (69 MPa) and in Cleveland it was 12 000 psi (83 MPa) in high strength. But in Seattle, where our practice is based, we got the concrete strength up to 20 000 psi (138 MPa)."

Seattle, not Chicago, is now the home of high strength concrete. "People in Chicago always made a big deal about their high strength concrete. I think what we did totally shocked them" adds Magnussen. He makes the point that the shorter the building, the lower the strength of concrete required. For instance, for an 11-storey building 69 MPa is sufficient.

The use of high strength concrete for buildings is likely to remain restricted to sites that have easy access to good ready-mixed batching plant. Although the cost of high strength is three times as much as that of normal strength concrete, it is still fourteen times cheaper than a client would pay for steel to carry the same load. One word of caution that Magnussen is quick to point out: "The formulae for structural design in codes and building regulations were based on tests made with low grade concrete, and just do not apply to HSC. What we have done at SWMB is to conduct our own tests for high strength concrete and evaluate the properties that are critical in design."

So will the double strength steel–concrete mega column structures catch on? "I don't know anyone outside this practice who is doing this kind of work except Les Robertson who designed the Bank of China in Hong Kong along similar lines. But then he was a former SWMB partner!"

Concrete-filled steel tubes

(taken from a paper written by J. Webb and J.J. Peyton of Connell Wagner in 1990 for a Structural Engineering Conference in Adelaide, Australia)

The choice between whether to design concrete or steel columns for a multi-storey building has never been an obvious one. The final decision has generally been driven by local conditions, contractor's preference, and the price of the materials at the time of tendering. Steel columns, although they are more expensive than their reinforced concrete counterparts, have considerable advantages. The desire to adopt a steelwork solution is driven by the growing trend to prefabricate building components off site, and the ability to work simultaneously on a number of levels without obstruction of formwork and back propping associated with concrete construction. When the Casselden Place building was being tendered, further bias was given to steel because formwork prices were very high.

These factors were considered by the contractors Baulderstone Hornibrook when they submitted their alternative design for the original all-concrete scheme proposed for the 46-floor project. The alternative design retained the concrete core and substituted a steel composite floor with steel tubes filled with high strength concrete for the columns (Fig. 1.16). This approach essentially left the structure formwork free except for the central core and increased the efficiency

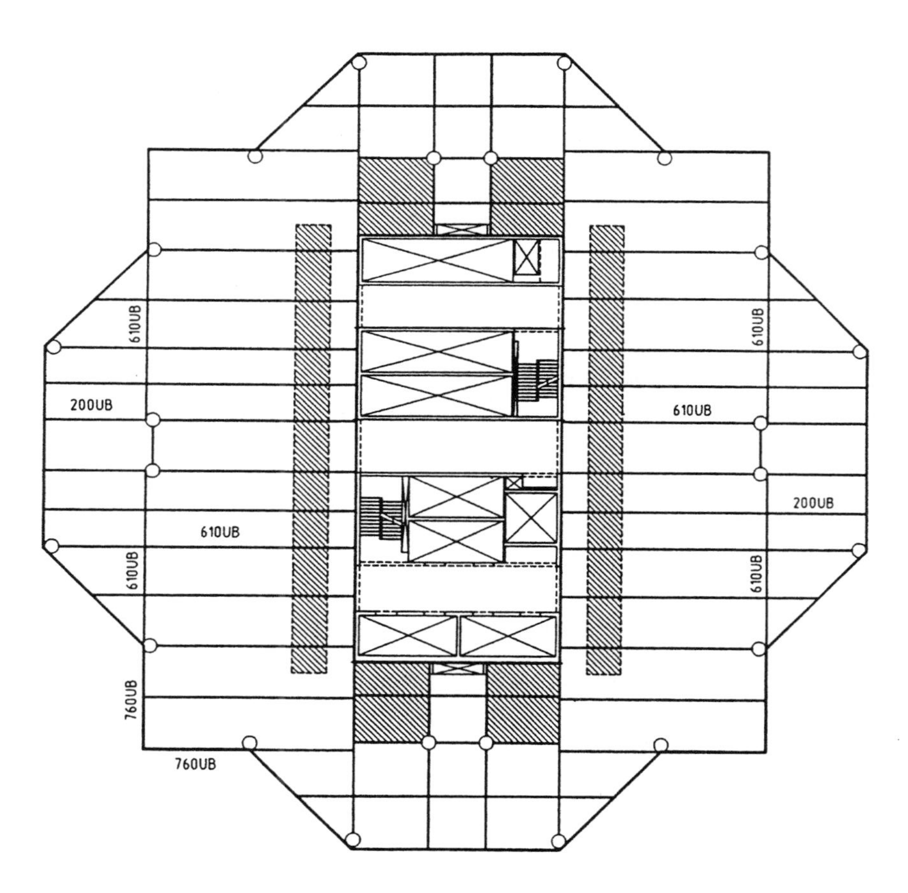

Fig. 1.16. Casselden Place: typical floor plan

of the floor construction cycle. The steel tube column section alone was robust enough to support one floor of main steel and metal deck, allowing the steel tube to be concreted with the floor deck in a later operation.

To validate the alternative design a detailed cost study was made of the comparative column options, and how best to integrate a hybrid column like this into the project. It soon became apparent that if reinforcement could be omitted further construction gains could be made. Fabrication and placing of rebar within the tube slows down the construction cycle, involves more site-based operations and leads to congestion as the beam rebar has to connect to the inner core of the column. The big issue, however, was fire protection.

Relative costs

The study of the relative cost of the six column types that could support a ten- and a thirty-storey building showed that reinforced columns were the cheapest and were used as the basis for comparative cost (Table 1.4). The competitiveness of an unreinforced concrete-filled steel tube was more marked for the taller building. It can be argued that for buildings higher than 30 storeys the concrete-filled tube may be the most cost-effective option.

The steel tube is fabricated from mild steel plate. The major proportion of the axial load is resisted by the high strength concrete. Confinement of the concrete by the tube enhances the capacity of the concrete to resist loads. Connections to steel floor beams are very simple. Plate web cleats were used for all except the moment-resisting joints. Fabrication cost for this type of connections was very competitive.

Main advantages

The big advantages of the concrete-filled tubular columns are rapid speed of erection, reduced site operations and facility to work on more than one floor level. The main points are:

1. With the lack of formwork and reinforcement, the column is erected quickly with minimum crane time.
2. The tubular steel column can support one or more floors without the column being concreted, thus keeping the concreting of the column off the critical path.
3. The ability to be able to pour up to four floors in one operation reduces columns activity compared to conventional reinforced concrete construction.

Table 1.4. Column comparison study

Number of storeys	Type 1 Conventional RC column	Type 2 Concrete column with steel erection column	Type 3 Concrete-encased steel strut	Type 4 Un-fireproofed tube filled with reinf. concrete	Type 5 Unreinforced steel tube filled with concrete	Type 6 Full steel column
	R 10–200 mm links	R 10–200 mm links		R 10–200 mm links Mild steel tube	Mild steel tube with fire spray	Fire spray to steel section
10 levels	450 × 450 8 T 32	450 × 450 8 T 32 203 × 203 UC 46	410 × 410 305 × 305 UC 118	500 × 6.4 CHS 6 T 20	500 × 6.4 CHS	305 × 305 UC 240
Relative cost	1.0	1.22	1.53	1.14	1.10	2.271
30 levels	750 × 750 20 T 36	750 × 750 12 T 36 254 × 254 UC 89	570 × 570 Plate girder 400 × 50 flanges 360 × 25 web	800 × 10 CHS 6 T 32	800 × 10 CHS	Plate girder 500 × 60 flanges 460 × 40 web
Relative cost	1.0	1.13	1.85	1.11	1.02	2.61

Column loaded by 8.4 × 8.4 m bay, steel framed. Concrete f_{cu} 60 MPa. Steel grade 50.

Main disadvantages

The main disadvantage of the unreinforced tube is the requirement for full strength butt welds on site, connecting the sections. However, this operation is considered not to be on the critical path of the floor cycle. A typical splice detail is shown in Fig. 1.17. Adoption of a convex handrail section as a backing strip to site joints will assist in the alignment of the upper tube section as well as providing the backing for a full strength butt weld on site. A series of shear studs is welded to the inner tube wall on either side of the joint to aid the transfer of load from the tube wall to the heart of the concrete column.

Fabrication

The steel tubes on Casselden Place were fabricated in 7.5 m lengths; this was long enough to stretch two floors. The tubes were produced from mild steel plate which was rolled into cylinder form and longitudinally seam welded. The steelwork fabricator was capable of making columns ranging from 488 mm up to 2500 mm in diameter, with a

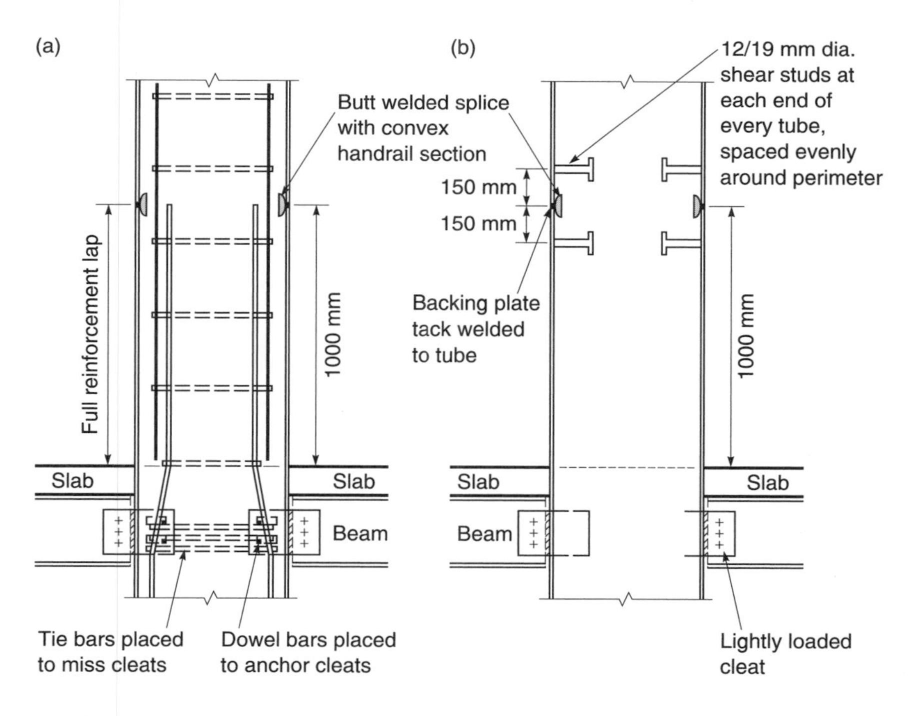

Fig. 1.17. Typical column splice details: (a) bare steel tube filled with reinforced concrete; (b) fire-protected steel tube filled with unreinforced concrete

plate thickness of between 6 mm and 28 mm and in continuous lengths of 12 m if required. Often the maximum fabricated length of steelwork is dictated by road width, bridge height restriction and lorry trailer dimensions.

Design analysis

For the unreinforced tube a method of analysis needs to be formulated, as Australian codes of practice could not be directly applied. A design method was adopted based on Eurocode EC4. The ultimate strength of the section was determined by rigorous strain compatibility analysis. To determine the forces in the concrete the Desai–Khrishnan stress/strain curve was used, while an elasto-plastic bilinear stress/strain function with E taken as 200 GPa was evaluated for analysing the forces in the steel tube and any steel reinforcing bars. The forces in the steel tube and concrete were then reduced by partial reduction factors, using 0.6 for concrete and 0.8 for steel.

The use of the stress/strain curve method for calculating concrete strength, rather than a rectangular stress block, substantially improves the strength of the columns in compression. By taking advantage of the confining effects of the tube on the concrete further enhancement of column strength is achieved (Fig. 1.18). These factors combine to give a substantial boost to the axial capacity of the composite column of around 30% over a conventionally reinforced concrete column.

Several other factors must be taken into account in the design strength of the tube under temporary and permanent load conditions including local stability, load transfer between shell and concrete, slenderness ratio and the maximum allowable eccentricity of load.

Fire resistance

For 2 h fire resistance, the column at the end of the exposure period must be capable of sustaining a load of 1.1 DL (dead load) + 0.4 LL (live load). The filled steel column was fire checked using a heat flow, finite element program. This analysis established the temperature gradient across the column section during the fire exposure period. At the end of the two-hour period the analysis showed that the steel tube was ineffective, reinforcement remained only 87% effective and the outer 36 mm of the column should be disregarded when determining the load capacity of the column. For an 800 mm × 10 mm thick CHS (circular hollow section) column, with 50 MPa concrete and 1% rebar, the analysis shows that the fire strength is 58% of normal strength. This is not critical for typical multi-storey buildings such as Casselden Place where the dead load is at least 1.6 times the reduced live load. However, for small columns where the loss of the outer 36 mm is sig-

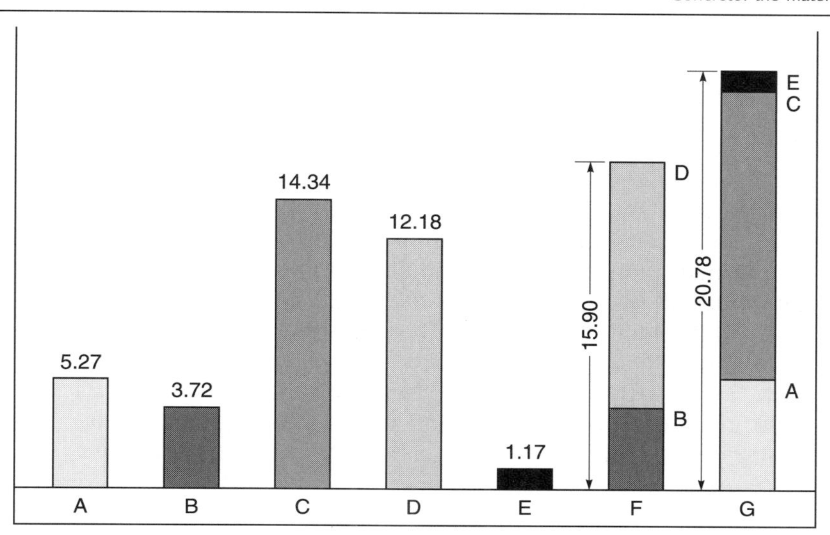

Fig. 1.18. Factors contributing to column strength: A, tube capacity factor of 0.85; B, equivalent amount of reinforcement with 0.6 capacity reduction factor; C, concrete strength (determined by rigorous analysis; D, concrete strength (based on rectangular stress block); E, confinement enhancement; F, conventionally designed RC column; G, tube column designed by the Connell–Wagner method (30% enhancement of squash load)

nificant, fire design becomes critical. It was found that the cost of fire protecting and cladding the columns on Casselden Place was more than compensated by the omission of reinforcement, the simplification of the beam joints and the speed of construction.

Conclusion

This form of column construction will be successfully used on future high rise projects. The tube column exploits the construction advantages of the all-steel column, whilst enjoying the economies of reinforced concrete. Further areas for potential development include the use of composite columns for jump-starting the construction of the suspended floor above ground level, in top-down construction.

Pyrament

Pyrament was developed by Lone Star Industries, America's largest cement manufacturer, for creating high strength, rapid-curing concrete that can support impact loads from commercial aircraft after just 4 h curing. According to Lone Star, Pyrament is the only cement that offers high strength, durable, rapid-curing concrete without relying on admixtures or accelerators of any kind.

In the development of Pyrament over a ten-year period, Lone Star came up with a cement that behaved in a different way to other cements. It did not behave like an OP cement with additives, because the chemistry was actually different. It is technically described as a high performance Type 1P blended hydraulic cement, meeting the requirements of ASTM C595 and modified with 'a proprietary interground functional addition' complying with ASTM C688. If we are none the wiser for that piece of jargon, it is because Lone Star wants to keep the formula confidential. However, Lone Star can confirm that no chlorides are added to Pyrament in the manufacturing process, but the company is reticent about disclosing any more information about its production.

How does Pyrament compare with Portland cement concrete? An ASTM Type 1P concrete fails to achieve sufficient flexural strength for highway, bridge or airport runway loadings in 12 h; while a Type 3, the fastest setting and curing concrete, can satisfy this condition only after 12 h curing. Pyrament exceeds the requirement for highway loading in just 4 h and for bridges and runways in only 6 h. As for comparison of compressive strength, Pyrament concrete achieves the 12 h strength of Type I concrete in 2 h.

Though it is twice the price of OPC, Pyrament is seen as being cost-effective in reducing long-term maintenance cost and for rapid repairs to airport runways and bridge decks where the cost of closure and loss of landing fees is significant. On the basis of accelerated wear tests carried out by Lone Star, it has been claimed that a highway built with 250 mm of Pyrament concrete will last four times longer than one built with 250 mm of Portland cement concrete. From 1989 cost figures, a 1.6 km new dual carriageway in the US costs about $32 million, and incorporating Pyrament instead of OPC would add only $160 000 to the cost.

The reduced water content of Pyrament concrete means that the concrete achieves freeze/thaw resistance without air entrainment; it is also resistant to sulphate attack and there is a significant reduction in drying shrinkage. Another useful characteristic is that it can be placed in temperatures below −2°C without impairing full strength gain. Generally, in mixing, handling and placing it behaves in the same way as an OPC concrete except that its hydration is much quicker and its wet 'mobile' phase is shorter.

An example of runway repair in freezing conditions took place in 1989 at Griffiths Airforce base in the US. This is a strategic air command centre on the alert 24 h a day with a large number of B52 bombers and intercept aircraft. Two Pyrament concrete mixes were tested on sections of an operational apron and primary taxiway. The air temperature at the time of laying the concrete was $-2°C$ and falling. Once the Pyrament concrete was laid it was covered with a plastic sheet and 4 h later the concrete was sufficiently tough and durable to take aircraft loading.

Self-compacting concrete

Structural concrete relies on full compaction, after placing, to achieve the required strength and durability. By compaction the wet concrete is consolidated, voids are removed to below a critical percentage and the finished concrete surface is left free from honeycombing and pronounced blowholes. Voids in the concrete are created by pockets of air that become entrapped between successive layers of placed concrete and by those formed during the drying of the excess water in the mix that has been added for workability. Self-compacting high performance concrete will not eliminate voids that form if the concrete is allowed to fall freely during placement, but it will minimise air voids induced during the mixing process and those formed because of the excess water required for workability (Bartos and Gauer, 1999).

Even before the introduction of superplasticisers and other admixtures, concrete had to be placed in conditions where compaction was impractical; for example, concreting in situ bored piles, underwater concreting, and pumping concrete into inaccessible spaces or openings. But such applications were isolated cases, where reinforcement was often absent. The introduction of superplasticisers permitted production of flowing concretes with high strength and minimum water content, but with risk of segregation for very high slumps and flowing concrete. Consequently highly workable but cohesive concrete was produced using admixtures, which still required compaction. Research in Japan in the 1990s on self-compacting concrete (SCC) based on the development of underwater concrete — where viscous agents were used to increase internal cohesion to prevent segregation and washout — was extended to normal applications (Okumura and Ozawa 1995). Initial use of SCC was confined to complex, thin-walled structures and heavily reinforced sections. More recently SCC has been used in major Japanese civil engineering structures such as the anchorages for suspension cables of the Akashi–Kaikyo bridge, which has the longest span in the world. The use of SCC eliminated the need for compaction and improved the rate of concrete placing to $1900\,m^3$/day, saving three months on anchorage construction time.

Viscous agents

Fresh concrete is susceptible to segregation because it is a composite material consisting of ingredients with different sizes and specific gravities. Since the risk of segregation increases with greater workability, there is a trade-off between high workability and resistance to segregation. Generally, stiffer mixes are prescribed to avoid the risk of segregation. By introducing a viscosity agent in the mix, the viscosity of the paste can be increased effectively to inhibit segregation. For cohesive underwater concrete segregation is inhibited by a high dosage of viscous agent. However, in air, a high viscosity underwater concrete may not release any air entrapped during placement and it may not easily pass through spaces congested with reinforcing bars. Consequently SCC produced using an underwater concrete mix must have a reduced viscosity. The balance between the amounts of workability admixture and viscosity agent is important for self-compaction. However, there is a limit on the amount of self-compactability that can be achieved by this method, because the admixtures are trying to negate the effects of one another. By modifying the coarse aggregate volume and mix proportions, further improvement in self-compaction can be achieved.

Limiting coarse aggregate volume

In conventional concrete, the larger the proportion of coarse aggregates of large diameter, the better the mechanical properties of the hardened concrete, assuming the total amount of aggregate is the same. However, for fresh concrete to pass around obstacles and through reinforcement, it is better to increase the proportion of fine aggregate and to reduce the coarse aggregate size. A surprising discovery made by Yoshida Tokujiro, when studying conventional concrete mixes, was that the optimum proportions of a concrete mix should be 1:1:2 to obtain the maximum strength using a particular set of materials. He noted that if the aggregate volume exceeded this ratio, the attainable maximum strength would fall drastically. Tokujiro also showed that when the ratio of cement to fine aggregate was set at 1:1.5, the ratio of cement to coarse aggregate should not be greater than 1:1.5 as this leads to a drastic loss in maximum attainable strength. The conventional mixes in this study all required careful and thorough compaction.

For practical necessity, the proportion of fine to coarse aggregate in SCC is kept at 1:1 by volume, which is a big reduction on the conventional coarse aggregate content, but the concrete does suffer from a drastic drop in ultimate strength. Also, with SCC, if the coarse aggregate content exceeds a certain limit, there is greater contact between the larger particles which increases interlocking and the risk of blockages on passing through spaces between reinforcing bars. The possibility of interlocking is negligible if the coarse aggregate fraction is lower than 50% of the total mix, provided adequate mortar is used. Smooth, rounded river gravels are generally preferred because this type of aggregate permits a larger coarse aggregate volume than an angular or rough-textured one.

Fine aggregate

Fine aggregate in SCC is defined as particles that are larger than 90 μm, anything smaller being defined as a powder. The amount of water confined by the fine aggregate is almost proportional to the volume of fine aggregate, so long as the fine aggregate proportion is around 20%. There is a drastic increase in the ratio of confined water above this percentage and this is believed to be due to increase in direct contact between fine particles. Tests have confirmed that for good self-compacting concrete the volume of fine aggregate should be set at 40% of the volume of mortar.

Powder materials

Selection of powder is critical because the properties affect self-compaction and govern the quality of the hardened concrete. SCC mixes contain higher than normal proportions of fine material smaller than 90 μm such as PFA, GGBS or limestone powder. One of the characteristic features of a powder is that a unit volume confines a large amount of water. There is an optimum water/powder ratio for imparting a viscosity to the mortar paste that is suitable for self-compacting concrete. Flow tests on paste show a linear relationship between the flow area and the volumetric water/powder ratio. Therefore the volumetric water/powder ratio at which the paste ceases to deform can be extrapolated. This ratio usually falls between 0.7 and 1.0, with a scatter depending on the grading and particle shape and other properties. The ratio of water confined by well-shaped powders such as PFA is extremely small, so the optimum water/powder ratio by volume is also small.

When the volume of fine aggregate has been specified the volumetric water/powder ratio and dosage of superplasticiser can be determined. High water/powder ratios not only cause deformation but also cause excessive reduction of paste viscosity, resulting in segregation of the coarse aggregate. Conversely, a low water/powder ratio leads to high viscosity, impairing the ability of the concrete to pass through small spaces. General methods have not yet been established for determining the optimum water/powder ratio and dosage of superplasticiser, over a broad range of materials, that will impart good workability and viscosity to the paste.

Hardened SCC

Properties of hardened SCC do not differ significantly from those of ordinary concrete of a similar basic composition. So for compressive strength compliance the standard concrete cube test will be adequate. However, verification testing of bond strength between concrete and reinforcement, from results to date carried out in the UK, suggest an improvement over ordinary concrete of the same strength. Although research in the UK on drying shrinkage and creep is still being carried out — the UK is well behind Japan — preliminary indications suggest that there are no abnormal increases. It is highly unlikely that the Japanese civil engineering industry would be specifying SCC for major bridge structures unless the properties of SCC were predictable, consistent and within acceptable limits.

Fresh properties of SCC

The behaviour and characteristics of fresh SCC put it well outside the scope of current standard tests for workability and other properties of fresh concrete. In essence, SCC must have the following characteristics in its fresh state; it is therefore these properties that need to be assessed:

(1) Flow and filling ability: it must be able to flow into all the spaces within the formwork under its own weight.
(2) Passing ability: it must flow through the small spaces between reinforcing bars under its own weight without blocking.
(3) Resistance to segregation: it must fulfil the requirements of (1) and (2) without segregating.

As with other types of new concrete, a number of non-standard tests have been developed for assessing the properties of fresh SCC. However, such test work will require substantial previous experience of handling SCC mixes to interpret the test results correctly. The most common method for assessing workability is the slump flow test, where the spread of a sample made in a slump cone is measured rather than the slump. The time taken for the concrete to spread 500 mm can also be taken with this method. The passing ability or 'blocking' test most commonly used is the L box or V funnel test. The L box test measures the time it takes the concrete to flow through an assembly of reinforcing bars into a trough. The V funnel test measures the speed at which a given volume of concrete placed in a cone flows from the outlet at the base of the cone. The outlet diameter is set at a proportion of the cone diameter at the top of the funnel. The funnel length is 650 mm. Many other tests exist, some showing potential to be adopted for possible standardisation. The long-term aim is to develop one or two simple yet inexpensive tests that give consistent results when carried out rapidly on site.

It is important that the properties of SCC are maintained for an adequate period of time, around 90 min or more, after completion of mixing when the concrete is being transported and placed. SCC differs from flowing superplasticised concrete, because it has to be resistant to segregation and must not cause blocking.

Summary

There is compelling evidence that self-compacting concrete can be produced satisfactorily and from many combinations of materials. The extra cost of the high cement content and special admixtures can be justified by the greater savings on the labour cost and time-consuming activity of vibrating concrete. Such value engineering exercises at present will probably discount the use of SCC on all but the largest projects. The elimination of compaction opens up opportunity for greater automation of concrete into the construction process. At present the general adoption of SCC is hindered by the lack of user-friendly guidance and standard test methods, and by the need for specialists to interpret the results from site testing.

*Reactive powder concrete**

The following is an edited version of an article in *Concrete*, January 1999, by Marcel Cheyrezy of Bouyges SA, France.

Over the past twenty years, significant improvements have been made in the development and improvement of high performance concrete. In seeking further improvements, two lines of research have been pursued — densified concrete (HSC) and macro defect free (MDF) polymer mortars. Densified concrete systems utilise a compact granular matrix of concrete, silica fume, certain reactive fly ashes with high-range water-reducing admixtures and ultra-hard aggregates. MDF mortar develops a very high tensile strength when mixed using high alumina cements.

Both alternatives have low ductility, which can be overcome by the introduction of steel fibres. In practice this involves filling formwork panels with bulk fibres and then injecting a fluid mortar to create a dense composite matrix.

* Now branded as 'Ductal' by Lafarge.

Recently an ultra-high-strength and ductile composite concrete has been developed by Bouyges. Because of the fineness and reactivity of the dry components, the material has been called reactive powder concrete or RPC. Compressive strengths range from 200 to 800 MPa.

Basic principles of RPC

RPC is based on the principle of formulating concrete mixes with minimum defects such as air voids, micro cracking etc. to achieve high ultimate load carrying capacity and enhanced durability. A research programme was set up at Bouyges to develop a concrete with the following characteristics:

- enhancement of homogeneity;
- enhancement of density;
- enhancement of microstructure;
- enhancement of ductility;
- ease of handling for normal site use.

By eliminating the coarse aggregate fraction, applying pressure to the mix before and after setting, using heat treatment after hardening and incorporating steel fibres, these characteristics were obtained. However, the cost and practicality of applying pressure to the mix and supplying heat treatment in the field must be assessed for each application of RPC.

The mix proportions of a typical 200 MPa RPC mix are shown in Table 1.5(a). The steel fibres used were 0.2 mm in diameter and 12 mm long, and were incorporated at a dosage of 2% by volume of the mix. The RPC test specimens were mixed in a conventional batching plant and placed in formwork without the application of pressure. The material was heat treated between 1 and 3 days after the concrete had hardened. The concrete was heated to 90°C and 'cooked' for two days.

In general, the test results (Fig. 1.19 and Table 1.5(b)) confirmed that RPC exhibited linear elastic behaviour up to first crack formation for standard beam specimens, which were subjected to the 3-point loading test. Thereafter the load capacity of RPC beams increased very gradually as the beam deflections grew larger until ultimate flexure load was reached, after which the load decreased. The ultimate bending stress of the RPC test beams was more than double the first crack stress and ten times greater than the ultimate strength of conventional mortar beam specimens.

RPC conceptual design

In RPC structures primary tensile stresses are resisted by prestressing, whereas secondary tensile stresses, shear stresses and all

Table 1.5. (a) Typical composition and (b) mechanical properties of reactive powder concrete

(a)

	Fine sand	Cement	Quartz powder	Silica fume	Superplasticiser	Total water	Fibres
Content (kg/m^3)	1020	710	215	230	10	140	160

(b)

Property	Mean value	Characteristic value
Compressive strength	200 MPa	180 MPa
Flexural strength	40 MPa	34 MPa
Fracture energy	30 kJ/m^2	–
Young's modulus	50 GPa	–

compressive stresses are taken by the concrete. The high tensile strength of RPC combined with its high ductility makes conventional reinforcement unnecessary. Extensive tests carried out on 10 m long 'T' and 15 m long 'X' shaped prestressed beams have demonstrated

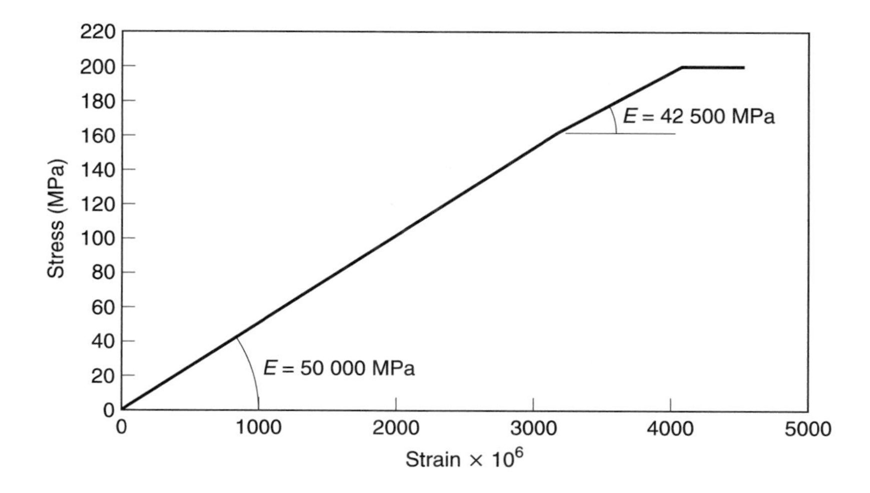

Fig. 1.19. Compressive behaviour of 200 MPa reactive powder concrete

the validity of these design principles. Extensive analysis has been conducted to validate the design rules and performance characteristics of RPC. Assurances on compressive strength and tensile strength can be given.

Regarding long-term behaviour, RPC exhibits very little sensitivity to creep and shrinkage. Moderate shrinkage is observed during heat treatment, after one to three days. Thereafter there is absolutely no further shrinkage movement. Basic creep strain is less than 10% of conventional concrete values. These properties of RPC eliminate most of the design problems associated with time-dependent strains in concrete. It is evident that, when RPC is fully developed, a new family of beam shapes for structural use will emerge which will closely match steel sections for economy of mass, reducing the self-weight of conventional prestressed concrete beams by a considerable margin.

RPC structures

A number of structures have been designed with RPC using the basic principles that have been outlined. Two of them are described.

Sherbrooke footbridge. The first major structure to use 200 MPa RPC is a 60 m span truss footbridge in Sherbrooke, Canada (Fig. 1.20). The 30 mm thick, 3.30 m wide RPC walkway slab acts as the stiffening top chord of the 3 m deep truss. The struts and tension ties that connect the top and bottom chord members are thin-walled stainless steel tubes filled with RPC. Confining RPC in external tubing enhances the compressive strength to 350 MPa. The bottom chord comprises twin RPC beams that are 320 mm wide × 380 mm deep.

The truss structure was fabricated in 10 m long segments and erected on falsework to complete the span. External post-tensioning tendons, placed longitudinally, pass under the top chord slab, whereas the 30 mm top slab itself was transversely prestressed. No rebar was used for any part of the structure, not even for distribution steel. The research and design development for Sherbrooke bridge was carried out at Sherbrooke University.

Large-span spherical dome. A dome with a clear span diameter of 120 m has been designed using 200 MPa RPC. The structure is supported on a series of post-tensioned arches springing from a prestressed ring beam. The roof canopy is made from plates of 30 mm thick RPC that span between the arch beams and overlap one another like roofing tiles. The beams are made in match-cast segments and placed on centring falsework before being post-tensioned. The overall depth of the roof structure, including the beams, is 100 mm.

(a)

PRE-STRESSED
REINFORCED CONCRETE
Ductal
STEEL

WEIGHT OF BEAMS (kg/linear meter)

(b) 140 112 467 530

Fig. 1.20. (a) Sherbrooke footbridge. (b) Beam sections: comparison of ordinary prestressed concrete, steel and RPC prestressed concrete beams

Compact reinforced composite

Compact reinforced composite (CRC) is a fibre-reinforced high performance concrete developed by Aalborg Portland in Denmark in 1986. CRC has been the subject of a number of research projects, and one of the properties that has been investigated is its bond strength. Due to its large content of micro silica and steel fibres, CRC has exceptional bond properties, which means that CRC can be used to glue together narrow joints between two reinforced concrete sections, similar in concept to welding steel sections.

CRC applications were first developed for precast construction, but the material has been modified for in situ applications and is commercially known as CRC JointCast. It is sold as a binder, with suitable local sands and aggregates blended in the mix for each application. As CRC JointCast is used in critical parts of the structure where quality control is very important, it is provided as a dry mortar comprising binder, sand and steel fibres. Water is the only ingredient added on site. Some of the tests carried out have been described in papers by Nielsen (1993), Nielsen *et al.* (1996) and Aarup and Jensen (1998), and some of the early applications by Jensen *et al.* (1995). However, as there are many current applications for CRC JointCast a list will be given in the following of the more recent activities.

Aalborg University

CRC JointCast has been used for two buildings at the university in Aalborg, each building covering about 7000 m^2 in floor plan area. Before each application, extensive testing was carried out as part of a development project sponsored by the Danish Ministry of Education. The tests included a fire resistance test which established that the joint could sustain a standard fire for more than 90 min. The joint used (Figs. 1.21–1.23) was 100 mm wide and was applied in the section with maximum moment. Straight 8 mm bars with a characteristic yield strength of 550 MPa were used.

Repair work

CRC JointCast has also been used for repairing the Øresund bridge, where a number of reinforcing bars were accidentally cut. Using CRC

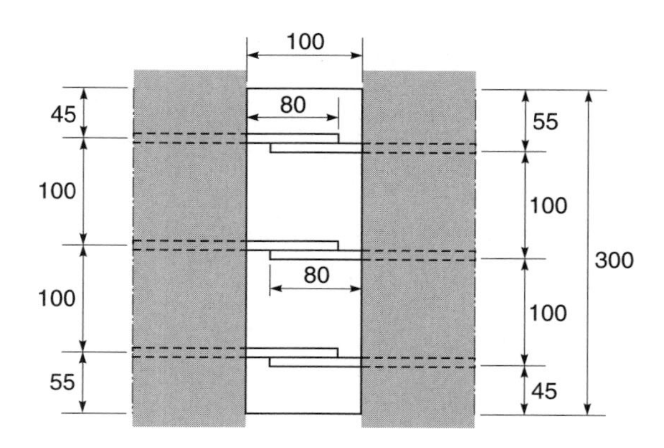

Fig. 1.21. Aalborg University: plan view of part of the slab connection

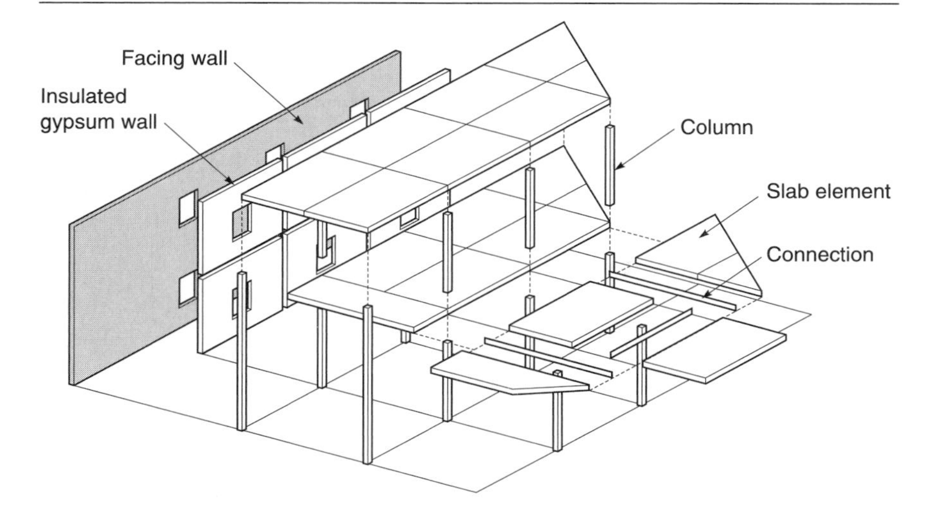

Fig. 1.22. Building system used at Aalborg University

Fig. 1.23. Slab joint prior to casting with compact reinforced concrete

JointCast, it was necessary only to cut out a relatively small area of the structure and introduce lap bars, whereas it would have been necessary to make a much larger repair patch if conventional concrete had been used.

Frame joints

Another type of application was used for a factory in Sdr. Felding, where CRC JointCast was used for connecting parts of the frame. The building had a floor area of 10 000 m^2 but, as the joints were quite small, only 4 m^3 of CRC JointCast was used for the entire project. A picture of the joint being cast is shown in Fig. 1.24.

Bridge joints

Together with Strängbetong, Sweden, an investigation is being carried out to verify whether CRC JointCast can be used with advantage to replace a rather cumbersome type of joint traditionally used in precast bridges. The difference between the traditional joint and the simpler CRC joint using straight bars is shown in Fig. 1.25. The corroborating tests were carried out at Chalmers University, Sweden, and included static as well as fatigue tests. The static tests were concluded in December 1998 and the fatigue tests were carried out during September and October 1999. The joint was typically stronger than the equivalent monolithic beam, except for the case that simulated

Fig. 1.24. Casting of frame joint using CRC

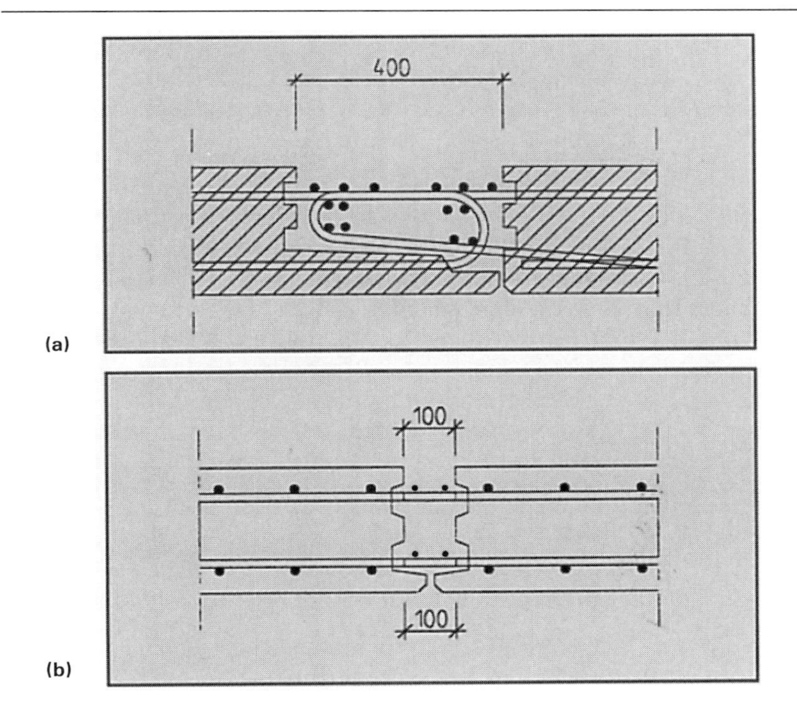

Fig. 1.25. (a) Traditional joint for precast concrete bridges and (b) an equivalent joint in CRC JointCast

the accidental omission of the transverse bars. In that case the failure mode was an anchorage failure. However, failure occurred after yielding of the rebar at approximately 96% of the ultimate load.

Testing

Beam tests, Japan. Shimizu, Japan, has carried out several tests on CRC JointCast. The first of these tests concerned a joint between two beam parts. The joint had a width of 250 mm and a lap length of 190 mm, using 19 mm diameter bars. When tested, the beam with a CRC joint had the strength and ductility corresponding to that of the monolithic beam.

Pull-out tests, Japan. Most of the pull-out tests for assessing the performance of reinforcing bars or prestressing strands have been carried out on diameters of about 20 mm. The reinforcement used in Japan consists typically of quite large diameter bars, owing to seismic design provisions. Shimizu has carried out a number of pull-out tests on bars of up to 51 mm diameter, in which the measured values of bond strength varied from 26.3 MPa for 51 mm bars to 79.5 MPa for 19 mm bars. The embedment lengths used were typically three times the bar diameter. It was observed that the calculated bond strength decreased with increasing embedment length.

Fatigue and other tests. Fatigue tests have previously been carried out using small test specimens with a low rebar cover. These tests are currently being complemented by a large study carried out at Aalborg University on larger specimens where the type and dimension of rebars are varied. Apart from these tests, CRC JointCast has been used in the BRE demonstration project at Cardington, England, and a number of tests on beam connections and pull-out specimens have been carried out at CBL, Aalborg Portland. These tests have included verification tests that joints can achieve adequate strength in only three maturity days as well as tests on the effects of lap length and of different types of British, Swiss and Danish rebar.

CRC has been used very recently to form thin precast balconies and staircases (Fig. 1.26).

Lightweight concrete

Foamed concrete

The idea of adding a foam to cement grout or mortar is not new, but the development of equipment to disperse foam into concrete mixes, in small or large quantities, is relatively recent. Foamed concrete is a lightweight material formed by entrapping or generating small bubbles of air into a Portland cement mix by mechanical or chemical means. Foamed concrete with densities from 400 to 1800 kg/m^3 can be produced at strengths in the range of 1 to 12 MPa. The cement content to produce this range of strength is generally between 300 and 400 kg/m^3 .

Foamed concrete is obviously lightweight; it is also durable, fire resistant, frost resistant and a good thermal and acoustic insulator. It is particularly successful as a single-shot trench fill for pipelines, conduits and sewers crossing roads and pathways. Unlike the traditional 'repetitive' repair process — where the surface has to be reinstated when the backfill has consolidated — foamed concrete backfill does not settle. It is easily pumped and poured into place, it needs no compaction, does not segregate and does not create a stiff beam like a ridge under the old road.

The main cause of damage to road pavements is from impact by heavy goods vehicles and opening up of the road surface by excavations carried out by private utilities. These excavations not only cause direct

(a)

(b)

Fig. 1.26. (a) Balcony slab, (b) staircase structure cast with CRC in a housing project.

damage to the road, but their reinstatement is often poor and is extremely vulnerable to the actions of vehicles and the weather. Settlement of the backfill and consequent realignment of adjacent trench walls means that surfacing is damaged and needs constant patching to maintain a weatherproof and even road surface.

In four pioneering projects carried out in 1988 by British Gas, foamed concrete was successfully used to overcome problems previously experienced in reinstating narrow (125–150 mm wide) trenches (Fig. 1.27). In the Netherlands foamed concrete has been developed as a porous slab, which is covered with an artificial playing surface for outdoor sports fields, and for self-compacting porous backfill with good drainage properties for roadside soakaways and drains.

Foamed concrete has a wide range of potential uses, the main ones being:

- ☐ trench reinstatement (Fig. 1.28);
- ☐ blinding layer;
- ☐ roof insulation and sound absorption;
- ☐ underfloor insulation;
- ☐ inert void filler;
- ☐ porous slab suitable for all-weather sports field.

The mechanical production of foamed concrete involves adding a foam into a base mortar and then discharging it into place. The foam

Fig. 1.27. Placing foamed concrete

Fig. 1.28. Trench reinstatement using foamed concrete

is produced by a foam generator in which a dilute foam concentrate is combined with compressed air and forced through a restriction to give a predetermined amount of foam. The generator can be hand-held like a gunite nozzle or fixed to discharge directly into a mixer. The density of foam produced in this way is typically between 25 and 80 g/l. The most commonly used foam concentrates are based on protein hydrolysates or synthetic surfactants. The foams are formulated to produce air bubbles that are able to resist the physical and chemical forces imposed on them during mixing, placing and hardening of the concrete. The majority of the foam bubbles are between 0.3 mm and 1.5 mm in diameter.

For batch production, a set amount of foam is added to the base mix, according to the density and compressive strength required of the hardened concrete. For batches of up to 6 m^3 a normal truck mixer or mini truck mixers can be used. For continuous or intermittent production a machine will produce a controlled output of foam to a known flow of the concrete base mix. Some machines will take the base mix from a ready-mixed truck, while others produce the base mix and foam, intermixing the two before discharge. Output can vary from 25 m^3/h down to 6 m^3/h. Foamed concrete can be pumped both horizontally or vertically or poured in place. If re-excavation is necessary, foamed concrete is simple to remove.

Foamed concretes with dry densities below 600 kg/m^3 usually consist of cement, foam and water, or cement, PFA, foam and water. Higher densities are produced by adding building sand or concreting sands with a maximum particle size of 5 mm. Research has indicated that a higher strength is obtained using fine sands with a maximum grain size of 2 mm, with 60–95% passing the 600 μm sieve. Coarse natural aggregates cannot be used to produce foamed concrete because they would segregate in the lightweight foam mix. It may be possible to use lightweight aggregates with a similar density to the foamed concrete. Ordinary Portland cement is used as the binder in most foamed concretes, but some manufacturers claim that rapid-hardening Portland cement gives higher strength for a given cement content, as well as a faster development of strength. Cement contents for most commonly used mixes are between 300 and 375 kg/m^3. For optimum results the water/cement ratio of the base mix should be kept fairly high to provide high workability lying between 0.5 and 0.6. A base mix that is too dry and stiff is liable to extract water from the foam and cause it to collapse.

The water used in the mix should always be clean and potable. This is particularly important when using protein-based foaming agents, as any organic contamination could have an adverse effect on the quality of the foam produced. A number of factors can affect the compressive strength of a foamed concrete mix. For example, in trench reinstatement work the aim is to design a mix to achieve an early strength of 1 MPa, as well as the specified 28-day strength, to allow early start to final surfacing. For a given mix, the density of the freshly placed foamed concrete will give a reliable indication of the expected strength.

Curing and testing of cube specimens for strength assurance is not quite as straightforward as normal concrete cube testing. Foamed concrete cubes stored in a fog room or under water will absorb a considerable amount of water which can raise the density by over 20%. It is best if test cubes are sealed in polythene for curing, to reduce the risk of absorbing excess moisture. The density of the cube as it is taken out of the test mould, known as the 'stripped density', is closer to the actual density of the foamed concrete. It is recommended that 150 mm cube moulds are used for making foamed concrete test specimens and that they are crushed in a compression testing machine with a calibrated load scale to BS 1610. The failure load of foamed concrete test cubes may be as low as 40 kN or even less.

Test cubes made on site should be carefully stored for up to 24 h, with a minimum of movement once they are made. Any movement, for example, gentle rocking or vibration while the material is still fluid, will affect the density of the sample. A 'wet density' test, where a known volume of foamed concrete is poured into a cylinder and weighed, would also give a good indication of the placed density of the foamed concrete, for site quality control.

Some typical properties of foamed concrete are given in Table 1.6.

Table 1.6. Typical properties of foamed concrete (Van Dijk, 1991)

Dry density: kg/m^3	Compressive strength: MPa	Thermal conductivity: W/(m K)	Modulus of elasticity: GPa	Drying shrinkage: %
400	0.5–1.0	0.10	0.8–1.0	0.3–0.35
600	1.0–1.5	0.11	1.0–1.5	0.22–0.25
800	1.5–2.0	0.17–0.23	2.0–2.5	0.20–0.22
1000	2.5–3.0	0.23–0.30	2.5–3.0	0.18–0.15
1200	4.5–5.5	0.38–0.42	2.5–4.0	0.11–0.09
1400	6.0–8.0	0.50–0.55	5.0–6.0	0.09–0.07
1600	7.5–10.0	0.62–0.66	10.0–12.0	0.07–0.06

Expanded shale granules with a promising future

Lightweight aggregate concrete has been a structural engineer's dream as a concept, but the practical reality of its varied handling characteristics, its relatively low strength, low elastic modulus and shear capacity has largely relegated it to concrete block making and composite slab toppings. The concept of low density, high strength concrete has immediate benefits in reducing floor slab depth and dead load. Such benefits have been exploited successfully in composite metal deck construction, but not in reinforced concrete floor construction. It is ironic that one of in situ concrete's major disadvantages, of high comparative dead weight, has been further disadvantaged by lightweight concrete.

It has been common knowledge that lightweight aggregate concrete using pellitised PFA, commercially known as Lytag, has achieved concrete strengths of 70 MPa, but it requires a substantial volume of cement with a hefty cost penalty. By packing a high volume of cement into the mix the benefits of a low density concrete are also lost. Moreover, there is little to be gained by designing lightweight concrete reinforced with high yield steel bars. Both the Euro code and BS 8110 impose reduction factors on slenderness ratio, deflection and shear for lightweight concrete structures on the basis of normal concrete design data. For deflection, the reduction is 0.85, for shear it is 0.8, and for torsion it is 0.7. Whilst lightweight concrete has superior fire resistance to normal concrete, in BS 8110 it has a lower classification on durability so that cover has to be increased by 10 mm over the stated values. It is possible to carry out particular tests to demonstrate that the code is unnecessarily conservative, but it is an uphill struggle to convince clients and their advisers who view such suggestions as high risk.

The net result of all this is that the reduction in dead load of 30% is negated by the increased section depth necessary to overcome the code requirements on shear, deflection and cover. There is little difference in dead load between lightweight and normal concrete, but somehow with lightweight concrete the slab depth is greater and so is the cost.

Now, if we consider the use of lightweight concrete with prestressing and post-tensioning, the arguments are very persuasive. Prestressed lightweight concrete does not require a reduction for elastic modulus, nor for torsion. Nor is the shear capacity reduction factor a criterion for increasing slab depth. What is critical in post-tensioned concrete is the shrinkage movement of the slab, the allowance for prestress losses, the tensile strain capacity and the compressive strength of the concrete. The amount of prestress that is applied to the slab is proportional to the total dead load carried, i.e. self-weight plus imposed static load. The load carried by the prestress maintains the slab at zero deflection. An allowance is then made for the imposed live load, which is carried by normal reinforcement. The longer the span, the greater is the dead load and the more efficient is the design.

The benefit of a low density concrete lies in reduced dead load, resulting in less prestress and thinner slab depth. As prestressing reduces slab depth by 20% over the equivalent reinforced concrete slab, so we now need to search for a lightweight aggregate concrete that can combine high strength with low density. In BS 8110, lightweight concrete is taken as having an oven-dry density of less than 2000 kg/m^3. Any concrete with a density greater than that is assumed as normal weight.

The gift of the Jurassic

Shale is a precious gift of nature, originating many millions of years ago as 'Lias' in the organic sediments of the Jurassic seas. This compacted fossil rock is extracted by open pit mining in Bavaria, Germany, between Nuremburg and Munich. The shale is quarried and then transferred by tipper truck to the processing plant some 6 km away. It is crushed, dried and milled into a powder of less than 250 μm. The shale powder is then conditioned, mixed with water, pelletised into small round granules and then coated with a fine limestone dust. The pellets are graded into sizes and then conveyed to a three-stage rotary kiln. This technique enables the amount of expansion of the pellets to be controlled and the density and size of the granules to be engineered and graded precisely. The coating of lime powder increases the amount of surface vitrification, forming a dense impermeable coating when the pellets are heated to 1200°C. The product is known as Liapor.

Liapor is like no other man-made aggregate. It is a lightweight shale aggregate with an aerated core and a high compressive strength, available in sizes from a sand to 4–8 mm and 8–16 mm diameter aggregates (Fig. 1.29). The density can be varied from 325 to 800 kg/m^3 according to the application and usage. It has an ideal spherical shape, with a closed surface skin that has a slightly roughened texture. It is completely frost resistant and can be stored in the open under any weather condition. Its impervious outer skin means that water absorption is not high, and it does not need special admixtures or vacuum soaking beforehand. In fact it can be treated like a good quality natural aggregate. Table 1.7 gives the sizes and densities of the various grades, while Table 1.8 shows their uses and properties. Structures incorporating Liapor aggregate are shown in Figs 1.30 and 1.31.

Concrete using a Liapor 8 graded aggregate, comprising 4–12 mm aggregates with a bulk density of 800 kg/m^3 and combined with natural sand, has achieved a compressive strength of 80 MPa at a design density of 1800 kg/m^3. For lower concrete density, the natural sand is replaced with a Liapor sand. A concrete density of 1400 kg/m^3 can be achieved with a Liapor grade 6 coarse aggregate and a Liapor sand, to give a 40 MPa compressive strength. Clearly high strength, low density concrete can be fine tuned and engineered to suit the structural characteristics of a particular project. Harnessed to post-tensioning, this must be the most economic and dynamic structural floor system of the future.

Fig. 1.29. Liapor expanded shale aggregate

Table 1.7. Sizes and densities of Liapor expanded shale aggregate

Class	Grain sizes: mm	Bulk density: kg/m^3	Grain density: g/cm^3
Liapor 3	4–8 and 8–16	325 ± 25	0.55–0.65
Liapor 4	4–8 and 8–16	400 ± 25	0.70–0.80
Liapor 5	4–8 and 8–16	500 ± 25	0.90–1.00
Liapor 6	4–8 and 8–12	600 ± 25	1.05–1.15
Liapor 7	4–8 and 8–12	700 ± 25	1.25–1.35
Liapor 8	4–8 and 8–12	800 ± 25	1.45–1.55
Liapor-sand L	0–4	600 ± 50	1.30–1.50
Liapor-sand K	0–4	700 ± 50	1.50–1.70

Note. Table shows standard production. Special products upon request.

Lightweight prestressed concrete—design notes

BS 8110 and the draft European code permit the use of lightweight aggregate concrete in prestressed concrete. On the whole the design refers to the code requirements for reinforced concrete, with changes that relate to the tensile strength of the concrete.

Reinforcement cover for durability is the same as for normal weight concrete, but with an additional 10 mm. For fire protection, the cover is reduced relative to normal weight concrete. Permissible stresses in tension are based on the tensile strength of concrete, expressed as a function of the square root of the compressive strength. For lightweight

Table 1.8. Uses and properties of Liapor lightweight aggregate concrete

Main uses	Density: kg/m^3	Thermal conductivity: W/(m K)	Compressive strength: MPa
Loose in bulk	300–800	0.10–0.20	–
No-fines lightweight concrete for concrete blocks of all kinds; levelling concretes	500–1000	0.15–0.30	2–10
Fully compacted lightweight concrete for the complete concrete range in building construction and civil engineering	1000–1800	0.40–1.50	10–80

Fig. 1.30. BMW building, Germany

Fig. 1.31. Offshore oil platform, Norway

concrete this value is multiplied by a reduction factor of 0.8 in BS 8110. In ACI 318, the US equivalent of BS 8110, no reduction factor is required.

Shear capacity is always a sticking point with lightweight concrete. The current UK code values are based on extensive tests carried out on lightweight concrete in the US in the 1960s. As a result a reduction factor of 0.8 was imposed in BS 8110 and that figure has not changed. Lightweight aggregate concrete has improved and advanced since the 1960s, with higher strength aggregates being used. Tests carried out by Clarke in 1987 on modern lightweight concrete, with densities ranging from 1700 to 2000 kg/m^3 and strengths from 30 to 55 N, have led to the conclusion that without shear reinforcement a reduction factor of 0.9 would be more appropriate. Clarke recommended that for beams with links, no reduction factor was necessary. Replacing the 0.8 value in the code with 0.9 would lead to an effective reduction factor of 0.94 with nominal shear reinforcements.

In the draft European code the shear capacity of a prestressed concrete beam is treated as a special case of shear in a reinforced beam. The total shear capacity is a function of the tensile capacity of the concrete and the average level of stress due to the prestress. Thus, for reasonable levels of prestress, the difference between the shear capacity of lightweight aggregate concrete and normal weight concrete beams will be small.

Prestress losses in changing from normal concrete to lightweight aggregate will affect only the elastic, creep and shrinkage deformation values. BS 8110 gives guidance on the elastic modulus to be used in the absence of actual test data, multiplying the normal weight value by $(w/2400)^2$ where w is the lightweight concrete density. BS 8110 gives no specific guidance on suitable values for creep and shrinkage, but Spratt suggests that the creep of structural lightweight concrete is only slightly greater than that of normal concrete. The inclusion of natural sands in the mix will reduce creep. For general design purposes Spratt suggests that creep should be taken as between 1.3 and 1.6 times that of normal weight concrete.

Coloured concrete

Synthetic colour pigments for concrete

With the advances in coloured concrete technology, it is now possible to supply ready-mixed architectural concrete with greater durability and better integral colour than ever before. Site-cast architectural concrete has been attempted over many years in the UK with mixed success. The continuing lack of interest in this method of construction amongst architects and contractors stems from the lack of confidence expressed in quality and consistency of a ready-mixed supply, the poor quality of site workmanship and the lack of knowledge about long-term performance. There are too many well publicised accounts of concrete failing as an exterior surface in architecture. The jobbing architect and specifier with no great understanding of concrete, no feel for its plastic and sculptural possibilities like Tadao Ando, David Chipperfield, Ricardo Bofil or Betrand Goldberg, for example, would perhaps be wiser to stay clear.

Generally, where an architectural finish for concrete is being considered, a precast finish has been specified. The fact and fiction about precast production and site workmanship are discussed later. Suffice to say it is possible to achieve a good standard, factory finish with site-cast concrete. But in order to do so the designer must appreciate the technology that leads to good finishes and recognise the mistakes that can lead to substandard work. Good concrete architecture is about good design and that means knowledge of material properties.

A concrete that is colour stable, consistent in quality and supplied by ready-mixed truck to a site can significantly reduce the construction cost, the weight penalty of factory-made precast concrete and long lead-in times. Two routes are available to the architect/designer when deciding on the specification for visual concrete. There is the choice between totally natural materials or synthetic pigments combined with a base concrete mix. For the record, white cement is not a natural material; it is a synthetic powdered rock manufactured just like Portland cement, except that all the iron oxide and magnesium oxide are removed in the processing. White cement is a rapid-hardening cement, much finer in particle size than Portland cement, with a faster setting time, and it costs twice the price of ordinary cement. All the white cement used in the UK is imported from Denmark.

To create a white concrete, one option could be to combine a white cement with white china clay sand and, say, a coarse Ballidon limestone. The other option could be to use a blastfurnace slag cement, a titanium dioxide pigment, a china clay sand and a light-coloured coarse gravel. The white cement route will generally give a whiter concrete surface finish immediately on exposure. The difference in whiteness may not be much, but in about two to three years' time it could be difficult to tell them apart as the grime and dirt begin to speckle and darken the original face. This assumes that the concrete surface is left with the original veneer of cement paste for the rest of its life. How wrong it would be to assume that this surface would not change and would not need surface treatment and cleaning, at regular intervals!

Whether there is an 'as struck' finish or an exposed aggregate finish to a concrete, it is important that the basic mix design and colour are correctly specified. The option of using synthetic pigments and a base concrete mix in site-cast production is reviewed.

What colours concrete?

The mechanism that gives colour to concrete is the light absorption qualities of the finest particles in the mix. For ordinary concrete mixes, the cement particles are the finest particles and therefore the colour of the cement will dominate. The final colour of the concrete is largely dependent, therefore, on the cement colour and to a lesser extent on the fine aggregate colour, if the surface is left with an 'as struck' finish.

If a pigment is introduced into the mix, the pigment colour will dominate the final colour because it is ten times finer than cement. The amount of pigment needed to colour a concrete will vary according to the cement content, the pigment type and the method of incorporating it into the mix. For instance, yellow pigments are more water absorbent than red pigments and have a higher water demand, thus increasing the dosage for the pigment. There are two types of pigment on the market — organic and synthetic types. What are they, and what are the differences between organic and synthetic pigment types?

Naturally occurring pigments are inert oxides and hydroxides of iron and titanium and copper complexes of phthalocyanine, found in mineral rocks. They range in colour through red oxide, brown oxide and yellow oxides. The full description of pigments specified for use in concrete or mortar is given in BS 1014. A pigment may be defined as a fine dry powder or an aqueous suspension or slurry of powder, virtually inert to the ingredients of concrete, intended to impart a specific colour to the finished product. The mineral rocks containing raw pigment deposits are quarried, heat treated, crushed and then ground to a flour consistency to create industrial pigments. And, like sands and coarse aggregates, pigments do have unique characteristics: some are needle shaped, some are spherical, while some, like the phthalocyanines, are hydrophobic.

Red oxide pigments from different sources may have the same particle size but their bulk density and water absorption may differ significantly. Some have a bulk density of 1500 kg/m^3 and water absorption of 20 ml/100 g and others a bulk density of 900 kg/m^3 and a water absorption of 35 ml/100 g. Yellow iron oxide pigments have needle-shaped particles that can vary in bulk density from 500 to 800 kg/m^3 and in water absorption from 50 to 90 ml/100 g. Green or blue phthalocyanines are hydrophobic and have a particle size ten times finer than a red oxide pigment, and a bulk density of

500 kg/m^3. The various pigment mixes of red, yellow and brown oxides are blended to create intermediate colours, and have to be carefully batched so that the bulk density and water absorption are known and can be adjusted in the mix.

It is for this reason that synthetic oxide pigments were developed by Bayer to create a more homogeneous pigment particle with a more uniform bulk density and water absorption. Synthetic pigments are more intense in colour than organic pigments, and have excellent long-term colour stability. The pigments are produced in primary colours — red, black, and yellow — by the aniline or Penniman–Zoph process. In the aniline process nitrobenzene is reduced to aniline in acid solution, using fine iron filings as the reducing agent. During this process the iron filings are oxidised to produce an iron oxide which eventually turns a black–grey colour. By controlling the oxidation it is possible to produce black and yellow slurries with a high tinting strength. After washing and filtration the slurry is dried out to produce black and yellow pigments or heat treated and calcined to produce red oxide pigments. In the Penniman–Zoph process, iron filings captured from scrap sheet metal are dissolved in acid solution in a hydrolysis process, involving the oxidation and hydrolysis of iron sulphate in the presence of metallic iron, to produce an iron oxide yellow pigment with needle-shaped particles. A range of brown pigments is blended from these three primary colours. Green and blue pigments are processed from copper oxides and cobalt deposits, and are very expensive. Synthetic pigments should be preferred for all architectural in situ concrete work.

Colour shades for concrete

Using synthetic oxide pigments in concrete with a high tinting strength means that colour saturation is usually achieved at 5% pigment/cement ratio. Any higher dosage will not increase the intensity of colour of the concrete. However, for yellow pigments the saturation level starts at a higher dosage, between 8 and 9%, when used with ordinary grey cements. If a light coloured cement such as slag or white cement were used it would need less pigment to reach colour saturation. A light coloured cement is important where pastel shades are required for concrete.

Colour intensity of pigments is also influenced by the method of incorporating pigments into the mix. With blending or ball mills to grind cement and pigments together, only half the concentration of pigment is needed in comparison with adding it to the mix separately. However, the high cost of installing ball mills and silos to store coloured cements, coupled with the relatively small demand for colour in concrete, has not yet made this option commercially viable in the UK. Thus, for most coloured concrete production, pigments are introduced into the mix by dispensing them with mixing water as a powder in water-soluble bags or through a plasticising admixture suspension.

One of the big headaches of coloured concrete and mortar production has been the problem of lime bloom or efflorescence caused by the carbonation of calcium hydroxide that migrates to the surface, forming white deposits. For pigmented and architectural concrete this condition could spoil the finish, and ways to minimise and control lime bloom should be seriously considered. In addition to a good concrete mix design with a low water/cement ratio, prevention of rapid drying out of concrete in the early days is the best way to eliminate secondary efflorescence, which can occur throughout the life of the concrete until it has fully carbonated. A surface coating of a transparent vapour-permeable membrane that is non-yellowing and does not break down under ultraviolet light would be beneficial. Primary efflorescence problems are more difficult to control and usually occur on the first day or so after the formwork has been removed. Keeping the exposed concrete surface wetted and at high humidity is likely to increase the risk of primary efflorescence, but doing so will reduce or eliminate a secondary efflorescence problem, which is the worse condition. The solution for coloured concrete is to use a non-staining, removable curing membrane as soon as the formwork is struck and to avoid damp curing. The curing of coloured concrete is discussed later.

Pigments suitable for concrete are usually dry batched and supplied in water-soluble bags or freeze-dried granules, in cubic metre dosage units. Where there is a large volume demand — as there is for ready-mixed pigmented cement mortars — pigments can be supplied in liquid suspension in large containers. The liquid suspension usually has a pigment concentration of 60–70%, the rest being made up of water, dispersing agent, preservative and other admixtures to ensure the pigment has a high viscosity and is well mixed. The pigment suspension is wet-metered by piston pumps or inductive flow meters and discharged directly into the ready-mixed truck. Quality control in the factory is maintained by frequent checks of batch samples, using a smear test. In addition, control cubes should be made in the test laboratory to check the finished colour of the concrete before starting ready-mixed production.

For any ready-mixed production of architectural concrete the following essentials are required:

- ☐ Cement content should not be less than 350 kg/m^3.

- ☐ Cement source and colour must remain the same throughout. Any change in cement colour will affect the finished colour.
- ☐ Water/cement ratio must be constant and not greater than 0.5.
- ☐ Aggregate/cement ratio must not exceed 6:1 for a good mix.
- ☐ Trial mixes to check final colour and surface finish should be made.

It goes without saying that if there is good control of mix proportions and principally the cement content, then good coloured concrete will result. Large variations in water content and water/cement ratio between batches of concrete will vary the pigment concentration and affect the finished colour. The higher the water content the lighter the shade, the lower the water content the darker the shade. The key to the tolerance threshold of colour variation and what can be detected by the eye was researched by Bayer some years ago when the company conducted colour trials on concrete block paving bricks. There is a clear relationship between the colour shades of concrete and the water/cement ratio—the higher the water cement ratio the lighter was the colour. A group of concrete paving blocks was batched for the test, each with a different colour pigment. Within each pigment group the water/cement ratio was varied while the cement content and pigment concentration were kept the same. The ratios tested were between 0.34 and 0.42. The conclusion drawn from the test confirmed that variations in w/c ratio between 0.34 and 0.38 had little influence on colour shade. There was a noticeable difference of colour surface between blocks made with a w/c ratio between 0.40 and 0.42 and those with a ratio between 0.34 and 0.36. There was not much difference between concretes with a 0.38 or 0.42 w/c ratio.

What does a fluctuation between 0.36 and 0.42 mean for a mix with a cement content of 300 kg/m^3? According to Bayer the water content will vary from 102 to 126 l/m^3. The question now is, can a ready-mixed company keep the water content between, say, 114 and 126 l/m^3 for every colour mix? The answer is emphatically yes if the target w/c ratio is 0.40 and the moisture content of the aggregates is constantly monitored.

Good weigh-batching and control of aggregate moisture content are essential for ready-mixed architectural concrete production. Whether the concrete is wet batched or truck mixed, the following steps should be followed:

1. Batch a minimum of 3 m^3 in a 6 m^3 mixer drum.
2. Add 50% of the water and ribbon-feed aggregates, cement and pigment.
3. Mix thoroughly before adding the remaining water.

4. For dry-batched concrete, truck mix for at least 15 min. For wet-batched concrete the time will be less as the truck mixer acts as an agitator.
5. Colour test the mix before discharge, by sampling from the front and back of the mixer, using a slump test.

Concrete handling

Workmanship is the 'Achilles' heel' of in situ concrete construction. No matter how well the concrete is mixed and kept at constant w/c ratio and colour consistency, this will be of academic interest if the same care and thought is not given to handling, compacting and curing the concrete on site. As ground floor slabs and flat surfaces are easier to work on than formed faces of vertical panels and walls, workmanship guidelines will centre on the more difficult application.

Wherever possible, cast vertical surfaces to their full height between designed construction joints. Tremie the concrete into position down the body of the formwork; avoid free-falling it from the top of the forms, or causing it to rebound against rebar and the formwork face on the way down. Changes in water/cement ratio can be caused by not observing these simple procedures.

Pour the concrete in definite pour planes not more than 0.5 m high. Place the first plane along the full length of the formwork, melting down any peaks lightly with a poker vibrator. The first pour plane should be roughly level all the way along. Now, using a constant amplitude, high frequency poker, vibrate and compact the depth of the first pour plane, keeping the poker equidistant away from each of the formwork faces and at an even distance from the previous position. The poker interval should be estimated by the frequency of vibration and radius effect of a poker in cohesive concrete.

Replacement aggregates for concrete

Recycled aggregate concrete

(with reference to BRE Digest 433)

Recycled aggregates from crushed concrete and brick masonry have been used in the UK for many years as general hardcore and fill material, usually for making temporary site roads or hardstanding platforms for piling rigs and other heavy machinery. Such aggregates have not been used as a replacement material for natural aggregates in concrete or road sub-bases because up to now there has been little quality control and testing carried out to show that the material complies with the required performance standards. Recent improvements in the grading, assessment and specification of recycled aggregate will enable greater confidence in specifying such materials for use in structural concrete and road base construction.

Recycled aggregates are defined as materials resulting from the crushing of demolition waste and reprocessing of construction waste created by over-ordering, sub-standard or damaged materials on site. Recycled materials are predominantly crushed concrete and brick masonry. They may contain significant quantities of deleterious materials such as wood, metal reinforcement, plastic and trace elements that are harmful to concrete. Crushers, separators and screens are used to remove the reinforcement, crush the concrete to size and grade the aggregates. The handling and processing cost of recycled aggregates will tend to make the material more expensive than locally available natural aggregates.

The environmental arguments for the greater use of recycled aggregates are not conclusive, apart from the obvious benefits of limiting the excavation and quarrying of natural aggregates on greenfield sites. Where natural aggregates are locally available, there appears to be no reduction in net CO_2 emissions by converting to recycled aggregate concrete. There is a clear advantage in using recycled aggregate concrete when natural materials have to be transported from some distant location. However, the angular shape and mechanical properties of recycled aggregates will increase the cement content for a given workability and concrete strength, in comparison with rounded or irregular natural aggregates. This will tend to increase the environmental impact of using recycled aggregate as a replacement for natural aggregates. In the long term, as quarrying and winning of land-based natural materials are restricted, recycled aggregate concrete will become more established.

Current regulations allow up to 60% replacement of natural aggregates using good quality demolition waste, which has been classified as RCA (II) in accordance with BRE Digest 433. The Digest has classified demolition waste into three grades of recycled aggregates (RCA):

- Class RCA (I) is the lowest grade and can contain up to 100% brick waste and a high level of impurities, and is not suitable for use in concrete.
- Class RCA (II) is the best grade, limits the brick waste content to not more than 10% and has a relatively low content of impurity. It is composed almost entirely of crushed concrete. The 10% fines value is usually in excess of 100 kN.

- Class RCA (III) is a mixed material somewhere between classes (I) and (II). It can contain up to 50% brick waste and may be accepted as coarse aggregate in concrete provided it does not exceed 20% of the total coarse aggregate fraction.

Quality control

Any recycled aggregate to be used in concrete must meet the requirements of BS 882. The maximum permitted levels of impurities (by weight) present in a Class (II) material are: wood 1%; foreign matter such as metal, plastic, asphalt, glass etc. 1%; sulphates 1% as acid-soluble SO_3. The chloride content should be checked if the concrete is to be reinforced or contain embedded metal. The maximum recommended grade of concrete depends on the nature and content of recycled aggregate (Table 1.9).

Each batch of unprocessed material arriving at the crushing plant should be inspected and placed in a different stockpile designated by the quality of the aggregate to be produced. Although waste transfer notes and input inspection may suffice for preliminary quality control, better quality materials benefit from inspection of the demolition sites and preparation of deconstruction plans for their recovery. The frequency of testing of crushed concrete aggregate will depend on the information available on the waste material at input.

Although a large proportion of natural aggregates used in concrete and construction are classed as 'normally reactive', some are not; therefore it is prudent to be cautious and to classify RCA as 'highly reactive' until more long-term test data become available. A calculation can be made according to BRE Digest 330, taking into account the measured alkali content of the RCA. It is likely that a high proportion of PFA or ground granulated blastfurnace slag will be required to reduce the total alkali content of concrete made with Class (II) crushed concrete aggregates. This assumes that all the conditions for alkali–silica reaction are prevalent, viz. highly reactive aggregate, high alkalinity and wetting and drying exposure conditions.

Table 1.9. Recommended maximum grade of RCA concrete

RCA (III)	C20
60% RCA (II)	C50
20% RCA (II) with 80% natural aggregates	No limit

RCA as a fine aggregate is not recommended for general use in concrete because of its adverse effect on water demand and increased levels of contamination.

Properties of RCA concrete

Frost resistance is not a problem with RCA concrete unless the source of the crushed concrete has suffered frost damage due to non-durable aggregates. But will the supplier know this and pass this information on to the customer? If in doubt, and where RCA concrete is specified under fairly severe exposure conditions, it is best to carry out freeze/thaw tests.

Due to the higher cement content of the mix, the elastic modulus will be reduced and shrinkage, creep and the coefficient of thermal expansion will slightly increase in comparison with concrete made with natural aggregates. For concrete with RCA replacing not more than 20% of the natural aggregates, the effect of these changes is marginal and can be ignored, but for any higher proportion the effects may need to be accounted for in the design.

Lunar concrete

MDF, DSF and lunar concrete

In February 1993 I wrote a paper for Jim Stevenson, then Chief Executive of the BCA, for a talk he was to give to the Civic Society in Cheltenham. I've never found out from Jim how it was received. Perhaps the ideas in it were too radical or just too incredulous to be believed and the talk was never given! Seven years on, what seemed a far-fetched idea then is now a serious contender for a lunar base structure. I am indebted to J.D. Birchall of ICI, J. Francis Young of the University of Illinois and T.D. Lin of PCA in Skokie for their inspirational thinking, illuminating papers and the brief discussion that I enjoyed with Lin some years ago.

"Familiarity breeds contempt and complacency leads to indifference. This applies so aptly to concrete the material, in the world of construction today. We walk on it, drive on it, and surround ourselves in it every day. And yet concrete has failed to capture the imagination of scientists, engineers and even those working in education." So says J. Francis Young, and who would disagree?

Concrete can be homogenised to be what you make it. Put in the hands of some clever scientists and innovative research engineers, it is surprising to see just how versatile and unique concrete can be made to be. All the developments that are highlighted in this essay have been pioneered for a reason—for instance, the commercial pressures to build more efficiently, the need to reduce energy requirements in the manufacture of building products and the need to develop materials of low weight but high strength to reduce payloads, especially if the material is being transported to the moon.

All of these products have stemmed from the unique properties of cement, in its dry powder form, its fluid state when mixed with water and aggregates, and in its hardened state as an engineered rock.

High performance, high strength

The commercial pressures to build tall buildings, or buildings with long clear spans to satisfy modern office space requirements, have resulted in columns dominating the space within a building. High strength concrete of 100 MPa was introduced to reduce column sizes without increasing the dead weight or the cost of column construction. In the case of bridge structures, whose dead weight controls the profile and depth of the span, increasing the concrete strength is beneficial in reducing the mass of the structure and the cost. High strength, high performance concrete has been used for many years for slip-forming the massive oil sea platforms that were launched from Scottish and Norwegian coastal sites. They regularly used concrete of 70 MPa, sometimes incorporating lightweight aggregates to reduce the dead weight, thus trimming down the section sizes of the platform structure.

Macro defect free (MDF) cement. A breakthrough in ultra-high strength concrete happened in the late 1970s when research chemists at ICI started looking for a wonder inorganic material that could replace organic polymers (plastics) and certain metals. The growing emphasis on non-combustibility of raw materials, concerns about depletion of the ozone layer, the rise in cost of hydrocarbons and high energy cost of production led researchers to look at the possibility of making MDF—macro defect free cement—as a replacement product. The major attraction of MDF was simply the energy cost savings in manufacture. To produce one cubic metre of cement, organic polymer or aluminium requires 10, 100 and 1000 gigajoules of energy, respectively. Clearly cement has an advantage in energy saving over aluminium and organic polymers.

Cement forms a rock-like structure with water at normal temperatures. There is no need to heat the process to high temperatures, as with ceramics, metals or organic polymers, for the chemical interaction to take place. However, cement has a low tensile and bending strength, and low fracture toughness compared to plastics and metals, which makes it unsound. The main reason for this is the large amount of voids in the material caused by air entrapped during mixing, and the pores and capillaries formed within the material when water is desiccated during cement hydration.

The removal of these large flaws from the 'cement paste' is made possible by better particle packing in the wet and dry state. Researchers found that, by introducing a small proportion of water-soluble polymers into the cement and water mix, inter-particle friction and surface tension were greatly reduced. This allowed particles to pack more closely, which significantly increased the tensile strength. Using such techniques, a deformable dough was made with a water/cement ratio of less than 0.2. By working the dough-like material with gentle compression, entrapped air bubbles are reduced further to a minute volume. The small residual voids that remain are filled by the chemical processes of hydration, so that they contract or collapse by desiccation. In this way it proved possible to reduce the voids in MDF cement to less than 0.1%.

Early results of MDF cement paste compare favourably with those of aluminium and exceed the properties of ordinary concrete by significant margins (Table 1.10). If MDF cement is reinforced by fibres such as glass, carbon or Kevlar, a composite material can be formed having a fracture toughness exceeding that of aluminium.

The processes used to fabricate products from MDF cement are quite unlike those used in conventional concrete technology. They are allied to those of inorganic polymer technology where MDF products are likely to be extruded or mould formed. Articles in MDF cement have been made that would normally be fabricated from metal or reinforced plastic, e.g. sheet metal and aluminium tubes, and include moulded forms such as door knobs, and even a suspension spring for a car (Fig. 1.32). But there is still more development work needed as the product can be unstable when exposed to moisture for prolonged periods.

Table 1.10. Properties of MDF cement paste compared with those of common materials

Material	Density: g/cm^3	Flexural strength: MPa	Young's modulus: GPa	Fracture energy: J/m^2
Ordinary cement paste	2.3	5–10	20–25	20
MDF cement	2.3–2.5	>150	40–45	300–1000
Aluminium	2.7	150–400	70	10^5
Glass	2.5	70	70	10
Wood	1.0	100	10	10^4

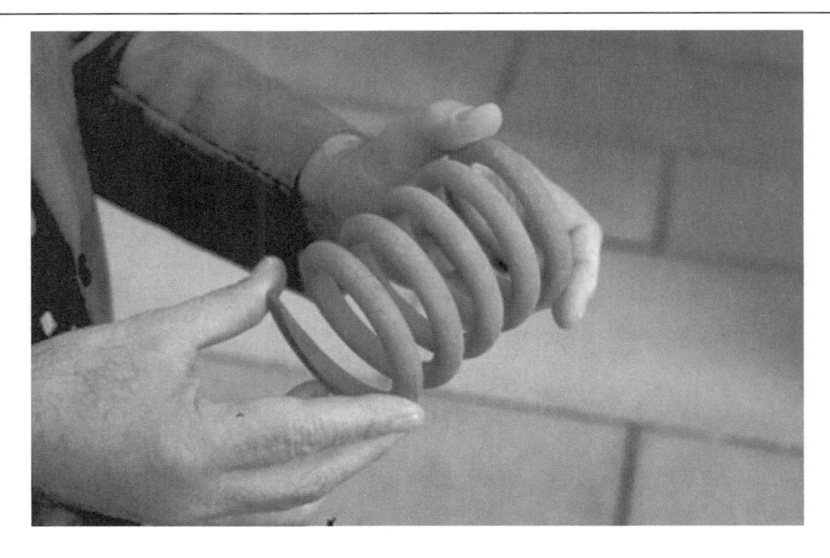

Fig. 1.32. Car suspension spring made from MDF cement

MDF cements are processed by creating a cement–polymer paste with the minimum of water and an earth-dry formulation that is mixed under high shear into a dough-like putty. The dough is then roll-mixed to eliminate entrapped air bubbles, and then moulded into desired shapes by extrusion, pressing or other conventional plastic pressing operations. The ratio of compressive strength to flexural strength strongly suggests that the polymer is not simply a processing aid, but is actively contributing to the engineering properties.

Dense silica particle (DSP) cement. DSP cements, which were conceived by Bache in 1981, are quite different, since they are processed to be workable as in ordinary concrete. This is done by control of the particle size distribution of the cement to minimise the void space between cement grains, which must be filled with water to make the system workable. Bache added a very finely divided silica (particle size 0.1 μm), a by-product of silicon and silicon alloy, to Portland cement in order to fill the voids left by the cement particles. The silica has the added advantage of reacting chemically with the cement paste to become an integral part of the cementitious matrix. A dispersion surfactant or superplasticiser is also necessary to achieve workability so that a DSP composite cement can be poured into forms and moulds. The properties of DSP are comparable with MDF in compression, but as it is much more brittle it has a very low tensile strength.

Future outlook. These new materials are both beginning to be used as useful substitutes for metals or reinforced plastics. MDF composites can be drilled, tapped or machined. They are being considered for ballistic protection and electromagnetic screening. DSP composites are being used for press tools and moulds for the aerospace and automotive industries, where they replace metals.

Building on the moon

It is clear that cement-based products can be engineered to provide potential structures with low mass and reduced cross-section thickness which can act as an airtight radiation shield. They can be processed to exhibit sufficient ductility and abrasion resistance to withstand meteorite impacts. Perhaps at present there is little known about the long-term properties of MDF and DSF composite cements, but it is likely that they will show good response to prolonged vacuum pressures, large temperature fluctuations and impact loading. It was this potential that inspired T.D. Lin to consider concrete as the material for lunar base construction.

It is not some far-flung fantasy straight out of a science fiction novel, but a planned reality, that designs for moon base structures are being finalised by NASA for construction at the turn of the century (Figs. 1.33 and 1.34). The permanent moon bases are required for a number of

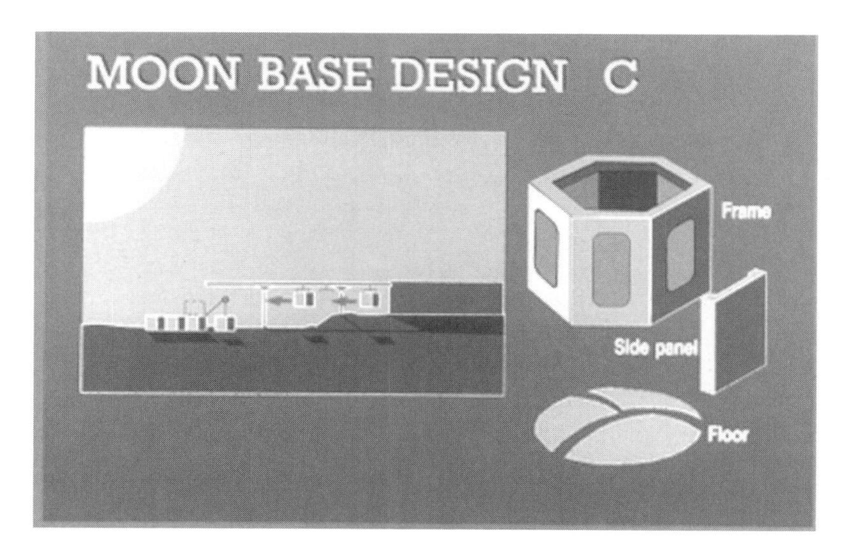

Fig. 1.33. Design for a moon base structure using MDF cement panels

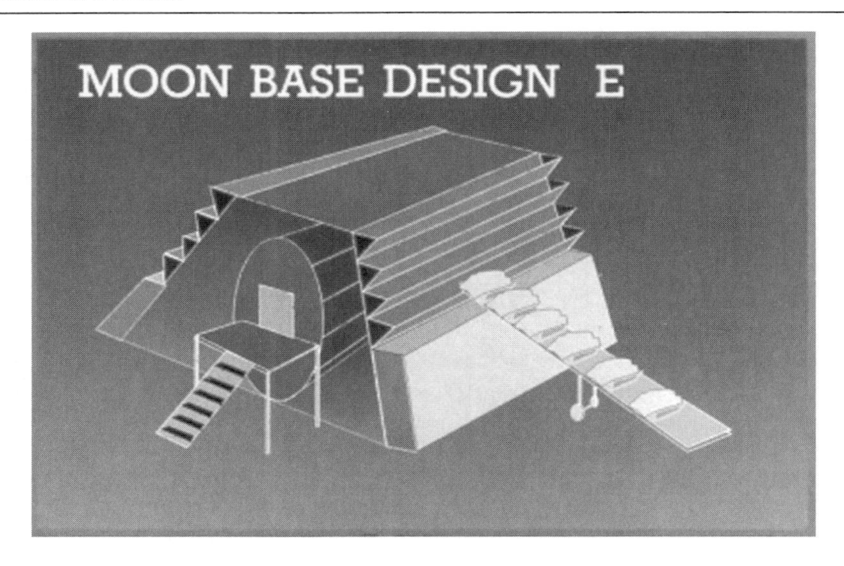

Fig. 1.34. Temporary moon base structure: the shell or canopy of the lunar module is being encased with concrete

reasons: to launch a mission to Mars, to carry out scientific work on interstellar space and to establish industrial production of, say, silicon chips, semiconductors and the like which would be cheaper to mass produce in the vacuum of the moon's atmosphere.

Fig. 1.35. Proposed concrete-encased lunar pressure vessels

Lunar designers all agree that they are designing pressure vessels as permanent structures on the moon, not airtight houses or igloos (Fig. 1.35). These structures must have the ability to shield the inhabitants from solar radiation and temperature extremes. Everything must be erected with the minimum of labour and with total reliability. While the temperature on the moon may vary from −150°C to +120°C, with gamma rays and X-rays bombarding the surface, there are other forces that do not have to be considered. No wind, no ice, little seismic action and only one-sixth the gravity to cope with. Also, while the outside temperature might vary, the internal temperature must remain constant at 20°C for a shirt-sleeve atmosphere.

The main advantages of concrete for lunar base structures can be summarised thus:

- ☐ Less energy is required in production than steel, aluminium or brick.
- ☐ Concrete is unaffected by temperature variations of +120°C to −150°C.
- ☐ Concrete is an excellent material for absorbing gamma ray energy.
- ☐ Under zero atmosphere and prolonged exposure to the lunar environment, the free moisture in the concrete will evaporate but the chemically bound water will not, so the material integrity will not be affected. However, as the concrete in this condition is not airtight, an epoxy coat will be applied to the internal surface of the concrete to ensure that it is.

Information from lunar rock and soil indicates that most lunar materials consist of sufficient amounts of silicate, alumina and calcium oxide for possible production of cementitious materials. Aggregate could be obtained by physical processing of lunar rocks and soil. To produce water, which does not exist on the moon, either hydrogen will be imported from earth and combined with oxygen extracted from lunar rock or hydrogen could be synthesised entirely on the moon. Whatever the option, water will be a scarce and expensive commodity. The basic building material for constructing internal compartments and rooms would be a low-grade concrete block. For the exterior skin of the structure a high-grade DSF type concrete, with better tensile strength than plain concrete, would be necessary. This material could be processed from DSF composite cements, which need a very small amount of water for hydration.

The proper hydration of concrete will need controlled conditions of humidity and temperature if it is to harden fully. This will have to be achieved by cocooning the structure under an airtight canopy and introducing moisture and carbon dioxide to maintain the right internal conditions for curing. Inorganic materials such as steel fibres can be processed from iron extracted from moon rock, but it might be more economical to import lightweight, high tensile reinforcement, such as carbon or Kevlar, from earth. The tensile strength and extremely low dead weight of these fibres offer significant advantages over steel.

Concrete lunar bases will be essential facilities for scientific research and manufacture of semiconductors etc. on the moon. Concrete is probably the ultimate material for the colonisation of the moon, the solar system and anywhere that is habitable in outer space.

References and further reading

High strength concrete

PARROTT, L J (1988). *A literature review of high strength concrete.* British Cement Association, Slough.
PRICE, W (1996). Stronger, bigger, better. *Concrete*, January.
WHITTLE, R (1997). Structural use of high strength concrete. *World of Concrete Europe '97*, Aberdeen, Seminar Paper 26–12.

Concrete-filled tubes

WEBB, J and PEYTON, J J (1990). Composite concrete filled steel tube columns. *Structural Engineering Conference*, Adelaide, Australia.

Pyrament

ANON (1989). Laid at night, land at dawn. *Construction Weekly*, 5 April.

Self-compacting concrete

BARTOS, P and GAUER, M (1999). Self-compacting concrete. *Concrete*, April.
OKUMURA, H and OZAWA, K (1995). *Mix design for self-compacting concrete.* Japan Society of Civil Engineers. Report No. 25.

Reactive powder concrete

CHEYREZY, M (1999). Structural applications of RPC. *Concrete*, January, pp. 20–23.

Compact reinforced composite

AARUP, B and JENSEN, B C (1998). Bond properties of high strength fiber reinforced concrete. *Bond and development of reinforcement.* ACI, Michigan. Publication SP-180.

NIELSEN, C V (1993). Presentation of cyclic load tests of rebars anchored in steel fibre reinforced high-strength composite. In *Fatigue of concrete structures.* Aalborg University, Aalborg, Denmark, pp. 65–71.
NIELSEN, C V, OLESEN, J F and AARUP, B (1996). Effect of fibres on the bond strength of high strength concrete. *BHP '96, Fourth International Symposium on utilization of high-strength/high-performance concrete*, Paris, 29–31 May.

Foamed concrete

VAN DIJK, S (1991). *Foamed concrete: composition and properties.* British Cement Association, Slough. BCA Reprint 2/91, publication 46.042.

Coloured concrete

LEVITT, M (1985). Pigments for concrete and mortar. Concrete Society, London. Cement Practice Sheet No. 99.

Recycled aggregate concrete

BUILDING RESEARCH ESTABLISHMENT (2000). *Recycled aggregate concrete.* BRE Digest 433.

MDF, DSF

ANON (1989). Cementing the future. *ACBM Publications* (USA), 1(2), Fall.
BIRCHALL, M (1984). *High strength concrete. ICT 12th Annual Convention*, Leeds University, 4–6 April.
ROBISON, R (1989). Building on the moon. *Civil Engineering* (USA), January, pp. 40–43.

2 Construction practice

Advances in construction technology — an introduction

In meeting the building challenge of the future, designers and constructors must first capitalise on the technological advances being made today. Continued reliance on traditional practices, heavily dependent upon a plentiful supply of building materials and labour, will not guarantee sustained growth of business in the future. Only by designing a structure for buildability will it be possible to reduce the cost and time of construction. If the financial cost of a project continues to be directly related to the period from commencement to receipt of rental income, then, clearly, speed of construction will continue to be the decisive factor in the choice of building options (Fig. 2.1).

The purpose of this chapter is to assist designers and constructors in evaluating the benefits of adopting alternative techniques to traditional in situ concrete construction. It reviews some of the techniques that have been developed to maximise the speed and economy of in situ concrete frame construction. Incorporation of some, if not all, of the construction techniques reviewed will ensure that in situ reinforced concrete and precast composite construction remain competitive building options for multi-storey structures, both now and in the future.

There are many ways to design an in situ concrete building. However, it must be recognised that clients will seldom give designers the luxury of unlimited time. More often than not an acceptable economic solution must be found under the most severe time constraints. Getting started, when the pages are blank, is the most difficult stage.

A useful reference document on general principles for the initial design of structures is *Economic concrete frame elements*, published by the Reinforced Concrete Council. It contains useful information on design loads, choice of basic structural form, stability requirements, fire resistance and durability, and provides simple and quick methods for sizing members of the superstructure on which estimates of cost and time for construction can be based.

For a design engineer seeking the optimum economic framing solution or a trade contractor bidding for a job, getting started requires a basic sense of construction logic. The first rule to remember is that construction economy begins at the design development stage and not on site. The second is to recognise that the most important influ-

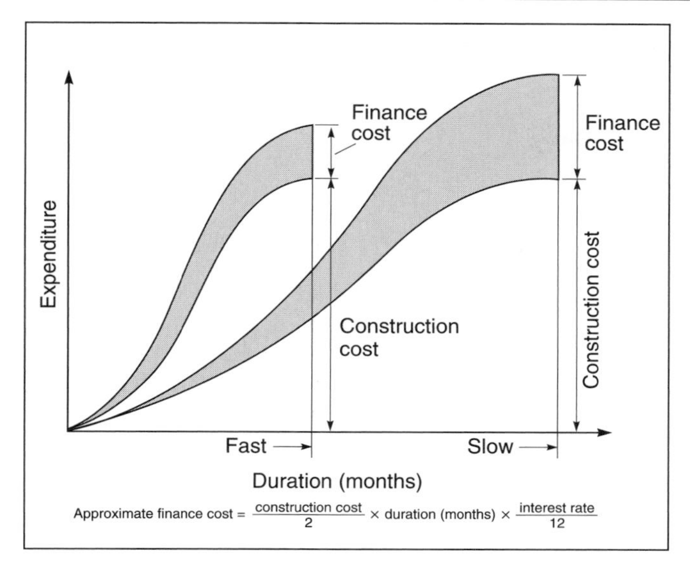

Fig. 2.1. Approximate relation between construction time and project cost

ence on concrete frame cost and project time is often the formwork. While formwork is not a tangible part of the finished structure, it can account for up to 50% of the frame cost. A typical breakdown of concrete construction cost is: pumping/placing 10–13%, concrete 14–16%, reinforcement 25–30%, and formwork 45–50% (Fig. 2.2(a)).

Selecting the most appropriate method from a wide choice of options requires a basic understanding of the construction process, the prevailing site conditions, labour demand and practicality. There is no mystique, no magic ingredient for making fast build or fast track a reinforced concrete reality; teamwork, design discipline, construction logic and honest toil are the key factors.

It follows that buildability, or making a structural frame or basement efficient and easier to build, must become a design objective and therefore the key element in all cost and time considerations. The opportunity to influence buildability is greatest in the design development stage and decreases rapidly thereafter (Fig. 2.2(b)).

Designing for speed

To develop faster building programmes, basic principles of buildability should be rigorously pursued at the conceptual design stage and followed through in all the design processes. Moreover, starting the design

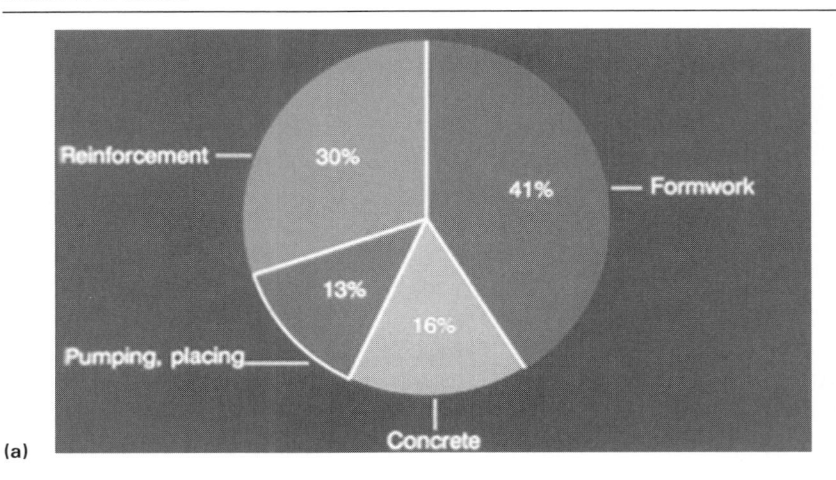

(a)

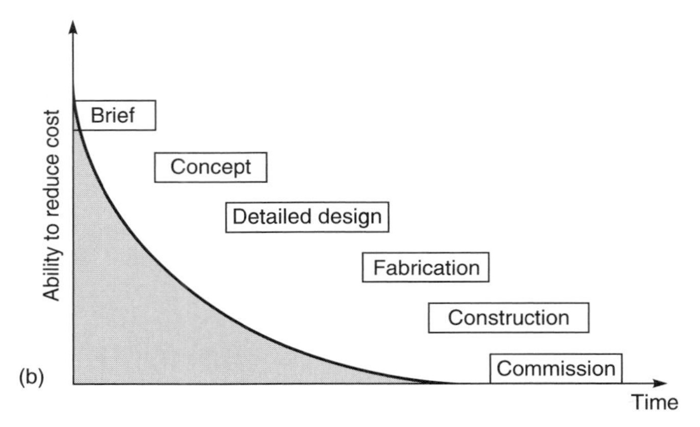

(b)

Fig. 2.2 (a) Typical breakdown of concrete construction cost. (b) Project stages and ability to reduce cost

with buildability in mind is more productive than modifying a design later.

Principles of buildability

Design for building repetition

Keep the floor layout and building grid repeating from bay to bay and from floor to floor. If work is repetitive and simple, labour and supervisory staff rapidly become familiar with it. The learning curve will be short and steep, and productivity should be higher.

Standardise member sizes

Keep to standard member sizes, maintain a constant depth of horizontal construction, constant storey height, and constant column dimensions from floor to floor. Construction and fabrication are easier, elements are simpler and interchangeable. It will usually be more cost-effective to increase concrete strength or the amount of reinforcement to accommodate differing loads and spans than to vary the size of the structural members.

Simplify construction

Simplicity reduces the chance of errors in the design and on site, and encourages faster construction. For example, in extreme cases core areas for lifts and stairs may require as much labour and resources as the rest of the floor. Economy of construction is achieved by eliminating as much complexity as possible from the core configuration.

Avoid mushroom or drop-head columns by using steel shearheads within the slab depth. Adopt flat slab designs, preferably solid or ribbed slabs, to minimise construction depth and encourage faster construction techniques using flying forms and table forms. Avoid waffle slab construction where possible because it can prevent the use of flying forms or table form systems, leading to slower construction.

Minimise sequential working

Reduce site operations by adopting kickerless construction techniques, cast-in screed rails and prefabricated stop-ends. Eliminate screeds by specifying direct finishes to floors. Design, where possible, for one pile per column to avoid the complexities of pile caps. Where there is a basement, consider building up and down at the same time. Slow activities, like core wall construction, should be kept off the critical path for the floor cycle by forming the core walls ahead of the frame; where feasible, this operation can be eliminated by designing the structure as an unbraced frame.

Draft economic specifications

Draft a specification that minimises cost and time spent. Write clear and unambiguous clauses that relate to the project, that can be priced, and that are performance-related. To encourage greater economy of formwork usage, allow early striking of forms based on the in situ strength of the concrete, rather than relying on tables which are over-conservative because of the use for which they are designed. Establish sensible dimensional tolerances for the structure, particularly floor surfaces for raised floors, which do not require a power-float finish.

Permit large pours

Permit large-area pours to give the contractor the opportunity to use flying or table forms economically. Improvements in the speed of in situ construction in recent years have largely resulted from specifiers and designers recognising the unnecessary restrictions caused by their specifications, particularly on pour size and striking times. These are the critical elements for faster in situ construction.

Encourage good quality control

The success of a project, whether fast or slow, can very much depend on a track record of good quality control and a minimum amount of defective work. This can be achieved by paying particular attention to getting cover right on site, compacting concrete correctly, providing adequate curing and protection to the immature concrete and maintaining good line and level of all finished work.

In coming to a final decision on the design of a frame, a number of formwork systems, construction techniques and framing solutions will be examined, each offering particular advantages and disadvantages in terms of speed, resources and time.

Formwork systems

Selecting the forming system for the frame is often the most critical decision affecting both speed of construction and building cost. In a building boom, with demand for construction outpacing supply, material and labour shortfalls are limiting factors. While many constructors feel comfortable keeping to the traditional forming methods, others will invest in more advanced construction hardware which is machine-driven and thus less labour-sensitive, for sustaining profitable growth in the future.

For example, in the USA today, the extensive use of prefabricated forming systems has given some concrete trade contractors an edge over their competitors. Speeds of up to three-day and even two-day floor cycles have been accomplished with forming systems available in the UK market. Those that can make a direct improvement on construction efficiency by saving time and money have been chosen for review.

Flying forms

Labour can represent up to 80% of traditional forming costs. The use of prefabricated formwork panels, popularly known as 'flying forms', can almost halve the labour content of forming and eliminate material wastage from cutting, damage and losses, which occurs with more traditional methods of construction.

Flying forms (Fig. 2.3) consist of a pair of steel or aluminium trusses topped with closely spaced aluminium or wood joists which support a plywood deck. In this way, panels of formwork measuring up to 100 m^2 can be decked out to form an entire bay of a suspended floor slab. Once the flying forms are assembled on the site, they are crane-handled into position and fixed to grid by a four-man crew. After striking, the forms are rolled out from under the newly formed slab on a set of tilt rollers placed under the trusses (Fig. 2.4) and then craned into the next position (Fig. 2.5). Typical daily outputs for a four-man crew, stripping, flying and resetting medium to large formwork panels, are around 700–1000 m^2. It is usual to begin using flying forms after casting the first floor above ground level, in order to get clear of obstructions.

While the building layout and the space external to the building must suit the use of flying forms, the formwork system is flexible enough to accommodate recurring cantilevers, overhangs, rib moulds and beam forms. Even edge beams up to 1m in depth can be accommodated under certain conditions. To use flying formwork efficiently and economically, there should be enough room to fly the entire system beyond the building line and back in again, and there must be at least seven or eight re-uses to amortise the equipment cost.

Fig. 2.3. A flying form

Fig. 2.4. Rolling out the flying form

Fig. 2.5. Craning the flying form into position

To benefit from this technique, the design of the structure must be sufficiently robust to allow early striking of formwork, and specifications should be written to encourage large-area concrete pours. Experienced constructors with a well-planned crane schedule have regularly achieved three-day floor cycles in the USA and Canada. A four-day cycle is becoming a routine production rate for multi-storey construction in the Far East and Australia, particularly with post-tensioned flat slabs. Flying forms are helping UK trade contractors speed in situ frame construction. A week per floor is now commonplace.

Column-mounted forms

In fast multi-storey construction the supporting floor is often not strong enough to carry the applied construction load at an early age, and therefore propping, through successive floor levels, may be necessary. Planning a fast construction cycle and scheduling the propping sequence to fit such a programme is time-consuming and labour-intensive, and is a critical activity. Using column-mounted forms (Fig. 2.6) can alleviate, if not eliminate, this problem. The system is essentially a flying form without the supporting trusses. Instead, two I-beams located on the sides of the forms carry the construction load onto brackets or jacks anchored to permanent concrete columns (Fig. 2.7). To strike and fly the panels, the forms are lowered onto special rollers using adjusting

Fig. 2.6. Installing column-mounted formwork

(a)

(b)

Fig. 2.7. Column-mounted formwork: (a) column-mounted bracket; (b) jacked-up form

screws on the column brackets. Then, with the help of mechanical pulleys, the formwork assembly can be moved out and lifted by crane onto column-mounted jacks located on the next floor. Columns are prepared with pre-drilled holes or cast-in openings at a predetermined height to accept the support jacks.

This technique has been used in the USA, and the makers claim that floor construction output for a four-man crew is about the same as for flying forms, around 700–1000 m^2 per day. The main benefit of this system lies in the reduction or elimination of all propping. Column-mounted form systems work best with flat slab or shallow beam and slab construction, and must have at least seven to eight re-uses to be economical. As with flying forms, building layout and column spacing must be constant, although the storey height can be varied without penalty. It is not essential to have a prepared surface at ground level in order to get started. Deep edge beams can be accommodated with this type of system.

Column-mounted forms were used to construct the floors of the 36-storey Westin Hotel in Boston, USA. The upper floors of the tower block were cycled every three days, and the building was topped out in less than 14 months. Florida's largest hotel and conference centre, the Marriot, was built using column-mounted forms. Building construction was complex because all floors were deliberately made non-typical to maintain the unique architectural appearance and end-user specification. Despite these complications and the Y-shaped geometry of the podium structure, the 27-storey tower block was cycled at the rate of 6000 m^2 of finished floor area every seven days. The flexibility and efficiency of column-mounted forms enabled the whole project to be completed three months ahead of programme.

Tunnel formwork

Tunnel formwork (Fig. 2.8) or, as it is sometimes called, 'apartment formwork' is a factory-made steel formwork system that enables walls and slabs to be cast monolithically on the same day to achieve a one-day construction cycle. It has been developed for rapid construction of room cells (i.e. walls and slabs) of varying sizes for housing projects, apartment blocks and hotels. The system has been used in the UK and is well established in Europe and the USA. Provided that the room layouts or apartments of a building project can be rationalised into one-way spanning slabs of not more than 7.5 m span, with a storey height of not more than 3.5 m, then a tunnel form system can be made to work efficiently. Production rates of around two apartments a day are usual, equivalent to 450 m^2 of total wall and slab construction.

(a)

(b)

Fig. 2.8. Tunnel formwork: (a) and (b) show the erection sequence using half tunnel sections

This can be achieved with a team of seven to nine workers stripping and erecting formwork, fixing steel, fixing plumbing and finally concreting the whole assembly on the evening of the same day.

The key to developing this fast rhythm of production is to keep the immature concrete floor slab propped the following morning, while removing a half-tunnel section of formwork. Minimum concrete striking strength is assured by tenting and heating the air space within the tunnel forms overnight.

The benefit of such a system is the reduction of site labour and the elimination of skilled finishing trades. Since the system is reliant on cranage to position and reset forms, labour is required only to wheel out and set down the forms, using simple hand-tools for jacking and clamping of the assembly. The accuracy of the wall forms and the exact positioning on successive lifts, using specially cast kicker plates, affords a high standard of surface finish and verticality, eliminating the need for plastering (Fig. 2.9).

The construction becomes a total building process with the minimum of following trades, offering significant savings in time and cost. However, tunnel formwork equipment is not cheap, and it usually requires about 40 to 50 re-uses to justify the cost. Alternatively, hire arrangements or 'buy back' agreements can be negotiated. The architectural form and structural layout, within the concept of bearing walls and one-way spanning slabs, does not limit the scope for imaginative and dramatic building design using tunnel forms. Major improvements in speed of construction can be achieved over more traditional methods. Tunnel forms helped to reduce the construction

Fig. 2.9. Tunnel formwork system in use (courtesy: Outinord)

time of a 96-room hotel building in Luton, England, from 12 weeks to 5 weeks.

Adjustable column forms

The speed of construction and the ability to modify the column section size are the main advantages of adjustable column forms (Fig. 2.10). The basic formwork modules enable variations to be made in square and rectangular column sizes from 150 mm to 1000 mm, in increments of 25 mm. Modification of the column dimensions within this range can easily be made at any level of the structure (Fig. 2.11).

The proprietary metal forms come in two standard heights, 2.7 m and 3.0 m, but it is possible to obtain other heights. The real benefits of the column form, apart from the safety features of integral working platform, guard rail and access ladders, are the speed, accuracy and quality of the construction. The rigidity of metal forms makes tie bolts and tie holes unnecessary. Only six wing-nuts need to be removed to strip the forms. Since the column formwork does not require a kicker,

Fig. 2.10. Adjustable column forms

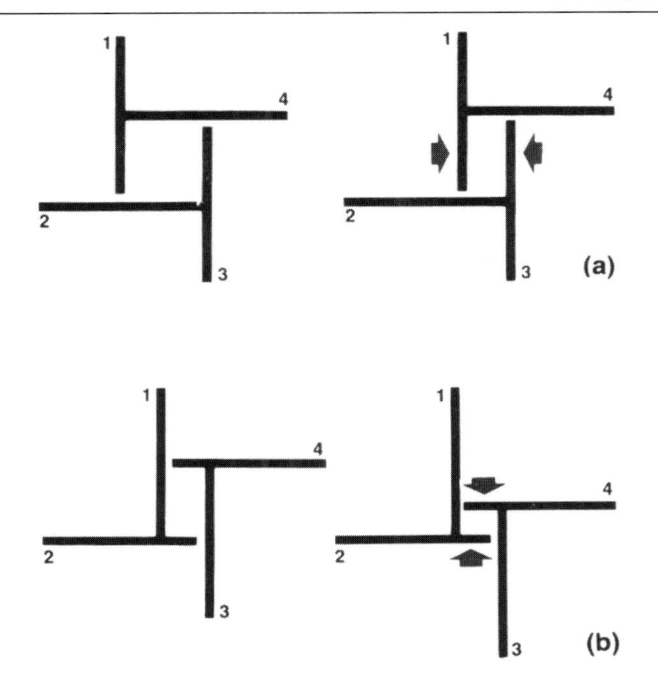

Fig. 2.11. Varying the column size with adjustable column forms: (a) panels 1 and 4, and 2 and 3 are coupled; (b) panels 1 and 2, and 3 and 4 are coupled

it is ready to be concreted once it has been positioned and plumbed. To give an example of the speed of construction, it took only half an hour for two men to strip and reposition a 7 m high adjustable column form on the Broadgate development in London. There is no way in which the same operation, using traditional timber formwork, could compete with this.

Fibreglass column forms

Fibreglass column forms are extensively used in the USA for circular column construction (Fig. 2.12). Moulded from fibre-reinforced polyester, these cylindrical, one-piece column forms are lightweight and can easily be handled by two workers with the minimum of crane time. The forms come in various standard sizes from 300 mm to 1000 mm in diameter, and in lengths from 1.5 m to 4 m. Other diameters up to 1500 mm can be made to order.

The one-piece column form makes for fast erection and quick stripping (Fig. 2.13). When handled carefully, the forms may be used many

Fig. 2.12. Fibreglass column form

times and are very cost-effective. A step-by-step installation procedure will illustrate the simplicity and speed of construction (Fig. 2.14):

1. A template constructed from timber batons or purpose-made steel collar is fixed to the base concrete with shot-fired nails.
2. The column reinforcement is then fixed in position.
3. The base of the column form is placed in the template, and the column forms are sprung open by hand (or by using hooks) to fit around the column reinforcement cage. Long column forms of large diameter, which may be difficult to manhandle, can be pulled into position by a rope fixed to the top of the form and secured over the top of the reinforcing cage. Several timber spreaders are used to keep the form open, and are then removed to allow the form to fit around the reinforcement as it is pulled into position.
4. The two flanges of the formwork are then locked together with special slotted pins and wedges.
5. A steel bracing collar is then clamped around the top of the form and attached to the lateral bracing legs for the alignment and plumbing of the column. The column is aligned and plumbed

Fig. 2.13. Stripping a fibreglass column form

using the adjustable turnbuckles located on the ends of each of the bracing legs (Fig. 2.14). The column is now ready to be constructed.

6. Once the concrete has set, the lateral supports and the steel bracing collar are removed. Starting at the base and working up, the slotted pins and wedges are then removed. Finally, the column is sprung open, working from the top down. The forms should not be sprung open more than their diameter as this could cause crazing and damage to the fibreglass, which could lead to failure.

Fibreglass column forms have been designed for a maximum liquid pressure of 115 kN/m^2. However, external vibrations or revibration of the concrete is not recommended and column forms must not be exposed to temperatures exceeding 60°C, as this may result in permanent deformation.

Disposable formwork

A number of manufacturers now offer disposable circular column formwork made from spirally wound rigid paper tube that is impermeable and resistant to water absorption (Fig. 2.15). The tube is internally lined with a smooth-faced plastic release sheet for good surface finish and appearance so there is no downtime on site spent applying release agent to the inside of the tube. The tubes must, however, be protected

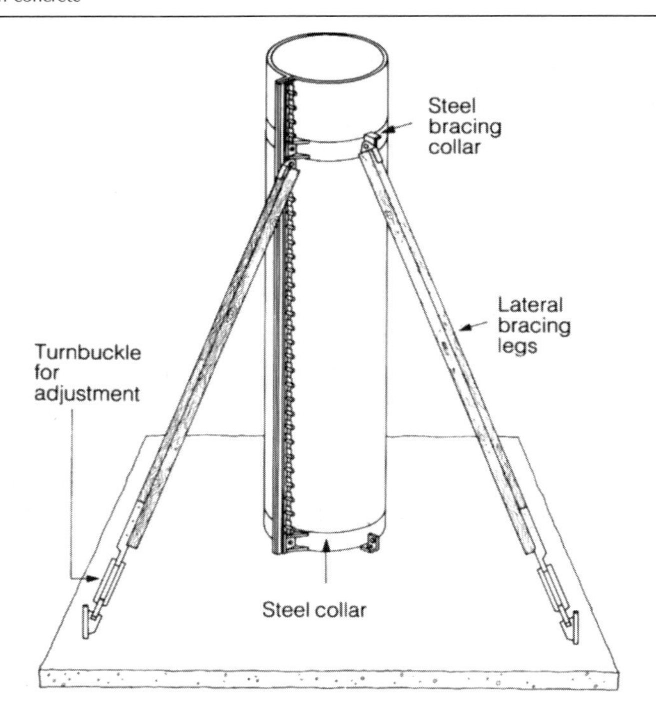

Fig. 2.14. Fibreglass column form: support details

from rain during storage and when in use. A good size tarpaulin that totally covers the tubes in storage would usually give sufficient short-term protection from the weather, provided the tubes are stacked off the ground on pallets. The disposable tubes are supplied to the required length and diameter. If tubes need to be shortened on site, they should be cut from the bottom of the tube using a fine tooth blade. This will ensure that the tear-off strip action is not impaired. Re-seal the cut rim of the liner with a suitable formwork tape.

Tubes can be supplied in diameters ranging from 150 mm to 1200 mm and in lengths to suit the project. To complement the standard circular form, the inside of the tube can be fitted with a moulded liner that can give square, hexagonal, octagonal, fluted, oval, L-shaped, rectangular and more column shapes, plus surface patterned finishes.

Installation

1. Ensure that the rebar cage is reasonably plumb before sleeving the tube into position. Always erect the tube with the arrows pointing upwards, thus ensuring correct operation of the tear-off strip.

Fig. 2.15. Disposable column formwork (courtesy: Rapidobat)

2. The base of the tube can sit on the concrete without a kicker, provided the concrete is level. A square timber template is fixed to the concrete to maintain the tube in position (Fig. 2.16(a)).
3. The top of the tube is restrained with a square timber frame for props to be used to plumb the column.
4. As the tube is quite lightweight, under a concrete pressure head it is liable to lift. Hold down the tubular column by wedging the top of the tube to the rebar cage or tying it back to the concrete slab (Fig. 2.16(b)).

Concreting

For columns up to 4 m high, use a skip or bucket to place the concrete. For columns higher than 4 m, it is better to pump or tremie the

(a) 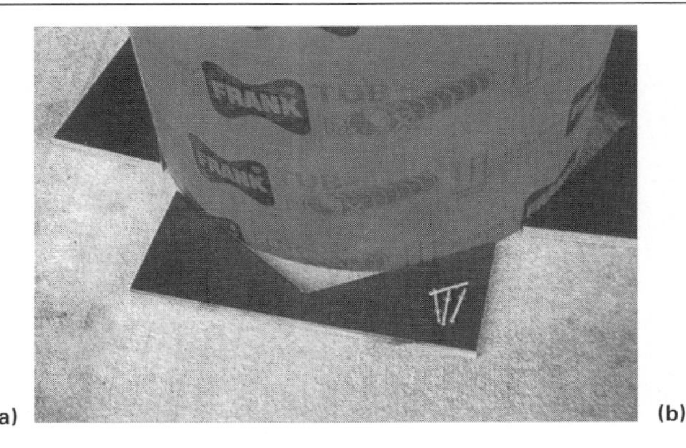

(b) 

Fig. 2.16. Restraining disposable column formwork: (a) at the bottom; (b) at the top (courtesy: Creteco)

concrete into place. Place concrete slowly to minimise trapped air and vibrate the concrete only from inside the tube.

Removal

Do not strip the disposable form until 48 h after casting. Pull the tear-off strip down the full length of the tube. Flap open the tube and ease it off the concrete. Then remove the liner off the column and check the surface finish. After completing any minor repairs, use the tube to protect the new concrete.

Core wall construction

Core walls and shear cores enclosing lifts, stairwells and service shafts can represent a significant element of the forming cycle and in some cases can take longer to form than the entire floor slab. Where possible, the layout of the core wall should be simplified to minimise the volume of construction work. Avoid boxed-in sections by standardising wall dimensions and considering robust L- or T-shaped wall layouts with infill to create a box section using dry lining or blockwork. On the other hand, where an enclosed box section cannot be avoided for stability reasons, the use of climbing or slipform construction can take this work off the critical path of the floor cycle by advancing it ahead of the frame.

Climbing formwork

Standard wall formwork panels are connected to climbing brackets, which allows the complete unit to be moved in one operation by crane (Fig. 2.17). Formwork panels are made in 2.4 m and 1.2 m lengths together with special panels to suit the specific lengths required. The panels are designed to suit the wall plan configuration and can be supplied complete with all fixings and sundry formwork items.

The formwork assembly is fixed to the carriage on the climbing bracket, which allows the formwork to be retracted from the concrete face without using the crane (Fig. 2.18). Plumbing equipment is fixed to the formwork and the climbing equipment for lining and levelling the units. The formwork and working platform are then lifted in one operation by the crane and located on to the next anchor position, using the steel climbing brackets. For internal shafts the formwork units are supported on purpose-made platforms and located on folding brackets fixed to the wall.

After the forms are struck, the complete formwork and access assembly is craned into position for the next lift. The lower working platforms

Fig. 2.17. Climbing formwork

Fig. 2.18. Climbing formwork: fixing a working platform

are fixed from the main working platforms for making-good operations and retrieving climbing anchors without the need for cranage. The main benefits of this system can be summarised as follows:

1. Formwork and working platforms are lifted in one operation, which is faster and saves crane time.
2. The retractable face on the external formwork gives safe access for reinforcement operations and cleaning.
3. The formwork units do not require storage space on site between uses, unlike other traditional systems.
4. Climbing formwork enables integration with the general floor construction cycle.
5. The system uses standard equipment which can be hired or purchased.

It is usual to put together the climbing form assembly ahead of the floor and to cast the walls two floors ahead of the level of horizontal construction, i.e. the floor and column supports.

Slipforming

Slipforming is a method of constructing a box-shaped core using specially designed formwork which constantly climbs, leaving behind the required core wall shape cast in the concrete (Fig. 2.19). The system can

Fig. 2.19. Slipforming the building core

be worked on a 24 h continuous process or a 12 h stop–go arrangement. Concrete placement is continuous as the formwork slowly climbs the full height of the structure. Multiple hydraulic jacks fitted to the forms climb special rods positioned inside the walls of the structure being formed. The rods have a trailing sheath to give them clearance within the concrete so that the rods can be removed when the climb has reached its full height.

The forms are usually 1.2 m high and are fixed so that there is a slight taper from top to bottom to reduce the drag on the concrete surface as it is slipped. All the climbing jacks are interlinked to a central power unit, which means that individual jacks can be controlled by the amount of travel and the pressure applied. Control stops can also be planned for crane fixing ties etc. to ensure that the slipform is kept on alignment and within vertical tolerance. Vertical tolerance is 50 mm for structures over 30 m high. The slipform assembly will include provision for an upper working platform and a lower platform for cleaning out and inspections.

The sliding speed is determined by the rate of setting and hardening of the freshly placed concrete and the rate at which concrete, reinforcement steel and timber blockouts can be supplied and fixed. When OPC concrete is used the average sliding speed is in the region of 150 mm to 300 mm per hour and it should not drop below 150 mm/h as the concrete may be harmed through premature setting and cracking when the forms are being jacked up. Concrete is usually site-batched for a 24 h operation. It is placed in the forms from the working platforms in 150 mm layers and poker-vibrated, with the poker lowered into the previous layer to ensure a good bond between successive layers.

The main advantage of slipforming over conventional methods is speed. It allows difficult tall structures to be formed easily because, once the initial slipform equipment is assembled, it is used throughout the construction period. However, because it is a continuous site operation it is usual in the UK to slipform the core in its entirety before any other work commences (Fig. 2.20). This makes the rest of the floors and columns sequential activities and may add time to the overall completion of the frame. Working practice in North America suggests that

Fig. 2.20. Slipformed core and lift slab construction

slipform cores can be advanced four or five floors ahead of the main structure to integrate construction operations and to reduce the overall time for completion.

The important points to consider when using slip forming are:

1. It is difficult to accommodate design changes once a core is being slipformed.
2. Slipforming calls for a higher level of management and supervision during operations.
3. Pre-planning for box-outs and voids is critical.
4. Any delay in material deliveries or shortage in labour can seriously affect slipform progress.
5. Rebar has to be specially designed for slipform requirements. Vertical bars are supplied in 6 m lengths, while horizontal bars have to be detailed for placement between slipform jack rods.

Construction techniques

Simple, efficient and proven construction techniques and practices that reduce labour content and save time and money should be used, no matter how small or large the project. Regrettably, the use of many of these practices is almost non-existent on UK sites, with few exceptions, emphasising the gap in technology that still yawns between the UK on the one side, and the USA and the rest of Europe on the other.

Kickerless construction

Eliminating the traditional kicker in walls and columns can reduce construction time, save materials and other resources and improve workmanship (Figs. 2.21 and 2.22). The two most popular methods for forming kickerless joints are to use 'shot-fired' plywood battens, or adjustable metal T-bar spacers. The advantages over the traditional kicker are:

1. *Faster construction.* Columns and walls can be concreted the day after casting the floor slabs or base, saving a day in the construction cycle. On multi-storey buildings, this can result in a significant saving in total construction time.
2. *Lower labour requirements.* It takes two trades, working part-time over a two-day period, to form the traditional kicker. Kickerless construction requires one trade for one simple operation taking less than one hour.
3. *Low cost.* In addition to saving labour and construction time, kickerless construction uses items which are inexpensive to buy

or which can be made on site. This makes it considerably cheaper than forming the traditional kicker.

There are a number of products and techniques for forming kickerless joints, most of which have been developed on building sites in America and Europe. The accuracy of this method can be better than

(a)

(b)

Fig. 2.21. Kickerless column construction: (a) shot-fired plywood battens; (b) fixing formwork clamps

(c)

Fig. 2.21 (continued). (c) Kickerless joint

Fig. 2.22. Kickerless wall construction

that achieved with the traditional kicker, because the T-bar spacers and the shot-fired battens are fixed on the setting-out line and any discrepancy can be easily spotted and rectified. Additionally, some types of T-bar spacer can be adjusted after fixing. To gain the maximum construction benefit from this technique, kickerless construction should be specified in the contract documents and the wall and column reinforcement detailed appropriately on the drawings.

Prefabricated stop-ends

Traditionally, stop-ends are formed with timber, which can be time-consuming, particularly on large-area pours. Further delays can and often do occur while awaiting the removal of the stop-end formwork and in preparing the construction joint before the adjacent bay can be cast.

The use of permanent formwork, such as galvanised expanded metal, for forming stop-ends can speed up the pour cycle by eliminating stripping times and joint preparation associated with timber stop-ends.

Most concretes, including pumped mixes, can be used with this type of stop-end, although precautions may need to be taken with lightweight aggregate mixes and very workable concretes which tend to 'bleed' during vibration. Excess 'bleed' water brought to the surface during vibration should be directed away from the construction joint, as it may wash away the small amount of concrete that extrudes through the expanded metal and assists the structural key.

Continuity strips

To simplify and speed the forming of construction joints, starter bar assemblies or continuity strips can be used for walls and slabs. The strip consists of a protective housing made from plastic, metal or other suitable material, with the projecting part of the starter bar bent down inside. The strip is fixed to the face of the formwork and concreted in to form the construction joint (Fig. 2.23). There is then no need to drill the forms to allow the bars to project through.

After the forms are struck the protective housing is removed, exposing the reinforcement, which is carefully straightened into position. The housing may be left in or removed, according to the maker's instructions.

This technique is particularly suitable for both mild steel and high-yield deformed bars up to 16 mm in diameter. For high-yield deformed Type 2 bars, it is advisable to check that the specified product accommodates the bent portion of the bar within the protective housing or covering.

Fig. 2.23. Rear view of a continuity strip fixed to inside of formwork for a vertical wall joint

Continuity strips help to reduce damage to concrete and starter bars and ensure that the joint is properly formed. They cut out unnecessary formwork preparation and labour and eliminate reinforcement projecting through formwork (Fig. 2.24). However, care in selecting the appropriate continuity strip is important since rebending of reinforcement can lead to brittleness. A German code of practice, *Bending back reinforcing steel*, provides useful guidance on possible problems with product selection and improper use on site. It also includes recommendations to prevent premature corrosion resulting from inadequate cover to metal housings if these are left in.

When correctly used, the performance of continuity strips will be comparable to that of similar cast-in bars. Note that rebent bars may not be suitable for applications where the applied loading is predominantly dynamic, for example, bridge decks, heavy machine bases, etc.

Wide-bay construction

In the early 1970s, a concerted effort was made to rationalise and improve the method of ground-floor slab construction for industrial buildings in the UK. Since then, long-strip, wide-floor construction

Fig. 2.24. Continuity strips used for a staircase

has become well established, making it feasible to place and finish over 1000 m^2 of concrete with a power-trowelled surface in a working day.

Similar improvements in the cost of placing and finishing, with a reduction in labour time, can be achieved on multi-storey floor construction by adopting this technique. Instead of using temporary metal screed rails and working on a long-bay construction plan, shallow precast concrete rails, typically 50 mm deep, are positioned 5–9 m apart within the depth of the suspended slab, and the slab cast continuously across the floor plan. The precast concrete rail is supported on adjustable chairs and kept rigidly in position by securing it to the main reinforcement. This gives an accurate and rigid control of level for the placed concrete, which is compacted and levelled in one operation using a lightweight, triangular-framed vibrating beam (Fig. 2.25).

The benefits of this method of construction over other more traditional methods, using temporary scaffold tubes and conventional poker vibration, are:

- ☐ greater productivity;
- ☐ elimination of separate labour operations for poker vibration and tamping;
- ☐ better level control;

Fig. 2.25. Levelling and compacting a wide-bay slab

- ☐ elimination of reworking of finished surfaces, hand floating or trowelling of areas needed to smooth out score marks or footprints made when retrieving scaffold tubes;
- ☐ a much smoother and more level surface, which is a distinct advantage when a raised pedestal floor is specified.

Precast construction

Spanlight floors

Spanlight is a precast concrete structural floor and frame system, developed by Tarmac Precast, of exceptionally shallow depth that does not require the use of extensive propping or any formwork (Fig. 2.26). Spanlight combines elements of proven technology in precasting and prestressing. The system is especially suitable for car park construction. The benefits of the system are:

- ☐ shallow construction depth;
- ☐ clear working areas;
- ☐ superior fire resistance;
- ☐ continuity;
- ☐ rigidity;
- ☐ reduced vibration;
- ☐ fast erection.

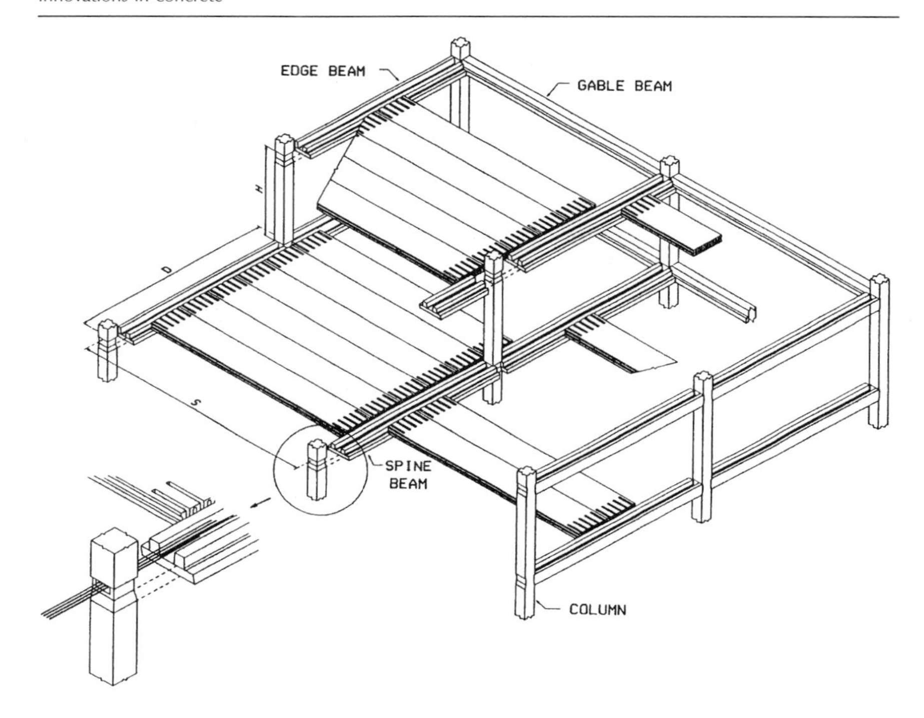

Fig. 2.26. The Spanlight system

Shallow construction depth

Because the main beams and flooring units are prestressed and have structural continuity, the overall construction depth is extremely shallow. Clearly, the span-to-depth ratio varies with column spacing and floor loading, but the system is amongst the shallowest available solutions. This saving in depth allows easy integration of services and saves on the overall structural height. The span-to-depth ratios offered in the system are equivalent to those offered by in situ reinforced concrete and structural steel slab construction.

Clear working areas

Spanlight does not require formwork or heavy propping during construction. The only temporary work required is a removable support around the precast column to carry the spine and edge beams prior to stressing and concreting. This is then immediately removed, releasing the area below for the installation of suspended ceilings or services.

Superior fire resistance

Being prefabricated in precast concrete, the Spanlight system has built-in fire resistance of up to 1 h. Greater ratings, to 2 h and beyond, can be achieved if required.

Continuity

Once the prefabricated elements of the frame are connected on site they are stressed together and the hollowcore flooring is tied over the beams. The frame then has continuity, with moments being transferred into the columns. This provides a stiff overall frame with greater structural continuity.

Reduced vibration

The increased mass damping and rigidity of a concrete structure means less transmission of vibration.

Fast erection

The system is essentially prefabricated off site, to very high tolerance. It is then transported to site and readily erected in the sequence described below. This fast erection allows for rapid possession of the completed floors.

Construction sequence

The Spanlight system is based on four main component groups:

- ☐ precast, reinforced columns with beam connections;
- ☐ prestressed trough-shaped spine and edge beams;
- ☐ reinforced concrete gable beams;
- ☐ prestressed hollowcore flooring.

The columns are erected, and temporary supports are placed around the columns below the beam connection area. The trough-shaped spine beams are then placed onto the supports and the hollowcore flooring is then placed onto the spine beam. Prestressing strand is threaded along the trough, passed through the columns, and anchored using normal barrels and wedges.

The precast spine beams rest on steel sections introduced into holes in the columns. Bending moment across the beam column joint is carried by prestressing strand that passes through the section and is continuous in the beam of the adjacent spans.

Figures 2.27 and 2.28 show a mock-up of the Spanlight system and a test, at Westminster University, of a full-scale Spanlight floor.

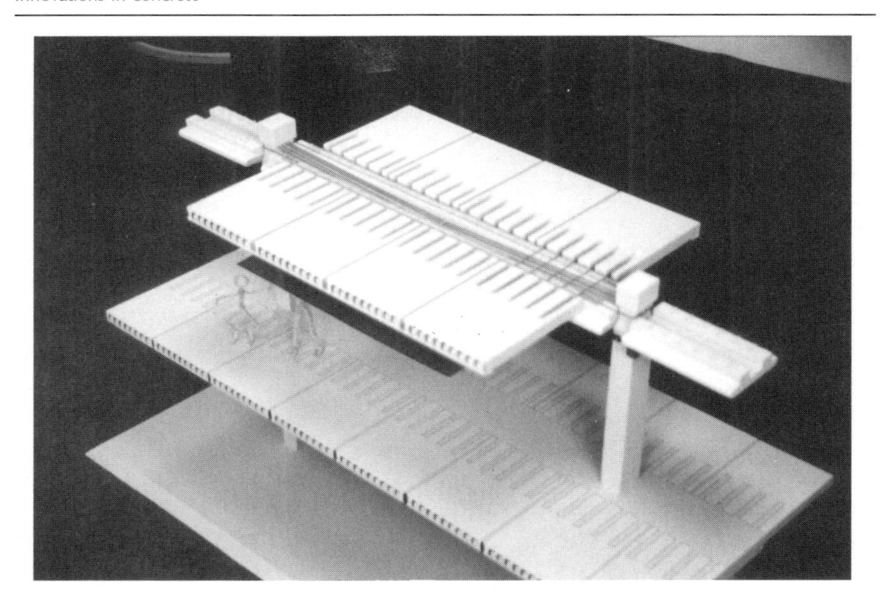

Fig. 2.27. Scale model of the Spanlight floor system.

Fig. 2.28. Full-scale Spanlight floor under test

TermoDeck flooring

Prestressed hollow core precast slabs can be the most economic lightweight flooring option for one-way spanning flat slabs, especially for longer spans from 9 m to 18 m. Conventional hollow core units are mass-produced in a factory; they are cheap and rough-cast and not a smooth, blemish-free product. For structural purposes they are adequate. The circular hollows in the units are formed primarily to save weight and reduce the volume of concrete required.

For some years the author searched unsuccessfully for a precast manufacturer in the UK who would be prepared to develop a hollow core unit that could work as a structural floor and a conduit for service pipework, to save construction depth. In the late 1980s, quite by coincidence, Strangbetong of Sweden launched TermoDeck, the first structural hollow core slab to be used for air distribution and thermal mass damping in indoor climate control. That meant a reduced cost of service installation, a dramatic reduction in depth of structure, and service zone provision as well as a cost-effective suspended floor. Since its introduction in the UK market, a number of highly acclaimed buildings using TermoDeck have been completed, quite a few of which have precast soffit slabs with smooth painted finishes that look aesthetically pleasing. TermoDeck UK is owned by Anglo American, having taken over Tarmac Building Products when Tarmac plc was restructuring its core business.

TermoDeck makes the ceiling and floor of a building into a mass storage sink for energy, with wide surfaces for cooling the air during summer and heating it during the winter (Figs. 2.29 and 2.30). This gives thermal stability to the internal space, reducing the peaks and troughs of external temperature variations and internal heat emissions by absorbing or radiating up to 70% of energy within the concrete slab. As hollow core units span up to 18 m, the soffit can cover a large open plan space or several rooms for a cellular floor layout. The units can therefore act as a heat exchanger between rooms, even when each room has a different temperature setting. Surplus heat from thermally 'loaded' rooms can be transferred to the 'unloaded' rooms via the concrete slabs. The hollow cores of the units are used to duct cool fresh air through the slabs in the hot summer months or convey preheated warm air in the winter. The conditioned air comes into the room via diffusers near the external walls or through the slab soffit at about the same temperature as the room space, which means that there are no draughts. The influx of air is quiet as the cores in the slabs act as efficient silencers against the noise of fans and dampers. The distribution air is generally circulated through three of the cores in a single 1200 mm wide

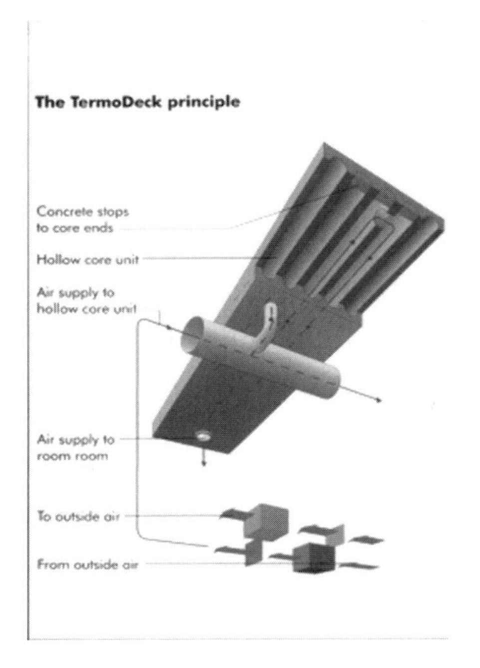

Fig. 2.29. The TermoDeck system

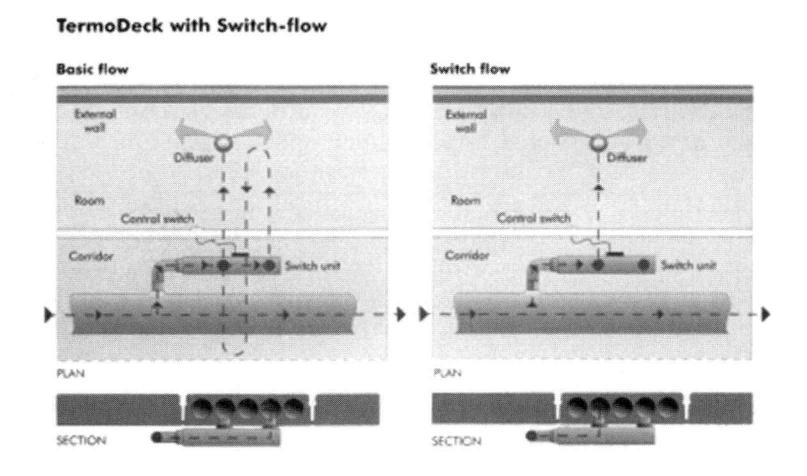

Fig. 2.30. TermoDeck with (left) displacement ventilation and (right) switch flow

hollow core unit. The remaining two cores can be used for electrical wiring or similar purposes.

The basic TermoDeck system consists of air conditioning units, distribution ducts with connections and dampers in corridors and diffusers in the ceiling. Offices with high internal heat emission can use a combination of TermoDeck and comfort cooling which can consist of a plain direct expansion unit. Cooled supply air is circulated in the hollow cores of the slab, absorbing the heat from the slab to reach the room at a temperature not more than 3°C lower than room temperature. The switch-flow option offers very high performance requirements for individual adjustment of room temperature. The system is based upon cooling the hollow core slab during non-working hours, coupled with direct cooling during working hours. The switch unit consists of a damper airflow meter and dampers for adjusting the basic and switch-flow conditions. There are other combinations of heating and cooling options available, including displaced ventilation and thermal radiation systems.

Overall, the need for mechanical services installations such as fans, cooling roofs, refrigeration machines, water accumulators and radiator systems is much reduced. This means there is a high degree of reliability during operation, coupled with a low cost of installation and maintenance.

The Elizabeth Fry building, University of East Anglia, Norwich
(Courtesy of Building Services Journal, April 1998)

This four-storey building (Fig. 2.31) has a gross floor area of 3250 m^2. The top two floors contain 50 cellular offices for about 70 staff. The School of Social Work is mainly located on the first floor, with the School of Health Policy and Practice on the second floor. The lower and ground floors contain lecture and seminar rooms. There are two dining rooms and a small catering kitchen on the second floor.

Well insulated, tightly sealed, triple-glazed windows were specified to meet the client's low energy criteria. An airtight, double-skin external blockwork wall with nylon wall ties, to reduce thermal bridging, encloses the internal concrete structure. All floors, including the top floor, have exposed concrete ceilings made of ventilated hollowcore slabs. The roof has 300 mm of insulation, and 100 mm is applied to the exposed floor soffits over the perimeter walkways on the upper ground floor.

The windows use argon-filled glazing with an inner sealed unit, a perforated metal venetian blind in the outer cavity and external protective single glazing. These were carefully detailed to minimise

Fig. 2.31. The Elizabeth Fry building

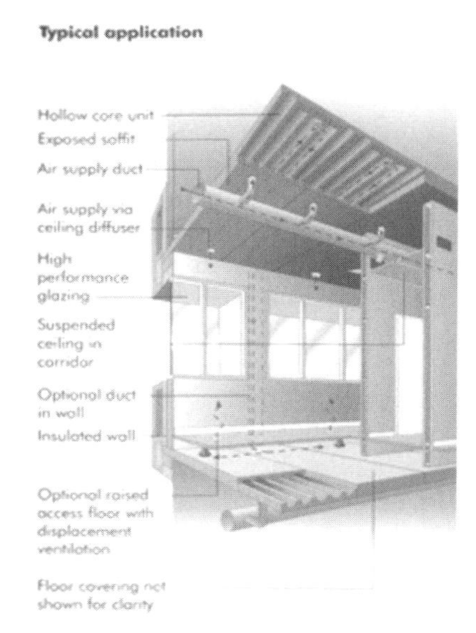

Fig. 2.32. TermoDeck system: typical application as used in the Elizabeth Fry building

cold bridging and any air leakage. Natural ventilation was considered, but the designer opted for the TermoDeck system (Fig. 2.32). By enhancing access to the thermal capacity of the structure, TermoDeck offers year-round tempering of incoming fresh air. In the offices and seminar rooms, air is supplied to the hollowcore units via stubs from ducts running above suspended ceilings in the corridors. After three passes through the hollowcore slab the air enters the room via annular soffit diffusers. Return air is extracted from behind the ceiling cornice and back to the air handling units via the corridor ceiling plenum.

Flow reversal and recirculation of air are achieved by using a set of fast-acting, mechanically linked dampers. The system claims to recover 85% of the available heat—a figure verified by monitoring when the building was occupied. Three 24 kW domestic condensing boilers with standby reserve are used for heating and hot water provision. During monitoring of the building performance, it was observed that even when the outside air temperature dropped to as low as 8°C, the boilers were not required all day. The standard Swedish specification of TermoDeck includes low-power perimeter heaters in each room. In view of the milder UK climate it was decided to omit them.

Key design lessons. Ventilated hollowcare slabs have produced very stable and comfortable temperatures during winter and summer, as good as or better than those in air-conditioned buildings. Using conventional heating and cooling services has helped to make this low-energy approach affordable within normal UK academic budgets.

Construction supervision by the client, the design team and the contractor, along with careful detailing and specification, ensured that the pioneering design was not compromised during construction. Special attention was paid to making the design requirements clear to the contractor, and critical details such as window reveals were inspected before concealment by wet trades.

Aftercare is important. Careful commissioning and handover training during the first two years of building occupancy have ensured that the design and operational performance have been fully achieved in practice (Fig. 2.33).

By any standards, the occupancy survey results conducted by the PROBE 14 team were excellent, one of the best ever recorded over the ten years that such statistics have been collected. The Elizabeth Fry building is likely to become a role model for future excellence in building design and management. It is one of the rare buildings where users give it unprompted praise: "I love it. It combines a sense of tranquillity with aesthetic delight."

Elizabeth Fry building: Project team

Client: University of East Anglia
Architect: John Miller & Partners
Services Consultant: Fulcrum Engineering Partnership
Structural Engineer: F.J. Samuely & Partners
Main Contractor: Willmott Dixon
Completed 1995

Tilt-up construction

(extracted from Southcott and Tovey, 1998)

Tilt-up is a construction method that allows design flexibility and encourages innovation. Walls can be produced as solid panels for use with added insulation or as sandwich panels where insulation is installed during production. The use of concrete offers the designer structural capacity, fire resistance, sound insulation, thermal capacity, aesthetic qualities and durability. Tilt-up panels commonly carry the roof load and provide transverse load resistance in shear, and thus may be used for both loadbearing and non-loadbearing walls in a variety of structures.

Fig. 2.33. The Elizabeth Fry building in use

The benefits

For a building method to be chosen in the highly competitive construction market it has to show that its benefits outweigh those of the alternatives.

Cost of construction

Extensive cost modelling has shown tilt-up to be highly competitive with traditional construction over a wide range of single-storey warehouse-type buildings.

Speed of construction

This is often a major reason for choosing tilt-up. After the floor slab is placed, the typical elapsed time from starting to form the panels until the building shell is completed may be only four to six weeks. Materials for wall panels are easily procured with minimal lead times, allowing a fast start to a near-factory assembly line process which progresses while any roof steel is fabricated.

Durability

The strength of concrete coupled with the uncomplicated method of construction offers reliable durability, as attested by buildings dating back to the 1940s.

Sound reduction

Buildings in a noisy area, such as near an airport or roads, can benefit from the sound reduction properties of concrete. The mass absorbs the sound rather than transmitting it as can occur with lighter forms of construction. Alternatively, noisy processes within buildings are more easily isolated.

Method of construction

Tilt-up construction is the on-site precasting of the walls of a building. In one method of construction, the perimeter foundations and internal ground slab are cast first, and then the wall panels are cast individually, contiguously or continuously on the slab. Alternatively, panels may be cast one on top of another, as stack-casting. After the panels have gained sufficient strength, they are tilted up and positioned around the perimeter. The internal frame is then constructed and the roof built. If desired, stack-casting the panels on a separate temporary casting bed enables the main building ground slab to be cast later, following erection of the roof. Thus tilt-up offers more than one method of construction programming.

A typical construction starts with the levelling of the site before foundations are dug and cast. The ground floor sub-base material is then rolled and accurately levelled; a membrane is laid and simple edge formwork fixed. This is typically laser-levelled for extreme accuracy. The main floor slab (Fig. 2.34(a)) may be laid by the long-strip method, typically 4 m wide by the full length, and finished by power float. But, increasingly, the slab may be laid in wide pours and finished by laser levelling and power float.

When the slab has gained sufficient strength, the tiltup wall panels are built upon it. The panels may be cut to size after long-strip casting, or more commonly are formed individually. The main floor slab or previous panels act as the panel's casting face, and completing the formwork requires only simple perimeter side forms. Climbing forms or full-depth formwork is used when the panels are stack-cast. Before each is cast, a bond breaker, form liner or other material is placed on the floor slab, or on top of the previous panel in the case of stack-casting.

Finally, the panel reinforcement and fixings for the roof are placed and the panel is then concreted and finished (Fig. 2.34(b)). Careful attention is given to the casting position of panels on the base slab in order to minimise crane movement and achieve the most efficient construction sequence. After, typically, two to seven days, when a panel has gained sufficient strength, props and lifting devices are attached. The panel is then gradually lifted or tilted up until it is upright (Fig. 2.34(c)). The flexural stresses during the lifting reach a maximum when the panels are at an angle of about 30°. At this point, the stresses are often greater than when the panel is in place, which can give the designer added assurance in the completed structure.

Particular economic benefits come from the methods used to lift the panels. It is common to use a multi-point lifting system so that the bending stresses are kept below the flexural tensile strength of the concrete, thus minimising and often eliminating the need to rely on reinforcement during lifting. This not only enables savings in reinforcement to be made but also allows thinner panels to be erected, with consequential savings in concrete and final wall thickness. Reinforcement is normally placed in a single central mat.

When the panel is in an upright position it is carefully swung to the perimeter where it is propped in place (Fig. 2.34(d)). Multi-point lifting can enable larger panels to be erected, thus saving time in construction. These larger panels require the use of appropriate cranage but since they are rapidly erected it means that heavy-capacity cranage is often needed for only a few days. For example, the entire external walls to a

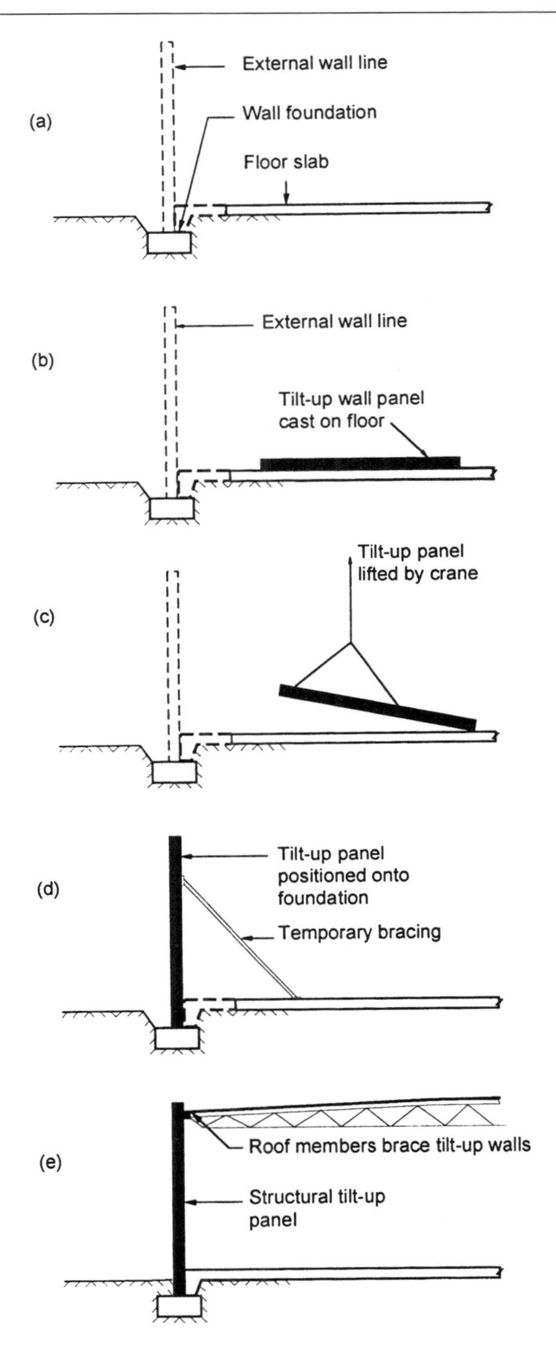

Fig. 2.34. Typical tilt-up construction sequence (Southcott and Tovey, 1998)

building covering an area of some 3000 m^2 can be erected in only a couple of days (Fig. 2.34(e)).

Bond breaker

Application of the bond breaker is one of the most important jobs on a tilt-up site. For best results, the compound must be applied evenly, at the correct rate per square metre, and must cover the surface totally. Whether application is by spray, brush or roller, it is a wise precaution to ensure that operatives know what they are doing, why they are doing it, how important it is and how to check that they are doing it correctly. When applying a bond breaker, it is essential that the manufacturer's instructions are followed carefully. The following points are offered as a general guide:

- Ensuring adequate bond breaking is one of the most important aspects of tilt-up construction.
- Bond breakers from reputable manufacturers will generally perform well if used in accordance with their instructions.
- If doubt exists regarding application rate, compatibility, procedures etc. the manufacturer should be consulted.
- If no other information is available, compatibility problems and procedures can be checked by casting small trial panels and lifting them.
- For best results, site personnel should be aware of the function, properties and limitations of the bond breaker and the importance of applying it correctly.

Panel erection

Before a panel may be lifted it must be confirmed that the concrete has reached the required design strength. This can be determined by use of temperature-matched cubes cured alongside the panels or assessed, where some previous experience exists, from time/temperature records. Evidence of adequate strength avoids premature lifting, which may cause cracked panels or actual panel failure as the concrete will be less resistant to flexural stresses; similarly, failure of the lifting inserts may also occur as the surrounding concrete also governs their strength.

A slightly longer curing period may be required for stack-cast panels, as the last and youngest panel in the stack must be lifted first. Speed of construction may dictate that special curing or a stronger mix may be justified, but this needs to be addressed at the planning stage. Before the lifting sequence is started, full consultation between the contractor and the crane driver will avoid misunderstanding and delays.

Erection sequence

It is important to ensure that the panels are erected in accordance with the specified sequence. Generally, panels should be erected progressively from one end of the building. The sequence of erection should, whenever possible, be designed to avoid multiple handling of panels. Generally, because of the increased risk of cracking, the designer should be notified before any lifted panels are lowered back for storing flat. In such circumstances the designer must determine the support system. However, panels should preferably be stored in the vertical position using temporary props, but great care is needed to ensure their safety in the temporary state. In some cases where continued activity is required on the panels (for example, cleaning of brick faces), A-frames may be needed in place of conventional bracing. In essence, the panels should be stored only in a position approved by the designer.

Rigging the panels

The design for lifting determines the number and position of lifting fixings. Lifting or rigging configurations are then determined so as to minimise the load on the inserts. Most manufacturers or suppliers of fixings can provide guidance on rigging configurations and may also provide rigging hardware. However, in most cases a specialist supplier will supply the rigging hardware.

Although the number of lifting inserts determines the amount of rigging, the actual rigging configuration used may vary. It is usual and far better for the rigging configurations to be designed to be self-equalising (Fig. 2.35). To facilitate this equalisation, the rigging system often includes the use of slings running through sheaves (pulleys) on spreader or lifting beams. The crane rigging cables must also be of sufficient length. Because of the extra loads imposed by their geometry, short cables may result in cracked panels or overloaded inserts. Cable of the largest sensible diameter is recommended to minimise tension stretch. Although thinner cables may be of sufficient strength, their 'springiness' may increase impact loading. Cables should be kept vertical and any side pulls should be executed only after prior consultation with the hardware supplier.

To avoid costly delays, rigging changes should be kept to a minimum. It may be more economical to add a few inexpensive inserts on the lighter panels for consistency than to constantly change rigging. Where possible, three rows of inserts should be avoided because of the complex rigging configurations required, but this is a design issue that must be addressed prior to the erection phase.

Fig. 2.35. Rigged panel being lifted

Some variations are possible on site, such as converting, say, a two-high rigging to four-high without removing existing rigging. However, this will require consultations with the hardware supplier.

Strongbacks

Panels that are oddly shaped, elongated, or with large or awkwardly located openings, are often strengthened for lifting by adding strongbacks (Fig. 2.36). This may be in addition to extra reinforcement in the panel.

The designer will normally have taken into account the size, shape and weight of the panel and whether strongbacks are to be used for lifting purposes. Potential clashes between strongbacks' roof/floor support angles, props and the rigging should be resolved at the planning stage. Any changes to the specified strongback system should be referred to the designer.

Lifting methods

A single crane generally erects tilt-up panels with the panel rigged for a typical multi-point lift (Fig. 2.37). However, there are other methods that can be employed for different panel shapes and erection requirements.

Tandem lift. Sometimes, for very large panels (typically in excess of 60 tonnes), two cranes are used in tandem. This will normally require

Fig. 2.36. Panel strengthened by strongbacks for lifting

the use of a spreader beam to connect each crane hook to a common lift point. Although this method is sometimes used it is preferable to use a single larger capacity crane where available.

Blind lifting. A reverse lift or blind lift is one where the crane operator is unable to see the upper face of the panel when the panel is lifted. This method is used occasionally but should be avoided whenever possible. However, a reverse lift is sometimes necessary, when, for example, the panels are erected from outside the building. It may also be unavoidable when erecting the last panel. The main danger is that the panel leans towards the crane and additional precautions may need to be taken. If a blind lift is going to occur it is important that this aspect is discussed at an early stage with the prospective crane company.

Top lift. This method is normally used only for small panels, typically not more than 4.5 m high. In this case the lift inserts are positioned in the top of the panel rather than on the face, and the panel will usually require the provision of reinforcement for lifting, which may be more than is required for in-service loadings. This is not an efficient panel design but it leaves an unmarked face as the panel hangs plumb. This method is also commonly used for factory-produced panels.

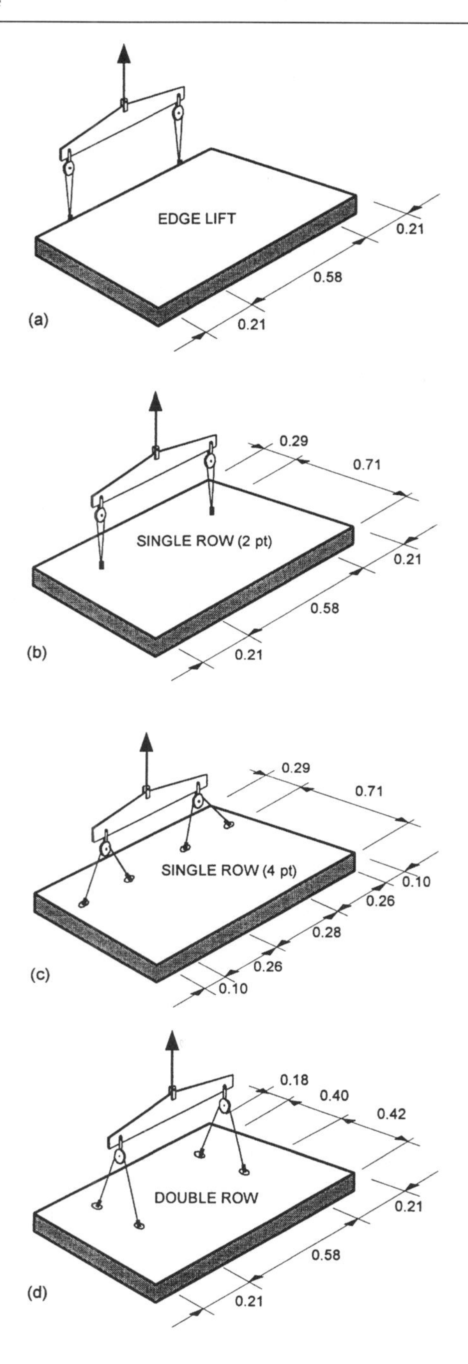

Fig. 2.37. Principles of lifting set-up

True vertical lift. The erection of panels on a boundary against an existing wall can often be difficult. Face-lifted panels will always hang slightly off vertical by 3–5°. Top-lifted panels will hang vertical. However, this lifting configuration is not economic for panels over about 4.5 m high. One solution for face-lifted panels is to use an extra set of inserts in the top edge. The load can be transferred to these using a second crane, or the panel could be temporarily propped off vertical and relifted off the top edge. Alternatively, offset lifting brackets can be used, so can trigger mechanisms, which lock the lifting ropes against the top of the panel when the panel is nearly vertical. These operations need to be planned beforehand and should be carried out only by an experienced crane operator.

Adjustments and tolerances

It is of the utmost importance that the specified panel and joint tolerances are realistic. Once established, they must be maintained. In general, panel variations lead to a growth in overall wall length. Depending on their size, joint details may be used to absorb these variations either progressively at each joint or collectively at one location, for example, at an oversail corner or doorway. If tilt-up panels are being used in conjunction with in situ construction, then the tolerances for tilt-up panels should not be used to absorb the construction errors of the in situ work.

The levelling pads should be compressed fibre sheet, in situ concrete or PVC shims of adequate strength to carry the loads. Steel shims should not be used since they can corrode, and can have too much frictional resistance, which can result in diagonal cracks near the ends of the panel as the panel shrinks. The tendency to crack can also be avoided by providing additional horizontal reinforcement at the bottom of the panel and continuous support from grouting under the panels. Concrete levelling pads should be the full thickness of the panel and at least 500 mm long. PVC shims should be at least 150 mm long. The length of the pads is dependent on the panel bearing stresses. The PVC shims have a high compressive strength (typically 55 MPa), sufficient to resist the weight of even the largest panels. Shims are available in a range of thicknesses (typically between 1.5 and 6.5 mm) and should be limited to a maximum combined thickness of 25 mm.

Bracing

Because a tilt-up panel is not a completed wall immediately upon lifting, temporary support in the form of bracing provides the necessary protection against the forces of wind (Fig. 2.38). Bracing gives the con-

Fig. 2.38. Example of braced panels

tractor time to turn isolated panels into one unified structure. A number of variables will affect the amount of force each brace must resist, including wind velocity, surface area of the panel, the presence or absence of openings, the dimensions of the panel and the angle of each brace relative to the panel. The number of braces is then found by dividing this force by the safe working load for the particular brace type being considered. A minimum of two per panel should generally be employed. The braces are normally attached to the panel by fixings incorporated during its fabrication. The foot of the brace is usually attached to the floor by special anchor bolts inserted into the holes drilled into the floor slab. Sometimes the braces may need to be strengthened by additional knee bracing.

Release of panel

The panel must not be released from the crane until the braces attached to the panel have been installed (including any knee, end or cross bracing) and the panel is approximately plumb. Some adjustment of the braces may be made after release (by turning the threaded portion), but the panel should be within about 100 mm of plumb before release. Before being released from the crane the panel must also be level so that the vertical joint between panels will be of a uniform width.

Final grouting

The gap between the bottom of the panel and the footing should be grouted or dry packed to transfer the load to the footings. This will normally need to be done before roof or other members are installed.

Panel finishes

Surface treatments

A major factor in the acceptance and increased use of tilt-up construction in Australia, the USA and New Zealand in recent years has been the improved appearance of the structures. A wide variety of aesthetic effects has been created by using treatments and finishes, most of which are simple to achieve (Fig. 2.39). In addition to the advent of specialised paints and coatings at reasonable cost, much of the improvement is due to the sensitive detailing of panels. Thoughtful use of grooves, textures and colour can break down the scale of a large flat wall to make it more appealing, and can highlight particular areas or features. Although several of these finishes can be achieved on the face-up side of a panel at casting, it is normal to apply them to the face-down surface in order to ensure consistent quality finishes, to avoid lifting fittings in the visible outer face, and to allow stack-casting.

Fig. 2.39. Grit-blasted ribbed panels were used on this Catholic church in Queensland, Australia

Grooves and relief

The creation of grooves in tilt-up panels at the time of casting is one of the easiest ways of providing visual interest to otherwise large flat areas of concrete. If the width of the groove is chosen to match a standard paintbrush or roller, it becomes a simple matter to apply a coloured coating in the groove that will contrast with the colour of the panel. Such grooves are easily formed in face-down casting by fixing strips of sealed timber or expanded polystyrene or polyethylene to the casting bed. The sides of the grooves should be tapered by a minimum of 15° to permit easy removal of the groove former and to avoid sharp corners, which may chip. Groove formers may be stuck to the floor slab using gun-applied mastic. Close attention to correct setting out and location is essential to ensure high quality finishes.

A similar technique using sheet material such as plywood or expanded polystyrene is possible for simulating windows, forming architectural features and providing relief on an otherwise flat surface. It is possible to use shaped formers to produce indented letters or logos to give the building a company identity.

Form liners are an alternative method for producing a series of grooves to give a ribbed finish. They are available in a variety of patterns. They can be made of rubber, plastics, timber or metal decking and are laid on the casting bed to form a profiled or textured surface against which to cast the concrete. In general, selective and sensitive use of form-lined texture/relief in relatively small areas is more effective than in larger areas, depending on the fineness of the relief produced. When using form liners, particular care must be taken to choose an appropriate bond breaker.

Exposed aggregates

Exposed aggregate finishes may be formed on either face-up or face-down surfaces, but quite different techniques are required. For face-up work, the traditional water-washing approach that is used for paving is appropriate. A chemical retarder is often used to create an exposed aggregate finish on panels cast face-down. Care must then be taken when placing the concrete not to abrade the retarder. Protect the casting surface during the initial placement of concrete by deflecting the pour with shovels, a wood baffle or similar equipment. Subsequent wet concrete should be placed onto previously placed material and then spread with rakes or other hand devices.

Sand embedment is the normal technique used for face-down casting. The selected larger aggregate or stone is spread in a single layer over a thin bed of sand and tamped into place. While pouring the concrete, care must be taken not to displace any aggregate from the sand bed. On lifting the panel, the sand is brushed off to leave the embedded aggregate showing on the surface.

Brick finish

A variety of methods is available to provide or simulate brick finishes to tilt-up panels. The Scott Brick system from the USA supplies brick slips with dovetail keys on their concealed face, which fit into a snap-together spacing system that simulates both stretcher and stack bonds (Fig. 2.40). The sacrificial spacing system replicates a lightly recessed jointing effect and the visible face of the brick slips is protected by wax against grout staining. Another method, successfully employed in the UK, uses bricks cropped in half along their long axis through holes formed during normal manufacture. However, the bricks are thicker than slips and require a spacing system in the mould used for their casting, which must be sequential, so compromising fast tiltup construction.

A further alternative is for tilt-up to be used as the loadbearing inner leaf of the building. This allows a brick outer to be laid later, off the critical path, and tied to the concrete panel across the cavity formed.

Surface coatings

Surface treatments are one of the easiest and most versatile ways of improving the appearance of smooth-finished surfaces and can vary

Fig. 2.40. Brick-faced panels enhance the external appearance of this YMCA/TWCA centre in Ontario, Canada. The inset shows brick slips laid face down in Scott Systems' Grid Snap™ system (information courtesy of CON-STEEL Tilt-up Systems)

from simple acrylic paints or high-build elastomeric coating systems to chemically bonded finishes or stains guaranteed for up to 25 years. The wide range of colours available makes it possible to choose an attractive colour scheme, whether it is for decoration or to give corporate identity to a building. Surface treatments can be applied to the total surface or used to highlight parts of it; they can be easily reinstated after damage and may be changed to give a new image following a change in ownership or tenancy.

For any surface treatment to be successful, the surface must be properly prepared to receive it. In tilt-up construction it is also advisable to check what effect the bond breaker or curing compound may have on any subsequent treatment. Some materials are incompatible and will impair coating adhesion unless physically removed.

Mock-ups and test panels

For many tilt-up buildings it is usual to erect a small test panel for all parties concerned to agree upon standards, particularly for finishes. Mock-ups such as these allow alterations before construction and easy reference during construction.

Rapid flooring cycles

Analysing shoring and reshoring in multi-storey buildings

(extracted from a paper written by M.K. Hurd and P.D. Courtois (1986))

Introduction

In multi-storey buildings, the shoring or props which support the freshly placed concrete on the top level are supported by lower floors which have been reshored or backpropped as they may not have attained their full strength, and may not have been designed to carry loads as great as those imposed during construction (Fig. 2.41). The additional load from fresh concrete has to be distributed over enough floors to minimise deflection and the risk of early-age cracking.

It is impossible to recommend a single general analytical procedure for shoring and reshoring multi-storey structures, since the inter-relationship of the variables such as design load, capacity of slab, allowance for construction load, dead load of concrete, actual construction live load, cycle time between floors, strength development of chosen concrete, span of slab, type of forming system and support points, will vary from project to project and differ widely.

Fig. 2.41. Building under construction, with many floors propped

Definition of shoring and reshoring

Shores are the vertical or included support members that are designed to carry weight from the floor above. Reshores are shores placed firmly under a stripped concrete slab after the original formwork has been removed. The stripped slab is allowed to deflect and expected to support its own weight and any light construction load posted to it before the formwork is struck. To ensure the reshore acts properly, it should be placed snugly under the slab but not wedged so tightly that it could pick up any slab load. The reshores are there to pick up additional construction load from the freshly placed concrete on the upper floors and to reduce deflection of the slab due to creep.

Actual practice on sites varies as to how reshores are placed. Sometimes they are wedged so tightly that they pick up load from the slab being reshored, instead of the correct procedure being followed. Because we cannot be certain how reshoring will be carried out, how the concrete will creep at early ages, or what local load redistribution will take place, a rigid precise theoretical solution seems impossible to justify. Therefore certain simplifications and basic rules need to be adopted to provide a robust yet valid empirical method of analysis.

Simplified approach

Research studies over 25 years have given validity to simplifying assumptions made for predicting the slab load distributions, and these have been corroborated by field measurement of actual loading. The simplifying assumptions in the floor cycling analysis are:

1. Shores and reshores are infinitely stiff, relative to the slab.
2. Slabs are interconnected by shores and therefore they deflect equally when a new load is added. They carry a share of the added load in proportion to their relative stiffness.
3. Slabs have equal stiffnesses and so added loads are shared equally by the interconnected slabs. This is not precisely true, but whether we assume this or the more accurate case that the stiffnesses are proportional to the various strength levels attained, it seems to make only a small difference in the results. We know this from the literature, as well as by trial and error analysis. So it seems both satisfactory and convenient to follow the simpler assumption.
4. Ground level or base support is rigid.

Floor load history: an illustrative example

In this example, to show the method of analysis, only two levels of shores are used and construction live load is ignored. The engineer is cautioned to make appropriate adjustments for actual site conditions and loading history in applying this method.

We refer to load as units of dead load *D*, per floor of structure, which is the weight of the slab ignoring the weight of formwork.

Step 1 (Refer to diagram in Table 2.1.) When the first floor is cast, the full load is carried by shores to the ground floor.

Step 2 When the second floor is poured, the full load is carried through the shores to the ground or base floor. We assume that shores are placed one above the other and aligned throughout the several storeys of construction. At this stage, neither the first- nor the second-floor slab is carrying any load and the props below the first floor are carrying the load of the two floors.

Step 3 When the concrete in the first-floor slab has gained sufficient strength to be self-supporting, remove the shores beneath the first floor. The first and second floors will deflect equally and carry equal loads—their own weight.

Step 4 Cast the third-floor slab. The load of this slab is distributed to the first- and second-floor slabs equally. Note the three columns of figures in Table 2.1. The first column shows load at the beginning of operation. The middle column shows load added during the operation and the third shows total slab loads.

The shore load between floors is determined by examining the number of floors (*D*) above each level of shores and subtracting the total load

Table 2.1. Simplified analysis of loads on shores and slabs of multi-storey structures. Two levels of shoring are considered (from Hurd and Courtois, 1986)

Step no.	Operation and remarks	Status of structure	Load carried by slab in multiples of D: At beginning of operation	Change during operation	Total at end of operation	Shore load at end of operation
1	Place level 1 concrete; full load transmitted by shores to ground	1	0	0	0	$1D$
2	Place level 2 concrete; all load goes through shores to ground, since slab 1 cannot deflect and pick up load	2	0	0	0	$1D$
		1	0	0	0	$2D$
3	Remove first-level forms and shores; load is divided equally between two slabs. In this case each slab carries its own weight	2	0	D	$1D$	0
		1	0	D	$1D$	
4	Place level 3 concrete. Its load is distributed equally to two interconnected levels below and added to the existing load in those slabs	3	0	0		$1D$
		2	$1D$	$+0.5D$	$1.5D$	$0.5D$
		1	$1D$	$+0.5D$	$1.5D$	

(continued)

Table 2.1 (continued)

Step no.	Operation and remarks	Status of structure	Load carried by slab in multiples of D			Shore load at end of operation
			At beginning of operation	Change during operation	Total at end of operation	
5	Remove forms and shores beneath level 2. Shore load is distributed equally to two interconnected slabs	3	0	$0.25D$	$0.25D$	
						$0.75D$
		2	$1.5D$	$0.25D$	$1.75D$	
		1	$1.5D$	$-0.5D$	$1.0D$	
6	Place concrete at level 4. Its load is divided equally between the two interconnected slabs	4				
						$1D$
		3	$0.25D$	$0.5D$	$0.75D$	
						$1.25D$
		2	$1.75D$	$0.5D$	$2.25D$	
		1				
7	Remove shores and forms beneath level 3. The shore load of $1.25D$ is divided equally between slabs 3 and 4	4	0	$0.62D$	$0.62D$	
						$0.38D$
		3	$0.75D$	$0.63D$	$1.38D$	
		2				
8	Place concrete at level 5. Divide the new load equally between slabs 3 and 4	5		0		
						$1D$
		4	$0.62D$	$0.5D$	$1.12D$	
						$0.88D$
		3	$1.38D$	$0.5D$	$1.88D$	

Step no.	Operation and remarks	Status of structure (level)	Load carried by slab in multiples of D: At beginning of operation	Change during operation	Total at end of operation	Shore load at end of operation
9	Remove shores and forms beneath level 4. Divide 0.88D shore load between levels 4 and 5	5	0	0.44D	0.44D	
						0.56D
		4	1.12D	0.44D	1.56D	
		3				
10	Place level 6 concrete. Load is distributed equally to slabs 5 and 4	6		0		
						1D
		5	0.44D	0.5D	0.94D	
						1.06D
		4	1.56D	0.5D	2.06D	

Freshly placed slab
Hardened slab
Storey of shores

D = weight of slab
No construction load included

carried by the slab (or slabs) above. In Step 4 the shore load is $1D$ between floors 2 and 3, while the shore load between floors 1 and 2 is $(2-1.5)D = 0.5D$. This technique is used throughout the remaining floors.

Step 5. Whenever a slab no longer has shore or reshore loads transferred to it, as for floor 1, the total slab load obviously is $1D$. This verifies that all calculations are correct to that point. The next step is to remove the props beneath the second floor. The load from these shores is divided equally between the second- and third-floor slabs. The half floor load carried by the shores just removed applies one-quarter floor load to each floor so that the second floor has $1.75D$ and the third has $0.25D$.

Step 6. Cast the fourth-floor slab. The second and third floors will deflect equally, each taking half the weight of the new floor. The second floor now has a load of $2.25D$.

Further sequences are shown in Steps 7 to 10 in Table 2.1. The maximum loads are always carried by the last slab in the system cast before the shores at ground or base level are removed. In this case it is the second-floor slab.

What carrying capacity will the floor have?

We need to determine whether a building can safely carry the loads we plan to impose on it for the given floor cycling times. We must calculate the probable carrying capacity of the slab at an early age, i.e. before it has gained its full design strength, to check whether the chosen floor cycle time is acceptable. According to the ACI Code, the floor slab must have an ultimate load capacity of 1.4 dead load (DL) plus 1.7 live load (LL). This is the capacity a slab would have when it reached its 28-day equivalent cube strength. It seems reasonable to assume that a slab less than 28 days old will have some fraction of the 28-day strength capacity, usually in proportion to its maturity. It is safe to assume that the modulus of elasticity, shear and tensile strength develop at the same rate or slightly quicker than the compressive strength in concrete, so that we can use compressive strength gain as the benchmark. For lightweight aggregate concrete these assumptions may not be valid, so further data must be considered.

To estimate the percentage strength gain with maturity, typical strength development curves can be plotted, based on actual strength development of the mix under different curing temperatures (see Fig. 2.42).

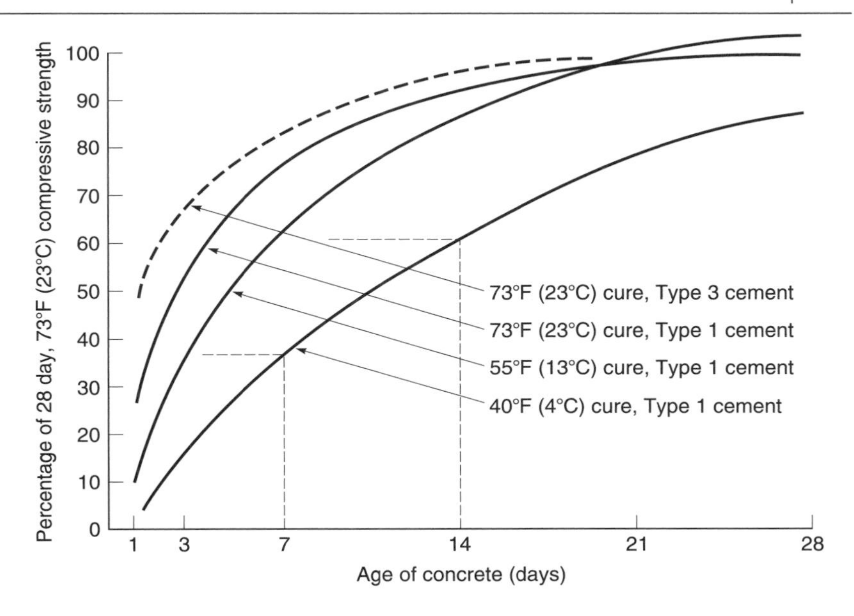

Fig. 2.42. Concrete strength development with age (data from Klieger)

Assessing slab strength for a four-day floor cycle

We wish to check the slab's capacity to support added loads for floor slabs cast every four days. We use the strength development curves to estimate the concrete strength at 4, 8 and 12 days. Divide this figure by an appropriate safety factor to minimise risk of collapse, and use it to check whether the slab has adequate capacity at the early age.

For illustration purposes, we assume a floor slab weighing 100 psf (lbf/ft^2) (4.8 kPa), which is designed to carry 50 psf (2.4 kPa) live load plus 15 psf (0.72 kPa) for services and partitions. Converting that to units of slab weight D, we have:

$$\text{dead load} = 1.15D$$

$$\text{live load} = 0.5D$$

ultimate load capacity of the slab at 28 days will be

$$1.4 \times 1.15\text{D} + 1.7 \times 0.5D = 2.46D$$

For the 4, 8 and 12-day strength we have to reduce this in proportion and give it a factor of safety. ACI 318 suggests a load factor of 1.3 should be used on construction loads; Hurd and Courtois (1986) prefer a more conservative 1.4. Table 2.2(a) shows allowable loads that the slab can support at 4, 8, 12, 16 and 24 days, using a factor of safety of 1.4.

Comparing, in Table 2.2(b), these allowable loads with the predicted loads from Step 8, we can see that the allowable load for slab 3 is

Table 2.2. (a) Effect of slab age on allowable load. (b) Comparison of estimated slab load and allowable load

(a)

Age of slab cured at 50°F (10°C) or above	Estimated percentage of 28-day compressive strength*	Ultimate load capacity at est. strength†	Allowable load at estimated strength (F.S. = 1.4)
4	42	1.03*D*	0.74*D*
8	66	1.62*D*	1.16*D*
12	79	1.94*D*	1.16*D*
16	90	2.21*D*	1.58*D*
24	98	2.41*D*	1.72*D*

* Based on data from Klieger.

† For slab assumed to have a 28-day ultimate capacity of 2.46*D*, where *D* is weight of slab.

(b)

Interconnected slabs	Estimated load	Age of slab: days	Allowable load on slab at age indicated (F.S. = 1.4)
Slab 5	0	0	0
Slab 4	0.91*D*	8	1.16*D*
Slab 3	1.75*D*	16	1.58*D*
Slab 2	1.34*D*	24	1.72*D*
	Total load 4.00*D*		

exceeded. Clearly, this plan would not be suitable for a building with a four-day floor cycle; either the propping regime would need to change, introducing another level of reshores and increasing concrete strength, or the floor cycle must be slowed.

Since this paper was written, refined mathematical analysis by Liu, Chen and Bowman, accounting for variable stiffness of floors, shores, lack of rigidity of support and other variables, has shown that in most cases a modification coefficient of between 1.05 and 1.10 would only be necessary to correct results obtained by this simplified method and convert them to more rigorous precision.

Early striking of formwork and forces in backprops (reshores)

(This section has been written with reference to the European Concrete Building Project.)

Research work carried out under the supervision of Professor A.W. Beeby of Leeds University in 1998, to predict the striking times for soffit formwork and the force in the shores and reshores in multi-storey flat slab construction, made the following observations:

1. Formulae are proposed for calculating the load that may be carried by immature slabs as a function of the concrete strength. The basic principle is that no part of the structure should be subjected to stresses, deformation or cracking greater than is implicit in the design for service conditions. To meet this criterion the following two inequalities have to be satisfied:

$$(w/w_{ser})(f_{cu}/f_c)^{0.6} \leq 1$$

$$(w/w_{ser}) \leq 1$$

   where w = construction loading, w_{ser} = design service loading, f_c = concrete strength under construction conditions, and f_{cu} = specified characteristic cube strength.
2. The forces in the props (shores) are significantly lower than is commonly assumed. On the Cardington test building they were generally about 1/3 of the prop forces rather than the commonly assumed 1/2. This difference has been shown to be due to two factors:

   (a) the stiffness of the props cannot be assumed to be infinite;
   (b) the relative arrangement of props and backprops (reshores) has significant effect on the force transmitted to the backprops.

   Because of these effects very little advantage was gained by backpropping through more than one level at Cardington.
3. The forces in the backprops may be calculated with sufficient accuracy on the assumption that the slabs remain elastic and uncracked during construction.
4. The use of more sophisticated methods of calculating floor slab behaviour during the construction cycle is of value only if adequate control is imposed on site operations to ensure that the assumptions used in the calculations remain valid.*

*The issue here is not just the accuracy of the calculation method but the level of control that is considered reasonable to apply over site operations. Some years ago the author asked Professor Beeby to investigate backpropping forces by monitoring them on a site. The results were so variable that they were effectively uninterpretable. He later suggested that one possible reason for this anomaly was that construction operatives may temporarily remove a backprop for a construction operation on the floor below and replace it later in the day. However, this does not usually happen on building sites with good supervision, so it is an exception. It is more likely that the props have been over-tightened by varying amounts, but this can be corrected on site by proper instruction. Up to now construction personnel have not been made fully aware of the critical nature of such operations.

5. The deflections measured in the test building at Cardington by precise levelling, up to an age of 300 days, may be predicted reasonably accurately on the assumption that the slabs have not cracked. An exception to this rule was on floor 3, where it seems likely that cracking had occurred.

Architectural concrete

Site-cast architectural concrete

An introduction

High quality, well finished site-cast concrete has been the ambition of many architects, bridge engineers and contractors the world over. Cast-in-place construction rather than precast concrete, so the theory goes, can also be a competitive cladding option as it offers greater construction efficiency and a reduction in cost. But the reality is that many architects and their clients believe that trying to achieve a precast finish using site-cast concrete is just not feasible. Why?

On the whole, it seems that not many architects are prepared to specify cast-in-place concrete because they do not have confidence in the building industry to produce good visual concrete. On many occasions the production of fair-face site-cast concrete has been a catalogue of remedial work, contractual losses and client dissatisfaction. Architects who have attempted to specify fair-face concrete have been appalled by the poor performance of the construction industry, more preoccupied with volume business rather than quality. Who is to blame? The architect for not understanding that concrete has a skin that breathes, soaks up moisture, attracts dirt and pollution and weathers just like stone, or the material technologists and the construction industry for not formulating additives and surface treatments that could minimise the effect of weathering?

Experienced contractors complain that architects are not giving them proper guidelines in the contract on how to achieve the fair-face work specified, nor the colour that is wanted, other than a very subjective description. More often than not, the contractor will price for something less than the architect has envisaged for the work, and when that happens the kango hammers and variation instruction start flying. Is there a solution?

In search of good news

It was a hot July day, and, with hat and camera, sitting half asleep, half awake, I was enjoying a slow burn on the deck of a river boat with a garbled commentary about Chicago's famous buildings, when suddenly ahead I saw a smooth, curved, grey-white building standing quite alone on the water's edge. It's been almost six years since I last saw Bertrand Goldberg's River City Plaza (Figs. 2.43 and 2.44). This architect had designed Marina City and unashamedly admired the work of Le Corbusier. The building facade looked crisp, clean and smart, no tell-tale rust runs or pockmarked infills to the fair-face concrete finish. It was ten years old and I was looking at the finest cast-in-place high-rise structure ever built in recent decades. So what's the story? How did Goldberg achieve what no one, apart from Tadao Ando or Peter Zumptor, has consistently done since?

The answer, according to Symons Corporation of Chicago, which supplied the formwork package, the form liners, the release agents and the upward pumping method on River City Plaza, was to ensure that the contractor adhered to the specification and workmanship techniques that were stipulated by Symons for architectural concrete. "Invariably, when we find there are problems on one of the jobs we have supplied with an architectural concrete package, it has been because the contractor has decided to skip or compromise on the pro-

Fig. 2.43. River City Plaza, Chicago (1986); Architect: Bertrand Goldberg

Fig. 2.44. River City Plaza: interior view

cedures that we have laid down'' says Ray Bartholomae, General Manager of Symons Corporation. ''We are able to offer a guarantee to the client and architect that a good visual concrete finish can be achieved every time if our procedures are followed. We have not lost a case against Symons on that point.''

The concrete for producing good architectural concrete, whether it is cast in place or precast, must, briefly, have the following characteristics:

☐ Cement content must be in excess of 350 kg/m^3.
☐ Aggregate/cement ratio must not be greater than 6:1.
☐ The sand should be a zone M to BS 882.
☐ Water/cement ratio should not be greater than 0.5.
☐ Formwork and release agent that will produce a blemish-free surface finish must be used.
☐ Release agent and any admixture in the concrete must be compatible.
☐ The compaction procedure for architectural concrete must be followed.

The right mix

To ensure that the concrete colour is consistent and uniform from one batch to the next, the cement content must remain constant, the source of the cement must not change and the water/cement ratio must remain tightly controlled with the minimum of variation. The colour of concrete is controlled by the finest particle in the mix, which is usually the cement, and any sand particles finer than 150 μm, unless a pigment is introduced which is finer. The paste that forms on the face of the concrete in contact with formwork is a mixture of cement and pigment if included, sand and water, with the cement and pigment dominating the colour. If, however, the water content changes, then in areas where the water content is higher the concrete tone will be lighter, and where it is lower, it will be darker in tone. Therefore, when writing a specification for architectural concrete, the source and the cement content of the approved mix must remain constant, and the water/cement ratio of the mix kept within a narrow margin of variability. Tests for conformity of concrete production should be based on cement content checks and water/cement ratio measurements and not cube strength. There are techniques and simple testing equipment for monitoring these properties quite quickly on site.

Form liner manufacturers such as Symons of USA, Noe of France and Reckli of Germany may not agree with Tadao Ando on the most cost-effective way of forming architectural concrete, but how Ando achieves good in situ concrete finishes (Fig. 2.45) is worth noting. "In civil engineering the concrete used is very hard, giving the impression of being very powerful and strong. I insist on using hard concrete, which is the final finish. We use plywood for formwork and then varnish all the panels to ensure a good surface to receive the concrete. In Japan we have a tradition of highly skilled carpenters and joiners and these craftsmen help to create aesthetically finished exposed concrete" explains Ando.

On River City Plaza, built in 1986, the whole structure was cast in place using a modular formwork system lined with an elastomeric form liner. The concrete was placed by upward pumping into the storey-high facade forms. The form liner and specially formulated neutral non-staining release agent ensured that the concrete was uniform in colour and did not suffer from surface discoloration during compaction and when the formwork was removed. Upward pumping avoids any trapped air getting into the mix when concrete is placed in the forms. The pumping action of the concrete also helps to remove any large air pockets in the mix. Frequency of vibration and poker size are checked for the radius effect of poker vibration to determine the ideal spacing of the poker and the depth of immersion into the concrete. This will ensure that vibration has the same intensity on the

Fig. 2.45. Children's Museum, Japan, designed by Tadao Ando

face of the forms. This is a critical procedure for architectural concrete and one that is often overlooked in most concrete specifications.

Chemically neutral non-staining release agents or specially formulated wax emulsions are important in good visual concrete production. Trial formwork panels should be cast to check the compatibility of the release agent and the forming surface. Very dense forming surfaces such as metal and plywood panels with a heavy-duty overlay, usually a bonded phenolic resin film, will give a shiny surface finish with a tendency to highlight and exaggerate small tonal changes in the colour of the concrete. The advantages of such systems is that they can be re-used fifty or more times. On the other hand, to produce a matt surface finish, it is best to use a plywood with a medium density overlay, usually a resin-impregnated film, which will reduce the incidence of blowholes and tonal variations but will have only ten re-uses. High density plastic form liners with a cheap formwork backing and the correct release agent, or a good birchwood ply coated with polyurethane paint and sanded to give a smooth matt surface finish, would be a viable alternative for multiple uses.

Tadao Ando suggests rubbing down the surface of the concrete after the formwork has been removed and when it has hardened sufficiently. This process will smooth over minor blemishes and take off any surface discoloration, but it may not be necessary if the correct formwork and release agent have been used. It is a hugely labour-intensive operation that will create a lot of dust and may not be appropriate for countries where labour is expensive.

It is vital to consider the long-term effects of weathering, moisture staining and lichen growth on concrete, which has a porous and absorbent surface skin. Coatings which blend well with the natural look and colour of concrete, that are not shiny nor turn straw colour or break down in ultraviolet light, should be prescribed to make the new concrete surface impermeable. Recently the National Theatre in London, that masterpiece of concrete architecture designed by Denys Lasdun in the 1960s, has had a much needed revamp by architects Stanton Williams. New bars, bookstalls and lifts have been installed, spaces within the building have been opened up and the dirty facade given a good clean at last (Figs. 2.46 and 2.47). It is a great building whose only flaw was to have left the exterior board-marked in situ concrete finish without a barrier coat to prevent moisture ingress and dirty staining. Now that has been put right, the National is going to find many more admirers amongst young architects.

Fig. 2.46. The National Theatre, London

Fig. 2.47. National Theatre: internal view

A guide to specifying visual concrete

These notes have been prepared from direct experience and research on visual concrete production, information collected from the projects visited by the author and some aspects on concrete finishes published by the Concrete Society in the UK in *Technical Report 52*. They have been drafted assuming British Standards.

The notes in *italics* give reasons why a particular aspect of visual concrete production has been specified, or highlight a case study to illustrate why guidance is required.

It is important that the ready-mixed concrete supplier is involved early in the discussion on the choice of concrete colour and finish, to check availability of cement types, availability of special aggregate or pigments, and to agree a practical and affordable concrete mix that can be supplied to the project. Consult with formwork, form liner and release agent suppliers to determine the best product(s) to achieve the required surface finish. Impart this information clearly and concisely in the specification to help the contractor to understand and price what is required.

Seek advice from the research and marketing division of the cement and concrete industry in your country. It's usually free.

The concrete mix

Concrete mix constituents shall be weigh-batched and truck-mixed generally in accordance with BS 5328 or its European equivalent.

The proposed concrete shall comply with the requirements for fair-face concrete work, viz. cement type and content, pigment, water/cement ratio, aggregate and sand content and not strength criteria. The concrete mix in combination with the formwork and selected release agent, when properly placed and compacted, will produce a blemish-free finish, free of blowholes, and give a uniform surface colour appearance.

The proposed concrete mix shall be cast in a sample panel to prove the integrity of the concrete mix, the formwork system and workmanship in meeting this requirement.

For general guidance the proposed concrete mix should comply with the following characteristic to satisfy uniformity of colour and surface finish:

1. The concrete workability shall be sufficiently cohesive for vibrator compaction, pump delivery, handling by conveyors and chutes on site, to free fall 2 m, and to be placed in vertical forms without segregation or causing excessive bleed water to rise to the surface.
2. The concrete must have a cement content not less than 350 kg/m^3 and must be taken from the same batch of cement, to eliminate possible changes in cement colour. *State what cement type is required—white cement, light grey GGBS cement or Portland cement—and describe what colour the concrete is required to be.*
3. The type of cement, cement content, water/cement ratio, fine aggregate content less than 150 μm, pigment concentration and any approved admixtures shall be fixed for all concrete supplied to the contract and must not be adjusted at any time during the contract. *Fine materials of particle size less than 150 μm control the surface colour of concrete. Pigments are the finest particles, followed by cement and a small proportion of sand that is smaller than 150 μm.*
4. The water/cement ratio shall not exceed 0.5. Once the ratio has been agreed by the architect, following successful trial mixes and panel construction, it must not be adjusted at any time during the contract, as any variation in the total water content will affect the surface finish colour.
5. The total aggregate/cement ratio shall not exceed 6.
6. The sand/cement ratio shall not exceed 2. The sand should be a zone M or similar type sand without too much fine dust. Sand content expressed as a percentage of the total aggregate by mass should not exceed 40%.

7. Coarse aggregate (20 – 5 mm): not more than 20% to pass a 10 mm sieve.
8. Any plasticiser, water-reducing admixture or pumping aid used in the mix must be stated and its compatibility with the release agent checked. Some admixtures can react with release agents to create gas bubbles and this could lead to entrapment of blowholes on the concrete surface.

Production quality control

Truck delivery tickets must show the batch weights of all mix constituents, including the total water content and total water/cement ratio, and show compliance with the approved mix constituents. They shall be given to the concrete contractor for recording and checking on arrival on site. If no delivery ticket is presented, the concrete shall be rejected.

Concrete shall be sampled from the truck mixer before discharge and slump tested to check the colour and uniformity of the mix. If any noticeable colour variation is evident, then the concrete shall be rejected.

Accurate weighbatching and control of aggregate moisture content is essential in the production of good visual concrete. Store aggregates under cover in bins to prevent rain wetting stockpiles. Monitor moisture content of aggregates regularly and adjust the free water content to maintain the correct water/cement ratio.

Wet-batched concrete is preferable to dry-batched concrete as it eliminates the variability and uncertainty of truck mixing efficiency. *Some trucks are not efficient concrete mixers; they can only agitate the mix because of the paddle configuration in the mixing drum.*

Concrete placing and compaction

Clean and wet concrete skips, conveyors and any other concrete handling plant before use. Remove all discharge of grout in the pumping line and excess water as the pipeline is primed.

Place the first layer of concrete into position across the whole length of the formwork for a wall pour and roughly tamp to level. If necessary, use the poker to melt the top of high points in the layer before beginning poker vibration (Fig. 2.48). The maximum depth of an uncompacted layer of concrete shall not exceed 500 mm (Fig. 2.49(a)). High frequency internal poker vibrators shall be used which do not drop speed when immersed in dense concrete, with the correct diameter to effectively vibrate the concrete to the full width of the form or the distance to the outside form face. The spaces between vibrating points shall ensure that the cone of vibration of the last position just overlaps the next

Fig. 2.48. Careful poker vibration is essential for architectural concrete

vibrating position (Fig. 2.49(b)). The rate of concrete placing shall be uniform and must exceed 2 m/h in vertical sections.

Prior to starting the work, the contractor shall be required to produce a sketch drawing of the pour planes and poker immersion points for a typical column and wall pour. This is necessary to show that the type and size of poker, and its radius effect of vibration, will adequately compact the concrete layers and maintain an even compaction force at the form face, thus ensuring a consistent concrete colour. *If the compaction on the form face varies greatly, this can inadvertently cause dark and light colour variations to appear on the finished surface.*

Use rubber linings over vibrating pokers for all compaction work to slab soffits.

Do not use the poker vibrator to move concrete into place.

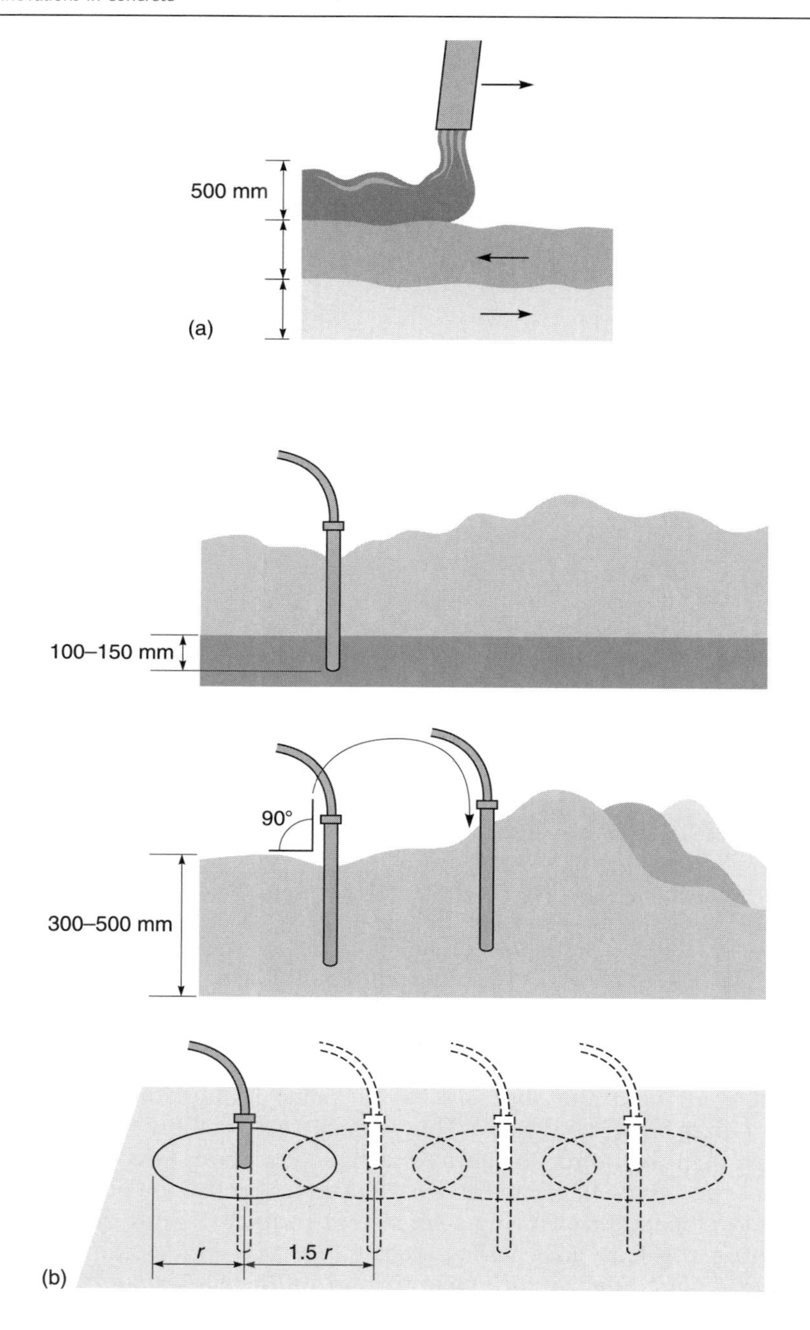

Fig. 2.49. (a) Placing of concrete in successive layers reduces segregation. (b) Guidelines for good poker technique.

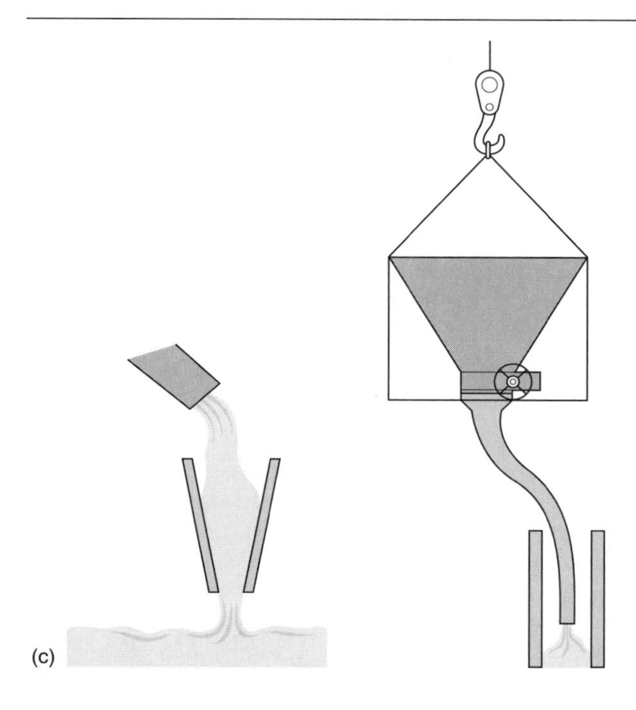

Fig. 2.49 (continued). (c) Use of chute or tremie will avoid blemishes due to segregation or splashing

Tremie or chute the concrete into positions for the vertical forms to avoid concrete splashes on the form face or segregation on the rebar before it reaches the bottom of the pour (Fig. 2.49(c)).

Spade the concrete near to the formwork face to release any air pockets. Do not place poker vibrators near the formwork face. This will avoid poker burns and noticeable variations in colour and aggregate density on the exposed concrete face.

Revibration

Revibrate concrete at the top of the vertical pour after 1 – 2 h to eliminate potential colour banding due to excess bleed water rising to the surface.

Surface tolerances

1. Abrupt changes to formed surfaces: 2 mm permissible abrupt changes between formwork panels.
2. Formed surface imperfections: blowholes, maximum size 3 mm, permissible; not more than ten in any square metre.

3. Floor surface finish must be smooth with no abrupt changes. Variation in level must not exceed 5 mm in 3 m, in any direction.

Formwork

Grout-tight joints. The formwork shall be grout tight and all joints between panels sealed with an appropriate sealant to eliminate grout loss. The formwork and tie bolts shall be designed to resist a full liquid head and high-amplitude vibration during compaction, and the deflection between tie bolts must not exceed 1/360 of the distance h between them. To prevent grout leakage at tie bolt holes, use plastic snap-tie cones with rubber gaskets on the tie.

Release agent. Use only neutral pH, non-staining release agents which, when applied to the formwork, will not wash off during rainy weather. The agent shall adhere to the formwork and debond the concrete effectively while the forms are in place (up to 36 h) without causing blemishes, discoloration or blowholes to form on the surface finish or later cause the surface to dust. Evidence that a proposed release agent complies with this requirement must be presented.

Construction joints and panel layout. Rebates shall be formed at concrete construction joints to form a neat edge to the construction joint. The position and detail of all construction joints shall be shown on the contractor's falsework drawing, together with a layout of formwork panels and tie-bolt hole positions for approval by the architect, prior to the start of concrete work.

Formwork striking times

Formwork should be stripped about 24 – 36 h after casting. This will give a consistent colour to the concrete and avoid exposing the forms to prolonged heat of hydration and chemical action of the concrete in breaking down the release agent. It is important to strip formwork to give the same equivalent maturity time throughout the project as even small variations in maturity time can cause colour variation between panels. Maturity time will be less in hot weather than in cold weather, and these times can be evaluated from cement content and actual concrete curing temperature.

Making good

No making good shall be permitted to exposed surface finishes indicated on the drawings. In all other areas making good may be permitted subject to approval by the architect and to a standard that matches the reference panel.

Reinforcement

Supply clean-cut reinforcement, with no rust marks, for all exposed concrete work.

Cover and spacers

Cover to slab and walls shall not be less than that appropriate to the durability requirements shown on the structural drawing. Spacers for slab and wall reinforcement shall be of plastic construction, rigid enough to maintain rebar cover without deformation and small in contact area with formwork to avoid marking the exposed concrete face.

Curing

Direct wet curing of concrete or wrapping concrete in polythene sheeting is not permitted as this will cause dark and light patches to form which may not fade in time. Maintain a small but definite air space between any impervious protection such as polythene or tarpaulins to avoid direct contact with concrete. Curing membranes are nor permitted as they will stain and prevent non-staining, transparent water-repellent coatings from bonding to the surface. It may be best to do nothing if, after removal of the formwork, the surface has hardened sufficiently and will not dry out prematurely and become dusty.

Reference panel

The reference panel should be a full storey high and of bay width. If form liners are used, the panel should be at least two liner panels wide to include the vertical joint. The finished result should be a true reflection of the quality of the workmanship, curing and construction method.

Formwork, form liners, mix design materials, release agent, concrete handling and compaction procedure and the actual operatives that will be deployed on the job should be involved in casting the panel (Fig. 2.50). Patch repairs of blowholes (should that be permitted), location and filling of tie-bolt holes and similar procedures should be evaluated and agreed at the same time. If the surface is to be rubbed, etched, mechanically abraded and/or coated with water-repellent covering to eliminate dirt staining or water absorbency, then this treatment shall be carried out. In this way the quality of the finish can be agreed, highlighting areas for possible improvement and setting the standard for the project. The surface finish may include small variations in tone even if the work has been done correctly, but these are likely to be distributed randomly over the whole surface rather than being concentrated in one spot.

Fig. 2.50. Casting a reference panel for architectural concrete

Consideration should be given to a sensible viewing distance for the reference panel, as scrutiny from close quarters would be unreasonable unless particular sections of work merit such inspection. In this event consideration should be given in the contract bill to "prepare and cement wash the concrete face to produce an even blemish-free finish", if that is what is desired. Some architects have preferred to leave the surface untouched, to show the natural grain of minor imperfections — the blowholes and blemishes — rather than cover them with a veneer of cement slurry or cement wash which make the surface look painted and plastic.

Quite often the finish tone of grey or white concrete can become an issue, because the architect or client is undecided on the final colour. There is no definition by which the quality and colour tone of the finish can be unequivocally described by specification alone. The same concrete mix poured into different formwork or placed in the same formwork at different temperatures, or retained in the forms for different lengths of time, will result in colour variations. The longer the concrete is cured or left in the forms, the darker the surface tone. The lower the water/cement ratio, the darker the tone. It would be prudent to conduct colour sample trials with the concrete supplier making concrete cubes, or small panels of concrete, prior to starting on an expensive, full-scale reference panel.

Compliance

Having verified to the architect or his representative that concreting procedures have been carried out correctly, the concrete supplied is to specification, formwork panels have been cleaned and prepared and removed diligently, every care taken to maintain the same quality of workmanship as the reference sample with no change of site operatives, then the resulting concrete finish shall be deemed to satisfy the specification provided that (a) blowholes are less than the specified diameter and number per square metre and match the reference panel, (b) there is no honeycombing from grout loss or the like, (c) the blemishes to the surface are no worse than those on the reference panel, (d) alignment and surface tolerance are within specification, and (e) making good, where permitted, brings the surface finish to within compliance standard. Otherwise the concrete shall be rejected and the work redone.

References and further reading

BEEBY, A W (1998). *Early striking of formwork and forces in back props.* European Concrete Building Project, University of Leeds.
BENNETT, D F H (1989). *Advances in concrete construction technology.* British Cement Association, Wexham Springs. Publication 97.309.
BENNETT, D F H (2001). *Concrete architecture: tone, texture, form.* Birkhauser, Berlin.
CLARKE, J L (1992). *Large area pours for suspended concrete slabs.* Reinforced Concrete Council and British Cement Association, Wexham Springs. Publication C/12.
GOODCHILD, C (1997). *Economic frame elements.* Reinforced Concrete Council and British Cement Association, Wexham Springs.
HURD, M K and COURTOIS, P D (1986). Method of analysis for shoring and reshoring in multistory buildings. *Proceedings of Second International Conference: Forming economic concrete buildings.* USA. Paper SP-90-8.
MONKS, W (1999). *Plain formed concrete finishes.* Concrete Society, London. Technical Report 52.
PUTTBACH, E (1986). *Pigments for colouring of concrete.* Bayer Technical Publication 9/86.
SOUTHCOTT, M F and TOVEY, A K (1998). *Tilt-up concrete buildings: design and construction guide.* British Cement Association, Crowthorne. Publication 97.366.
TERMODECK (2001). Technical brochure, April.

3 Reinforcement technology

Prefabrication of reinforcement

Prefabrication of reinforcement can lead to substantial savings in construction time. By enabling reinforcement fabrication to proceed ahead of construction, as much as 20% can be saved on construction time for the frame. Other benefits, such as a reduction in congestion at the workplace and the comparative ease of installation, help to improve overall construction efficiency.

Prefabrication is best carried out at the reinforcement supplier's factory, supplemented where necessary by on-site assembly at ground level. Simpler reinforcement cages and mats and welded fabric panels may be delivered to site ready for installation (Fig. 3.1). Units that may be too large for practical road transport are better assembled on site from a combination of smaller fabricated units.

In general, prefabricated panels, mats or cages for delivery by road should not exceed 12 m in length and 3.2 m in width. To maximise construction efficiency, the location of both the vehicle off-loading area and the site assembly area should enable prefabricated units to be crane-handled directly into position.

Design and detailing

Most pre-assembled arrangements of reinforcement may require a modification to the traditional reinforcement detailing in order to simplify and standardise the prefabrication process. Prefabrication of reinforcement requires a commitment at the design stage if the benefits of faster construction and earlier completion are to be fully exploited. Reinforcement should generally be detailed in accordance with the supplier's recommendations. Reinforcement suppliers provide a free advisory service on prefabrication detailing.

Reinforcement cages and mats

Prefabrication is possible for most structural reinforcement assemblies including those for foundations, loadbearing walls, columns, beams, spandrel panels, parapet and retaining walls and suspended floor slabs. By examining some specific examples, we can formulate a few

Fig. 3.1. Factory prefabrication of reinforcement

basic assembly rules and detailing tips that will help to rationalise the prefabrication process.

Loadbearing walls

In general, loadbearing reinforced concrete walls in building structures can be relatively lightly reinforced, but the fixing of the assembly in a vertical plane can be time consuming. This difficulty can be overcome by incorporating welded fabric sheets in the detailing of wall reinforcement. Fabric sheets designed for the purpose, with suitable lap lengths devoid of transverse steel, can accelerate wall construction (Fig. 3.2).

Suspended slabs

Most slabs are rectangular with an orthogonal grid of reinforcement; significant time savings can therefore be achieved if the reinforcement is placed in the form of prefabricated mats instead of loose bars assembled on the formwork. Most of the major fabric makers also produce tailor-made welded sheets, with reinforcement specifically arranged to give the most economical solution. In very simple cases it may be possible to pre-assemble entire panels, but generally,

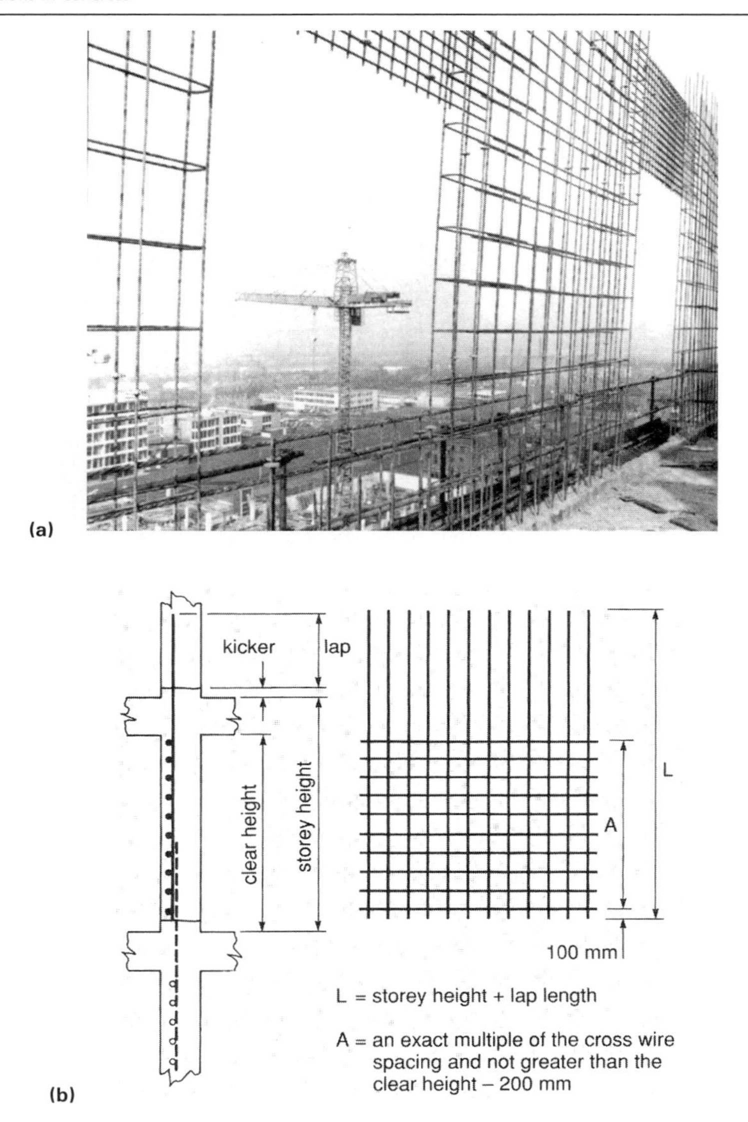

Fig. 3.2 (a) Welded fabric used in wall construction. (b) Detailing of wall fabric panel

because of interconnection with beam and column reinforcement, sub-panel assemblies are more practical. The example in Fig. 3.3(a) illustrates an effective prefabricated mat assembly, using bars, for a flat plate slab. The crane places the bulk of the main reinforcement directly into position. On the other hand, the same slab could be

(a)

(b)

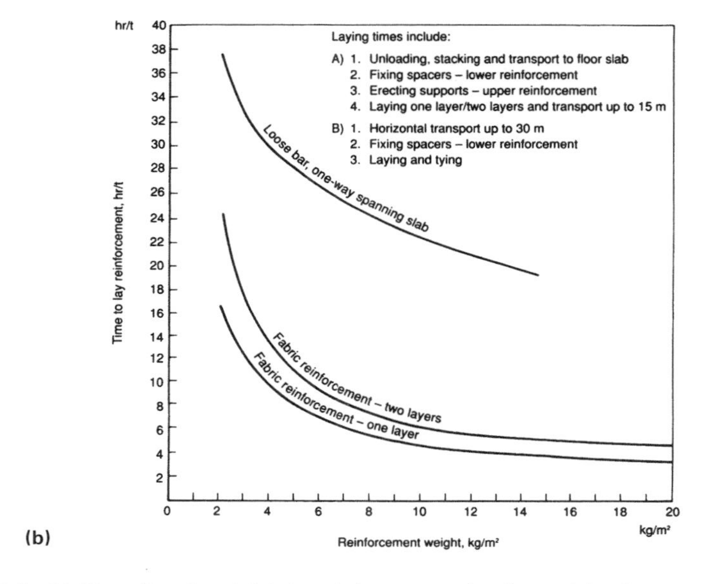

Fig. 3.3. (a) Two-directional fabric reinforcement for floor slabs (courtesy: RCI Ltd, Kettering). (b) Time taken to lay fabric reinforcement mats and loose bars in floor slabs (based on German studies)

detailed entirely for welded fabric mats, using two layers of mat with main bars at right angles. Fabric panels can be sized so that they are light enough to be manhandled into position. Thus crane time is reduced to lifting and placing the strips of fabric panels close to the work area.

Many other combinations of bar and fabric mat assemblies are feasible, making it possible to detail and schedule mats to suit a particular flat slab design and adapt them to the requirements of a particular site. Typical data for the rate of placing fabric mats and conventional bar reinforcement are shown in Fig. 3.3(b).

Column cages

Column cages are perhaps the best known example of pre-assembly. Prefabrication allows the work to be carried out at ground level with conveniently positioned components for effective tying and fixing. The vertical reinforcement for columns and for loadbearing walls should be detailed with all laps located immediately above or below the floor slabs or beams.

Beam cages

First assess the longitudinal reinforcement required for the beam. If it is a simply supported beam, then it is straightforward and very few modifications for prefabrication are necessary. If there is continuity over the supports, then the reinforcement requires some modification. The basic assembly should have all the bottom bars in place with a maximum length equal to the clear distance between the supports. Full shear reinforcement should be provided, with open-top stirrups to facilitate the placing of other bars (Fig. 3.4).

Nominally sized top reinforcement with a maximum length equal to the clear span distance between the supports should be provided to tie the tops of the stirrups. Further benefits may be gained by prefabricating the section of the beam cage between supports. This may be done using normal bar reinforcement, in which case temporary bracing to the outside of the cage may be necessary to prevent racking during handling and hoisting. An alternative is to prefabricate the cage, using fabric links.

If there are a number of beams of identical section and length, then standardisation of their reinforcement will enable faster and more accurate assembly. This aids interchangeability in the event of construction changes and is a cost-effective insurance against accidental misplacement.

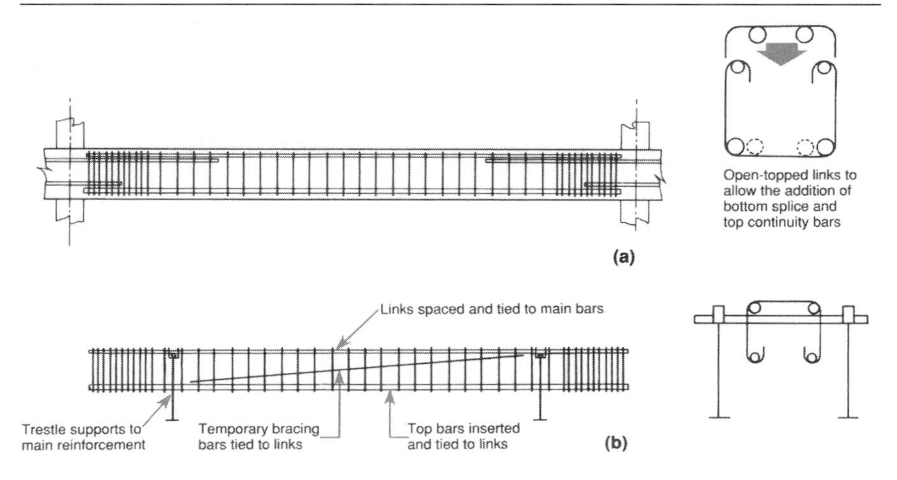

Fig. 3.4. Beam cages: (a) traditional detail, simplified for prefabrication; (b) prefabricated cage with bar links

Speed trials

Trials to measure the fixing time for different types of reinforcement assembly took place in the BCA's service yard at Wexham Springs in 1991. The structural elements were:

- ☐ flat slab;
- ☐ beam;
- ☐ column.

The aim of the trials was to compare the times taken to fix traditional loose bar assemblies with those for single-directional and two-way directional prefabricated reinforcement panels. The fixing gang was encouraged to have a few practice runs at fixing each assembly, to familiarise themselves with the method before the trial was timed.

Flat slab

The trial bay simulated an 8 m × 6 m internal panel of a 275 mm thick slab for a six-storey building. The floor was designed to take a typical imposed office loading of 5 kN/m^2. Speed trials were carried out for the following arrangements:

- ☐ a traditional loose bar layout;
- ☐ a hand-laid layout using single-directional reinforcement mats;
- ☐ a crane-handled layout using large two-directional reinforcement mats.

The crane was an 18 t capacity mobile crane with a jib height of 37 m and a maximum horizontal reach of 22 m when carrying a 1 tonne load. The rebar fixing team consisted of two fixers and a ganger. The ganger was used largely to identify and tag reinforcement and to help during crane lifting and not for rebar fixing. Prior to the trials, the bay area was marked out and column starter bars drilled into the concrete base at the column positions. The reinforcement was stored in a designated area outside the footprint of the mock building, with the crane positioned centrally.

Work was 'theoretically' carried out at level 5, some 15 m above ground level, by hoisting reinforcement from the storage area to a height of 18 m before landing it in the work area. The summer working day was regulated to follow site practice: start 8.30 am; tea break 10–10.15 am; lunch 12.30–1.30 pm; tea break 3–3.15 pm; finish not later than 6.30 pm.

The start time was noted with the laying of the spacer strips. Work was considered complete when the last bar was fixed in position and shear links placed over the column supports. Details of the reinforcement used and layouts for the speed trials are given in Figs 3.5 and 3.6.

Loose bar trials

The crane was used to pick up the reinforcement bundles from the storage area and place them near to the working zone. The time trial commenced after a dummy run to familiarise the team with identification and handling, so that the recorded trial run represents an efficient working cycle.

(a)

Mark	No	T/B
01	64	T
02	72	T
03	24	T
04	32	T
05	24	T
06	32	T
09	17	B
10	10	B
11	23	B
Total	298	

(b)

Sheet	Size (m)	No	Layer
B01	6.5 x 1.75	2	B
B02	6.5 x 2.75	2	B
B03	8.5 x 1.75	2	B
B04	8.5 x 1.5	2	B
T01	5.0 x 1.5	8	T
T02	5.0 x 1.5	4	T
T03	4.0 x 1.5	4	T
T04	3.6 x 1.5	8	T
T05	3.6 x 2.25	4	T
T06	3.45 x 2.0	4	T
	Total	40	

(c)

Sheet	Size (m)	No
BM01	9.0 x 3.0	1
BM02	9.0 x 3.0	1
BM03	9.0 x 3.0	1
TM01	9.0 x 3.0	2
TM02	6.0 x 3.0	2
Total		7

Fig. 3.5. Reinforcement schedules for flat slabs: (a) loose bar; (b) single-directional mat; (c) two-directional mat

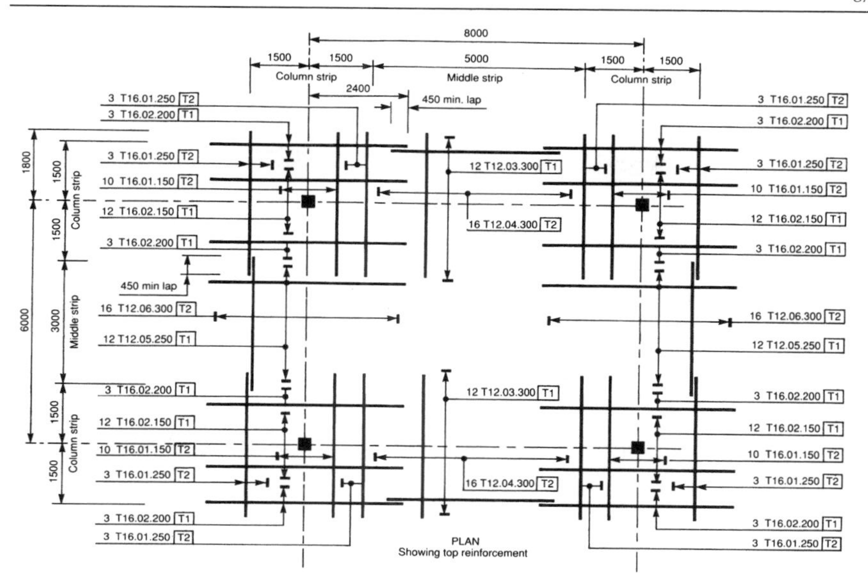

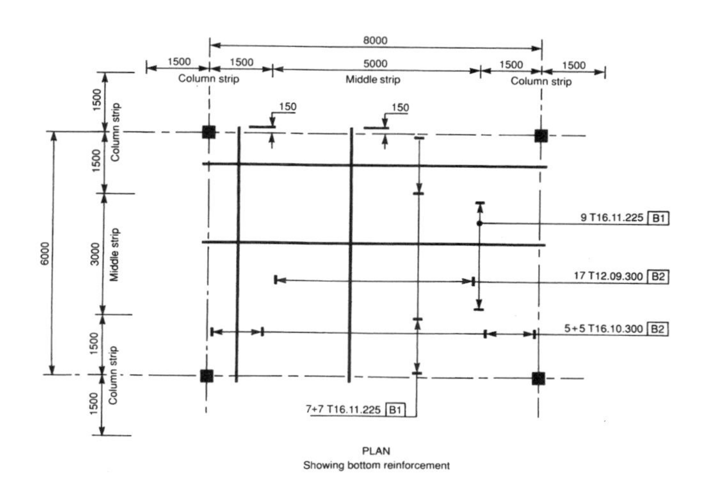

Fig. 3.6. Reinforcement layout: (a) loose bar

The work sequence for laying 298 single bars with a total weight of 1.8 t was as follows:

1. Cover spacer strips for the bottom reinforcement layer were laid out.
2. Bundles of bars were identified for the bay and craned onto the work area.

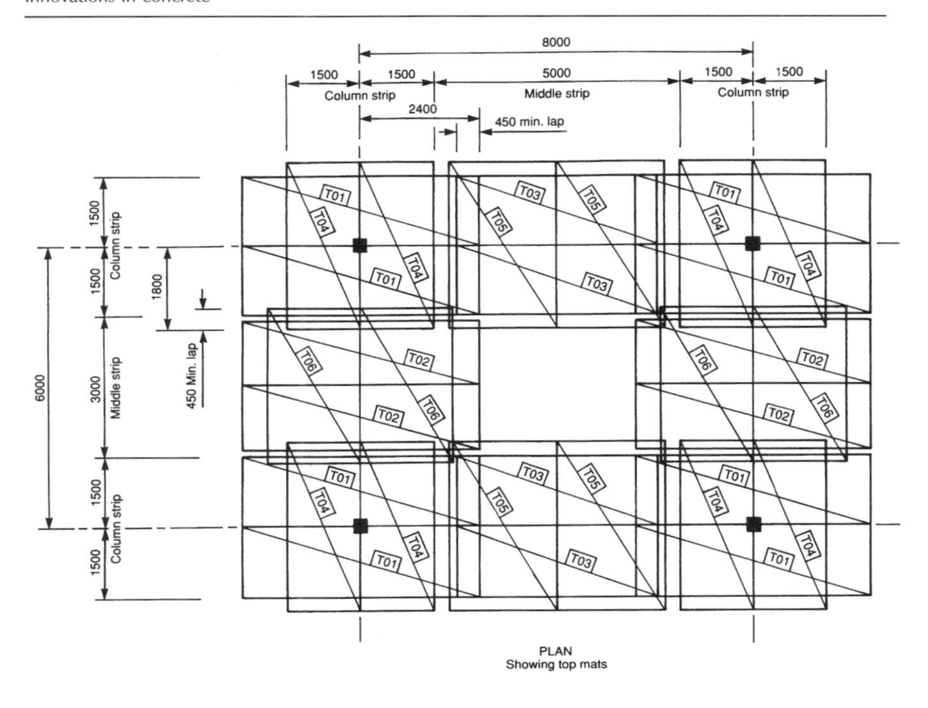

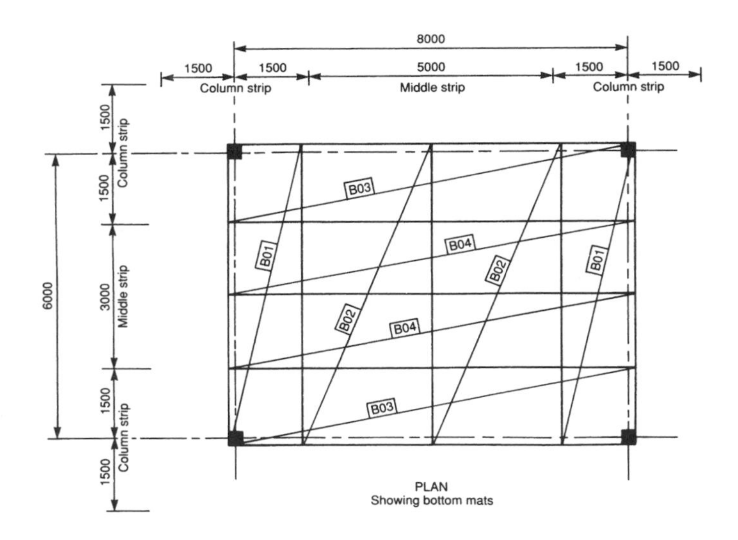

Fig. 3.6 (continued). (b) Single-directional mat

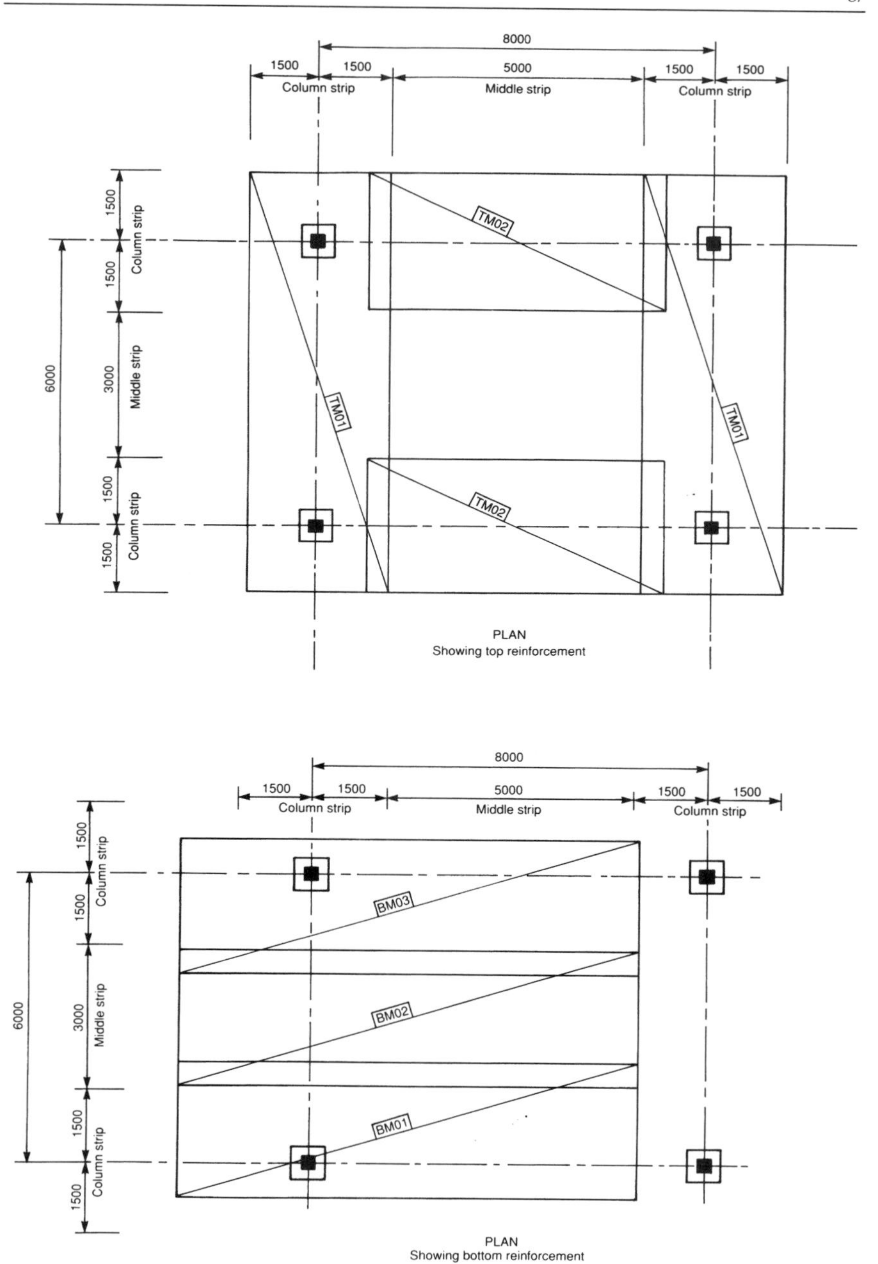

Fig. 3.6 (continued). (c) Two-directional mat

3. A single bar for the bottom 8 m span direction was positioned and the spacings for the other similar bars chalked out.
4. Bottom layer bars were laid out to the chalk-marked spacings (Fig. 3.7(a)).
5. A single bar in the opposite 6 m span direction was positioned and the centre for the top layer of bottom reinforcement chalked out.
6. The remaining bottom bars were located and fixed in position.
7. The bottom bars were tied at every other crossover to keep the assembly rigid (Fig. 3.7(b)).
8. Steel chairs to support the top reinforcement were placed over the bay and tied to the bottom reinforcement.
9. Bars for the lower layer of top reinforcement were identified, their positions chalked out and the bars placed and tied; bars for the upper layers were then extracted from the reinforcement bundle and positioned to finish the top reinforcement.
10. Loose bars were threaded through to complete the reinforcement over the columns.
11. Top reinforcement was tied together at every other crossover to fix the assembly.
12. Shear links were sleeved-in and tied.

Results. Fixing the loose bar reinforcement started at 8.30 am and was completed by 3.45 pm when the last shear link was in position.

Single-directional mats

For the second trial, traditional loose bar was replaced by single-directional fabric mats. The mats were light enough to be man-handled, with approximate panel sizes varying from 9 m × 3 m down to 3 m × 2 m. Each mat had the main reinforcement running in one direction with transverse wires welded on to keep the assembly rigid. A second mat was overlaid at right angles to provide the main reinforcement in the other direction. All mats were machine-fabricated and scheduled in accordance with sheet size, using standard dimensions. Loose bars were used for continuity over the columns and shear links. The mats were stored in the designated area and lifted by crane to the working area, with the bottom and top mats arranged in the required order for handling. The work sequence was as follows:

1. Spacer strips were laid for the bottom mats.
2. Bottom mats in the longitudinal, 8 m span direction were positioned by hand; four mats were laid.
3. Bottom mats in the transverse direction were placed over the top of the first layer and the two layers tied together.

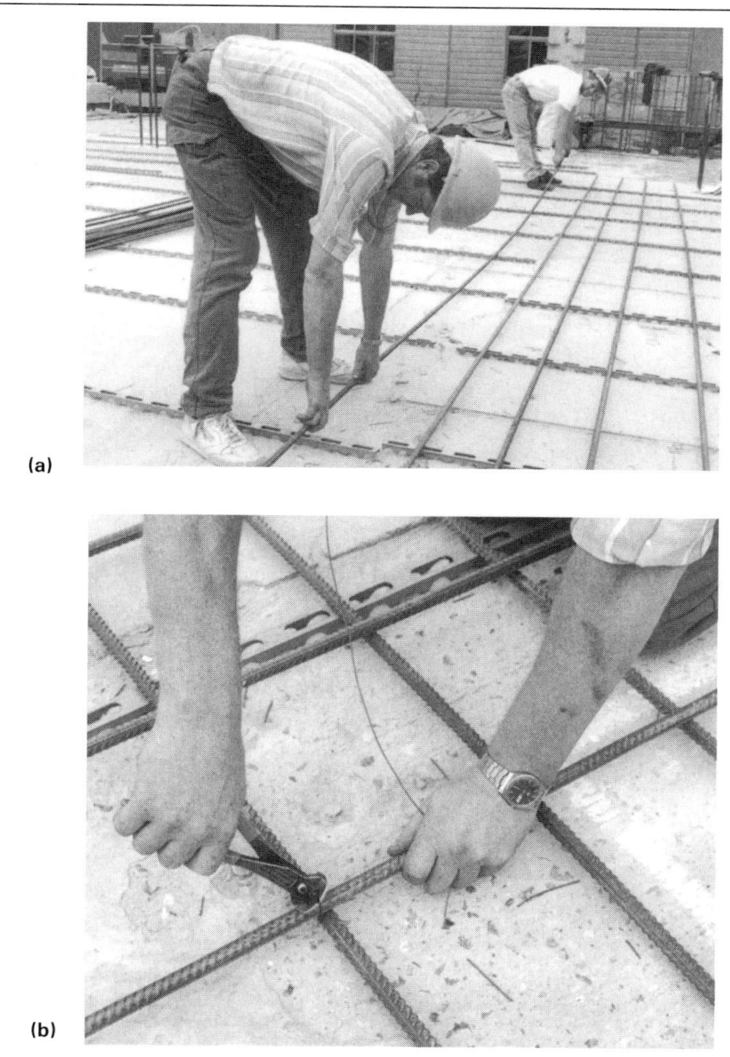

(a)

(b)

Fig. 3.7. Speed trials: (a) laying out loose bar reinforcement; (b) tying bottom slab reinforcement

4. Steel chairs to support the top reinforcement were positioned.
5. The first layer of top mats was placed over the support chairs in the 8 m direction (Fig. 3.8(a)).
6. The remaining top mats were then positioned and tied down to secure both layers (Fig. 3.8(b)).

(a)

(b)

Fig. 3.8. Speed trials: (a) laying bottom layer of single-directional mats for slabs; (b) laying top layer

7. Single bars were fixed over the column supports to provide continuity steel, and shear links positioned.

Results. The trial started at 8.30 am and finished by 10.00 am when the last shear link was fixed. In all, 40 reinforcement mats were fixed with a total weight of 2.0 t.

Two-directional mats

Two-directional crane-handled mats were substituted for single-directional mats. The mats were prefabricated in the largest possible sizes, limited in width only by the dimensions of the lorry trailer and the crane capacity. The bottom reinforcement was fabricated in three mats, each 9 m × 3 m, and the top layer fabricated in four mats, two of 9 m × 3 m and two of 6 m × 3 m. Reinforcement over and around the columns was provided by loose bars, as were the shear links.

The two-directional mats were stacked in the designated area and lifted by crane directly into position. The working sequence was as follows:

1. Spacer strips for the bottom mats were laid uniformly over the 8 m × 6 m bay area.
2. One mat for the bottom layer was craned directly into position (Fig. 3.9(a)); the remaining two mats were craned in together (Fig. 3.9(b)). The lower of the two mats was unhooked in position, then the upper mat was placed in position.
3. Support chairs were laid over the bottom mats and secured before the top reinforcement mats were landed.
4. The two 9 m × 3 m top mats were craned in together. The lower mat was unhooked before the crane lifted the top mat into position on the opposite side of the bay area.
5. The two smaller top mats, running in the 6 m direction, were then craned into position.
6. A minimum amount of tying was necessary only to secure the mats to the support chairs. The weight of the mats was sufficient to keep the assembly rigidly in place.
7. Continuity reinforcement using loose bars was positioned over the four columns and shear links were inserted and tied.

Results. The trial started at 12.15 pm, and by 12.45 pm the last mat for the top reinforcement was in position. By 12.51 pm all the shear links and continuity rebar over the columns had been fixed. Seven mats were fixed, with a total weight of 2.2 t.

Beam cages

Loose bars

An 8 m long beam cage having four 25 mm bottom bars, two 16 mm top bars and links to shape code 61 at 300 mm centres was assembled traditionally. The bottom bars were hung between trestles and the shear links fed through and tied before the top two bars were fixed into position. Total time taken for assembly was 20 min.

Fabric link

The 8 m long beam cage consisted of two sets of open-top fabric links, each 4 m in length. It was supported on trestles and the bottom 25 mm bars were fed through and tied. The open-top links were closed

Fig. 3.9. Speed trials: (a) the first two-directional mat being craned in for the slab; (b) two mats being craned in together

with a capping link to allow sleeving of the top reinforcement over the column positions. The assembly was then turned over and the remaining 16 mm bar fed through and tied to complete fabrication. Total time taken for completion was 10 min.

Column cages

Loose bar with single link cage

The column cage assembly consisted of four single bars, 25 mm in size and 4 m long with 20 T8 links, shape code 61 at 300 mm centres. Fixing trestles were set up and two T25s were hung between them. The links were then sleeved on and tied at the required centres. The whole assembly was then turned on the trestle and the remaining T25s fed through and tied to complete the column cage. Total time taken for assembly was 10 min.

Fabric with single link cage

Fabric link cages 4 m long were lifted onto the trestles and two T25 bars fed through and tied. The cage was then turned over and the remaining T25s fed through and tied into position. Total time taken for assembly was 4 min.

Loose bar with double link cage

The column cage consisted of four T25s and four T16s as main bars and a double link cage with two sets of T8 links. The cage was prefabricated in a manner similar to that for the single-column cage. The trestles were set up, the main bars hung, then two sets of links were fed through and tied before the whole cage was turned over and the remaining bars fixed. Total time taken for assembly was 20 min.

Fabric with double link cage

The two fabric cages were linked together as one cage before the main bars were sleeved into position and tied. The cage was turned over to allow the remaining top bars to be sleeved into position. Total time for completing the assembly was 10 min.

Overall results

Table 3.1 summarises the speed trial times for the flat slab, beam cages and column cages.

Rationalisation of reinforcement

Overview

Rationalisation of reinforcement has to be considered on many levels. It must begin in the design and detailing of reinforcement to ensure an efficient layout. In practice the subcontracting culture of the construction industry and the supply and demand of labour will often dictate the choice of reinforcement assembly. It does not follow that the most rational and material-efficient option will be the cheapest on price. Whilst reinforcement labour remains cheap and the price of rebar is based on tonnage irrespective of bar diameter, the value of total rebar prefabrication will never be realised. The current system of construc-

Table 3.1. Results of speed trials

	Time	Percentage of time for loose bars
Flat slab		
Loose bars (not including breaks)	5 h 45 min	100
Single-directional mats	1 h 30 min	26
Two-directional mats	36 min	10
Beam cages		
Loose bars	20 min	100
Fabric links	10 min	50
Column cages		
Single links — loose bars	10 min	100
Single links — fabric links	4 min	40
Double links— loose bars	20 min	100
Double links — fabric links	10 min	50

tion procurement has conditioned attitudes to such an extent that it makes it difficult to introduce change. Moreover, labour charges and material costs appear to bear no relationship to the process of manufacture and ease of assembly. They are both arbitrarily drawn up to satisfy tendering prices.

This study was carried out by the author for the Concrete Society in collaboration with Colin Gray of Reading University. It attempts to find an integrated approach to the problem of assessing the potential savings in assembly time and material cost due to rationalisation. The main findings of the study are:

- ☐ Rationalisation must begin with designers understanding the issues and offering the benefits to their client. Initial considerations based upon tonnage of steel give a misleading indication of the alternatives.
- ☐ Prefabrication is intrinsically expensive in materials, manufacture and transportation. There are large advantages to the installation process, but a cost structure based on weight will always penalise prefabrication.
- ☐ When in situ concrete is used primarily to achieve a customised solution, the opportunity for repetition and rationalisation is limited.
- ☐ Standardisation of components, at the level of bar diameters, length and cage construction, is very limited. Widely variable sections, depths, lengths and loading result in an infinitely variable range of requirements.

No single organisation has control of the complete value chain. The industry must decide where its best interests lie. If site work is to be significantly affected and reinforcement is to be rationalised, then a vertically integrated solution must be found. This can be achieved by design rationalisation.

Definition

Rationalising reinforcement means reducing the complexity of assembly and simplifying it to minimise the overall cost of construction. Rationalisation is defined as:

- ☐ simpler manufacture — by the use of fewer sizes, common shapes giving more repetition, more straight bars;
- ☐ reduced direct fixing time — by the use of fewer manhandled bars and large prefabricated units;
- ☐ reduced delays during direct fixing time — due to simpler reinforcement schedules;
- ☐ reduction in project duration.

Rationalisation can be done at various stages of the design: during initial design, detailed design, or when preparing the schedules. Rationalising the design causes an on-cost to the contractor because it involves the use of more materials than is necessary. However, it provides benefits in shorter fixing time, which can reduce the overall cost of construction to the client.

Scheme design stage

At this stage there are basic calculations of the amount of reinforcement required. The designers can put in additional effort, refining the design further in order to reduce the amount of steel. There is a trade-off between design effort and quantity of material. Putting in more design effort reduces the material cost but increases the complexity of the reinforcement assembly. There is a place for rationalising in the design at this stage, and this involves reducing design effort and providing a simple solution.

Engineering design stage

This stage involves taking the basic design and detailing it for later preparation of the reinforcement schedule. Similarly, if it were possible to simplify the detailing this would reduce design time. Less detailed layouts reduce complexity and would usually be simpler to fix, with fewer on-site delays. Less detailed designs can, in some instances, be prefabricated on site.

Site assembly

Fixing duration. This can be represented as a sum of the direct fixing task and any additional delays that may occur while performing the fixing task. The direct fixing time is a function of the quantity of material and the complexity of the detailed design. As the complexity of detailing increases, there will be a higher probability of delays occurring during the fixing process due to mistakes in locating the correct bars, fixing the bars correctly and unnecessarily repeating this process.

Complexity. This measure will be dependent on a number of factors: bar shape, lengths, bar location, number of repetitions of a reinforcement unit. The complexity measure will have to be developed separately for various structural sections — slabs, columns, stairs, etc. — and separately for different levels of rationalisation. There is a need for measurement on site in order to gain data on fixing time, complexity of detailing and material weights. It would also allow the establishment of the delays occurring on site caused by the reinforcement, and their effect on the project.

Benefit of early finish. If fixing is on the critical path, shorter fixing time could bring about shorter project duration. The benefit of a project finished early is less financial cost and reduced contract preliminaries.

Assessing reinforcement fixing rates

Bar reinforcement conforming to BS 4449 or BS 4481 is supplied cut and bent to various shape codes and in straight lengths, in bar diameters ranging from 8 mm to 40 mm. Current practice, using bills of quantity, calls off the weight of reinforcement as 'cut and bent', 'straight lengths' or 'fabric' without distinguishing the bar diameter. Although SMM7 requires that straight and bent bars and each bar size are shown separately, currently reinforcement suppliers are being asked to supply cut and bent bar of any bar diameter or length under a single price banner, with the exception of shape code 99 and helical bar. In the manufacturing process there is a real difference in the cost of making shape code 99 and a straight bar and between handling 6 mm and 16 mm bar. If a particular bar diameter and shape code are cheaper to manufacture, there is no scope under current practice to exploit the difference. Reinforcement has to be precisely scheduled if the economy of faster assembly time and lower manufacturing cost is to be realised.

Site assembly speed

The fixing times of reinforcement for a particular bar diameter, given in Table 3.2 and used in the following calculations, are based on an average time for a steel fixer to assemble a variety of shapes and lengths of bar. The rates are based on those commonly used by contractors when building up a price for reinforcement fixing.

Average fixing rate. The figures quoted in Table 3.2(a) are an 'average rate', irrespective of the location or whether it is for a floor slab, a beam, a wall or a pile cap. Average production figures will vary slightly according to the contractor and it would appear that outputs in general quoted by concrete frame contractors are a little higher than those quoted by concrete and steel fixing subcontractors. For the purposes of this study, the concrete frame contractors' rates for steel fixing are used as shown in Table 3.2(a).

A maximum output rate of 2.0 tonnes per man-day has been set for 25–40 mm diameter bars to account for the increased weight of bars and slower fixing rate. Conversely, because the smaller diameter bars are lighter in weight, they can be carried in bundles and hence their production rate is higher.

Flat slab fixing rate. The 'average' figures shown in Table 3.2(a) have been adjusted to provide typical rates for flat slab construction, which will generally have a higher percentage of straight bar lengths. A benchmark fixing rate for 12 mm and 16 mm bar, measured in the reinforcement assembly trials (Bennett and MacDonald, 1992), has been taken. From this the rates for smaller and larger bars have been derived according to their area ratio as shown in Table 3.2(b), except where the rate has been capped.

As the smaller bar diameters are used for links and the like, no increase in productivity above the 'average' value is shown for 10 mm to 6 mm bars. The enhanced output for larger diameter bars accounts for

Table 3.2. Fixing rates for reinforcement

Diameter (mm)	6	8	10	12	16	20	25	32	40
(a) Average fixing rate									
Tonnes/day[a]	0.30	0.35	0.50	0.75	1.30	1.75	2.0	2.0	2.0
Tonnes/day[b]	0.30	0.35	0.50	0.60	1.10	1.35	1.75	2.0	2.0
(b) Flat slab fixing rate									
Tonnes/day	0.30	0.35	0.50	1.00	1.80	2.50	2.50	2.50	2.50

Notes

[a] Data derived from estimates prepared by frame contractors (large contracts)

[b] Data derived from estimates prepared by concrete subcontractors (smaller contracts)

the ergonomic profile of a flat slab and the increase in percentage of straight bar lengths. These adjustments are based on estimates rather than measured values. Their derivation has been accepted by a panel of trade contractors as suitable for tender pricing.

Ribbed slab fixing rate. To transpose a 'ribbed slab' to an equivalent flat slab assembly, a reduction factor is applied to the bar fixing rates of a flat slab to account for changes in surface area and cross-section profile. A simple multiplying factor based on the area ratio of the surface profile relative to that of the flat slab is proposed.

A maximum and minimum reduction factor for ribbed slabs of between 0.90 and 0.75 is suggested. A wider rib spacing and a shallow beam depth gives the minimum reduction factor of 0.90, whilst a closer rib spacing and deeper beam depth gives the higher reduction factor of 0.75.

Although it is quicker to assemble a flatter rib profile, a deeper rib depth may result in less weight of reinforcement. Rationalisation will lead to the best compromise between speed of assembly and weight of material handled.

The reduction factor F_r is given by

$$F_r = \left\{1 + \frac{1}{1 + 2d_1/a}\right\}/2$$

where a is the rib spacing per metre run, and d_1 the rib depth below the slab soffit (see Fig. 3.10). The equation sums the average of the top surface area ratio and the bottom surface area ratio relative to a flat slab, where the top surface area ratio is always unity. The top slab area is a per metre run, and the bottom surface area b approximates to $a + 2d_1$ per metre run. Thus the average F_r is [(top area ratio $= a/a$) + (bottom area ratio $= a/(a + 2d_1)$)]/2, which reduces to the equation above.

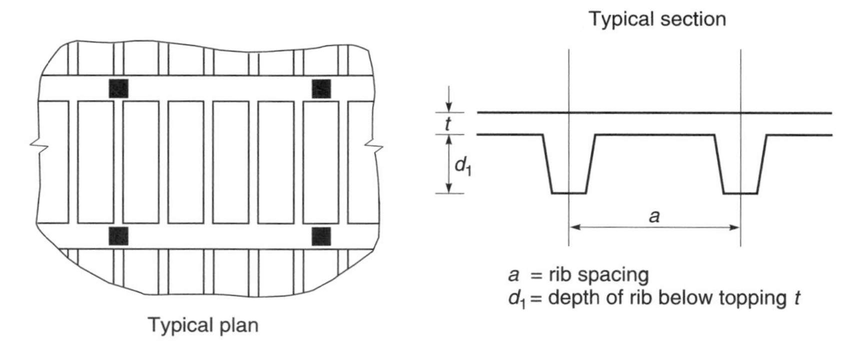

Fig. 3.10. Reduction factor for ribbed slabs

Ribbed slab reduction factor: worked examples

(i) Building 13 (see Fig. 3.11(a))
Shallow rib depth, wide rib spacing; $a = 1500\,\text{mm}$, $d_1 = 225\,\text{mm}$. Thus

$$F_r = [1 + 1/(1 + 450/1500)]/2 = [1 + 0.769]/2 = 0.88$$

(ii) Building 15 (see Fig 3.11(b))
Conventional rib spacing; average rib depth $a = 900\,\text{mm}$, $d_1 = 200\,\text{mm}$. Thus

$$F_r = [1 + 1/(1 + 400/900)]/2 = [1 + 0.692]/2 = 0.84$$

(iii) Building 21 (see Fig. 3.11(c))
Deep rib depth, conventional rib spacing; $a = 850\,\text{mm}$, $d_1 = 425\,\text{mm}$. Thus

$$F_r = [1 + 1/(1 + 850/850)]/2 = [1 + 0.5]/2 = 0.75$$

Waffle or coffered slab fixing rate (reduction factor F_w between 0.80 and 0.65). The area ratio rule assumes the equivalent vertical depth of ribs below the slab soffit. As there are four beams instead of just two for the ribbed slab, the reduction factor F_w (see Fig 3.12) becomes:

$$F_w = \left\{1 + \frac{1}{1 + 4d_1/a}\right\}/2$$

where, as before, a = rib spacing, d_1 = rib depth below the soffit. The top surface is flat and the ratio is unity.

Waffle slab reduction factor: worked examples

(i) Building 24 (see Fig. 3.13(a))
$a = 750\,\text{mm}$, $d_1 = 425\,\text{mm}$. Thus

$$F_w = [1 + 1/(1 + 1700/750)]/2 = [1 + 0.306]/2 = 0.66$$

(ii) Building 23 (see Fig. 3.13(b))
$a = 900\,\text{mm}$, $d_1 = 225\,\text{mm}$. Thus

$$F_w = [1 + 1/(1 + 900/900)]/2 = [1 + 0.5]/2 = 0.75$$

Composite reduction factor R_f. This applies to floor areas which may include a combination of flat slab, ribbed slab and/or waffle slab.

Having determined the structural grid and floor slab profile, an assessment is made of the whole floor by modifying a to account for the changing floor profile. Using the reduction factor calculated for each slab profile area, the composite factor for the entire floor area can be calculated.

Composite floor reduction factor: worked example

(See Fig. 3.14)

Ribbed slab $F_r = 0.75$, flat slab $F = 1.0$; typical bays are 9 m by 7 m. The ribbed slab area $= 36.5\,m^2$ and the flat slab area $= 26.5\,m^2$ out of total floor area $= 63.0\,m^2$. Thus

$$R_f = F_r(0.75)36.5/63 + F(1.00)26.5/63 = 0.43 + 0.42 = 0.85$$

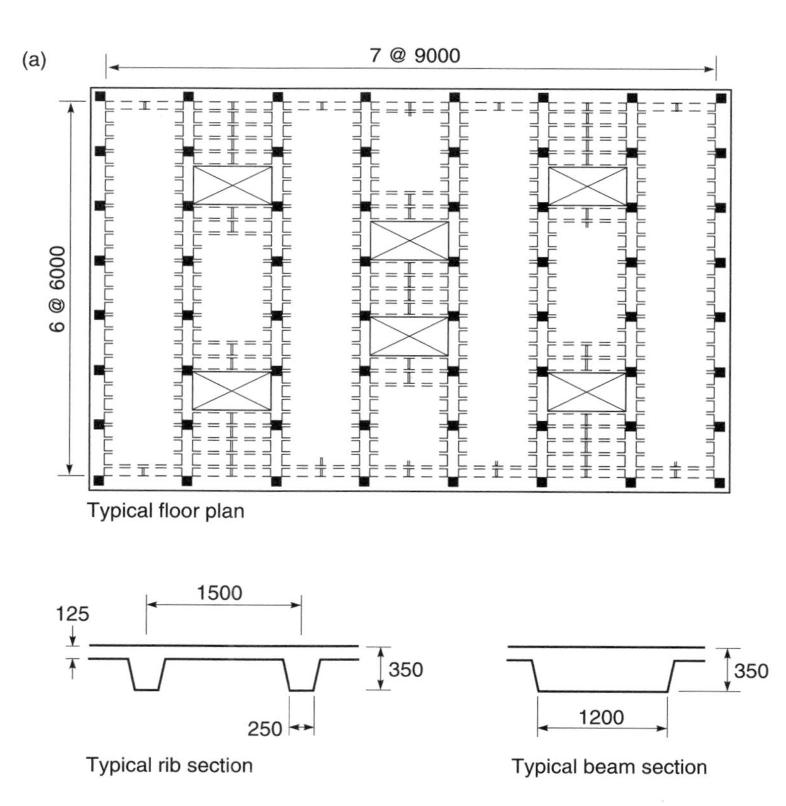

Fig. 3.11. Typical ribbed slab layouts: (a) Building 13

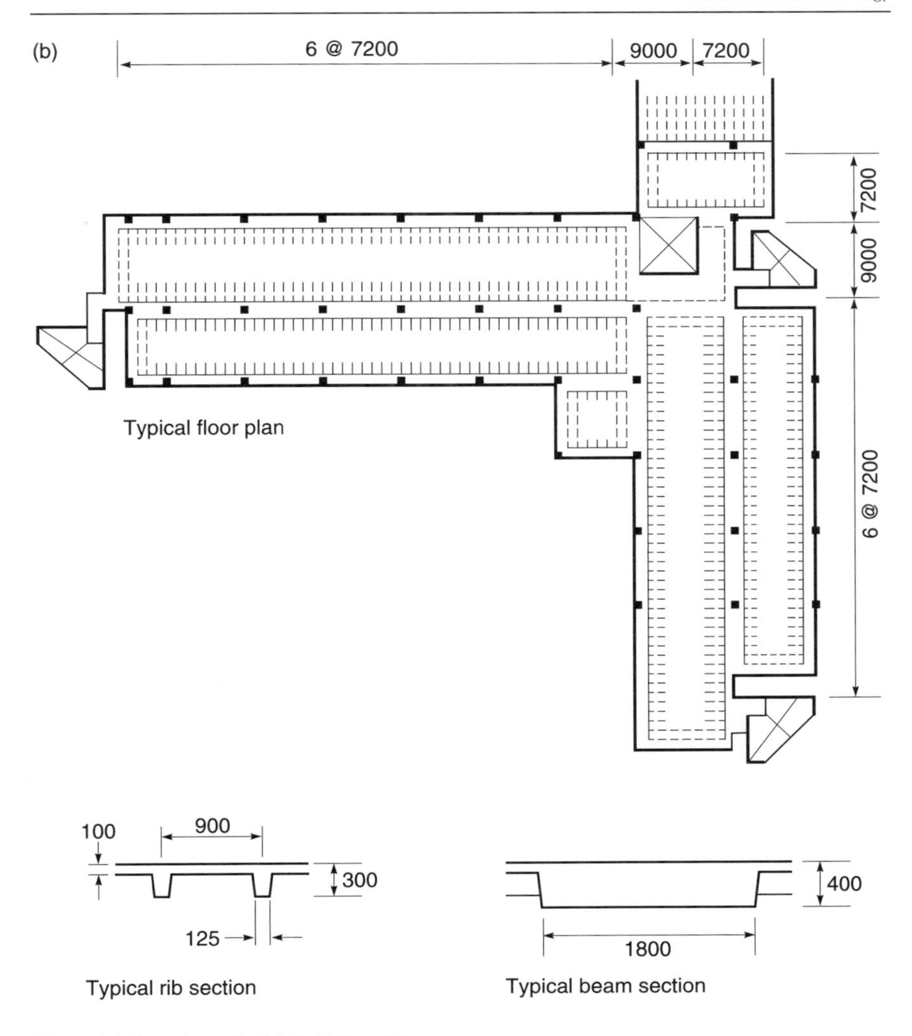

Fig. 3.11 (continued). (b) Building 15

Loose bar productivity

It is assumed that the smaller bar productivity rates remain unaltered irrespective of floor type and reduction factor, because they involve fixing links, stirrups and chairs. Therefore the composite reduction factor R_f is applied only to bar sizes from 12 mm to 40 mm.

Using the example from above with a composite reduction factor of 0.85, the *designed loose bar productivity* is found by applying this

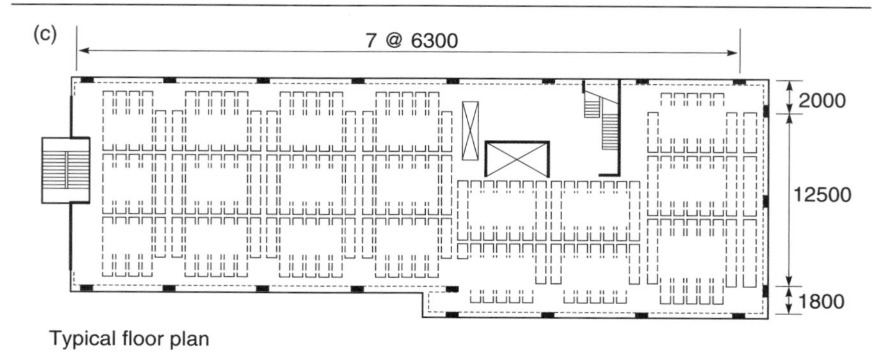

Typical floor plan

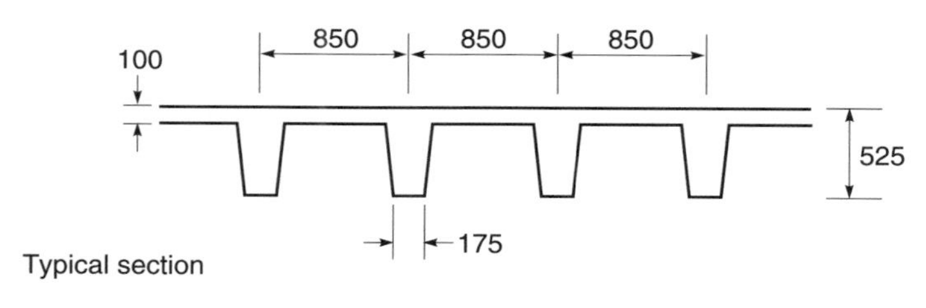

Typical section

Fig. 3.11 (continued). (c) Building 21

factor to the flat slab rates for 12 mm to 40 mm bar diameters. These outputs (Table 3.2(b)) can then be matched to the weights of the loose bar actually in the structure to find the time it will take to assemble the layout. A calculation for the vertical rebar can also be made.

Worked example

Using building No. 15 (see Fig. 3.11(b)), assume:

- ☐ output from Table 3.2(b) for the rates of rebar fixing;
- ☐ reinforcement density is 30 kg/m^2;
- ☐ 1000 m^2 of floor per pour area, with two pours per floor;
- ☐ all beam and column cages are fabricated on site;
- ☐ five floors in total, with a deduction of 10% for excluding vertical rebar.

Then loose bar productivity is:

Diameter (mm)	%	Tonnes	Rate	Time (days)
10	8	2.40	0.50	4.8
12	5	1.50	0.85	1.7
16	20	6.0	2.12	4.0
20	20	6.0	2.12	2.8
25	47	14.1	2.12	6.6
Totals	**100**	**30.0**		**20.0 days**

Rationalised loose bar productivity

Assume that fixing time can be reduced by minimising the tonnage of smaller diameter bars and replacing them with the larger diameter bars: 30% of 10 mm bars are converted to 12 mm bars, while 50% of 16 mm bars are converted to 20 mm bars. The reduction in time is the difference between the original fixing time, including conventional loose bar detailing, and the new arrangement:

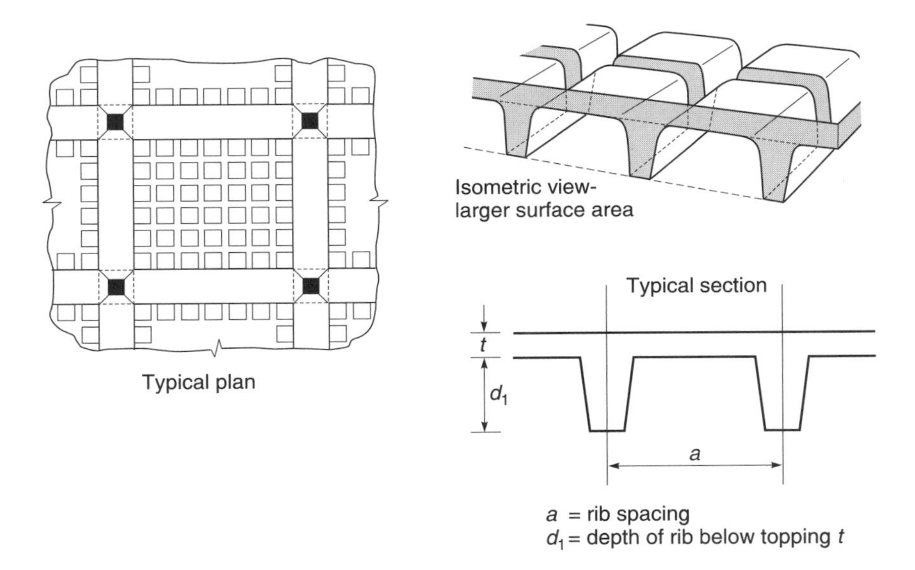

Fig. 3.12. Reduction factor for waffle slabs

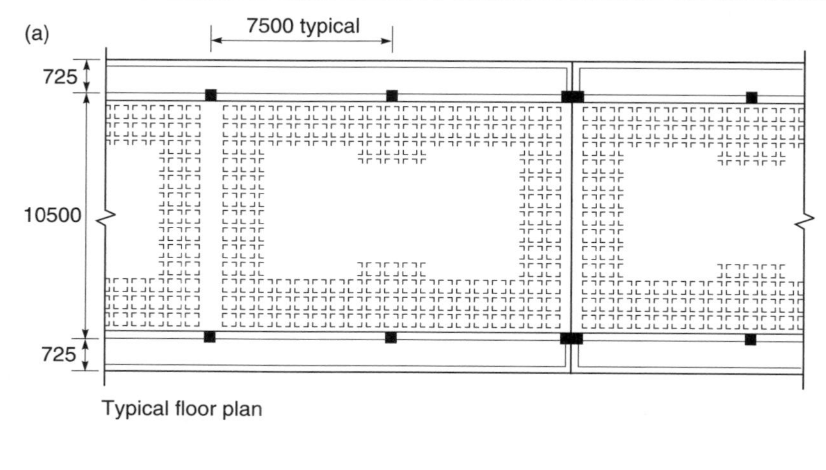

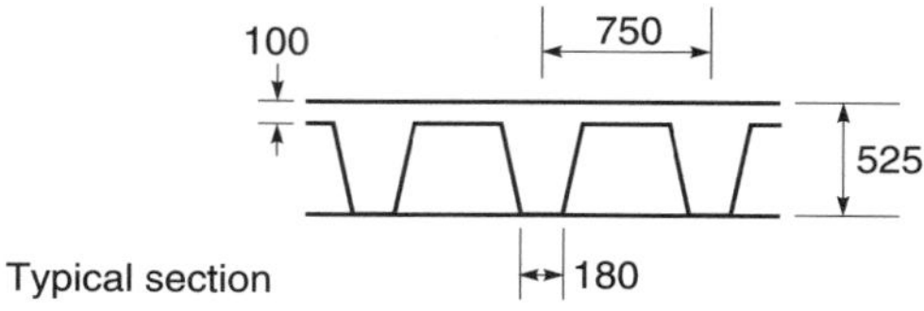

Fig. 3.13. Typical waffle slab layouts: (a) Building 24

Original time for fixing	10 mm bars =	4.8 days
	16 mm bars =	4.0 days
	Total	**8.8 days**
New time for fixing	50% 10 mm bars =	2.4 days
	50% 16 mm bars =	2.0 days
	1.2 t 12 mm bars =	1.4 days
	3.0 t 20 mm bars =	1.4 days
	Total	**7.2 days**

Rationalisation for speed of loose bar assembly can effect savings in the time of assembly in the order of half a man-day per 1000 m^2 of rebar placed. This is not as significant a saving in time as the option presented by prefabricated reinforcement assembly.

Prefabricated mesh panel assembly

Standard meshes to BS 4449 are not ideally sized to suit common building grids based on modules of 1.5 and 3.0 m; neither are they sufficiently heavy in reinforcement weight to cater for structural floor slabs. They evolved for use primarily in ground-supported slabs and for

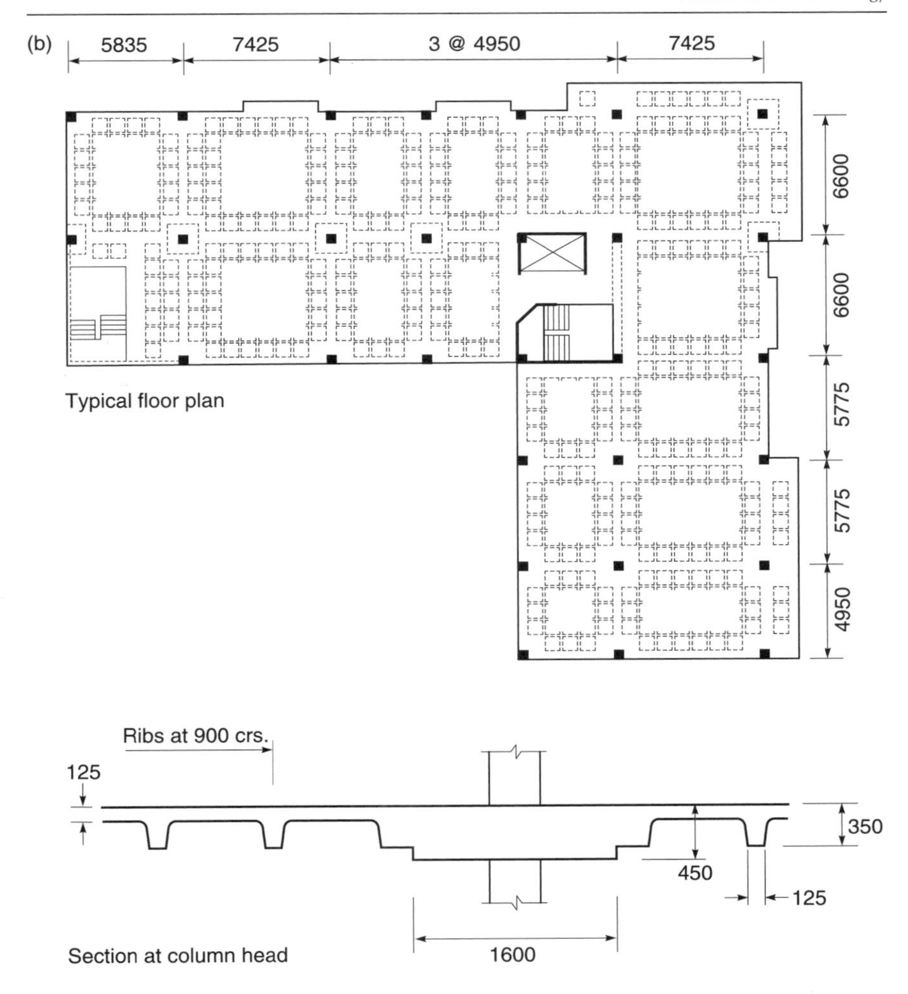

Fig. 3.13 (continued). (b) Building 23

road construction, which suited a fixed 2.4 m by 4.8 m panel size. Moreover, the ends of the BS mesh are left with a free length of half the bar spacing, which is not enough for a full tension or compression lap. For reinforcement continuity, between panels, loose bars have to lap with the mesh bars.

For these reasons, designed mesh panels are preferred for prefabricated panel assembly. It is possible to fabricate BS meshes with flying ends as a standard product which will not attract a cost premium. Meshes with flying ends are the preferred option if BS meshes are

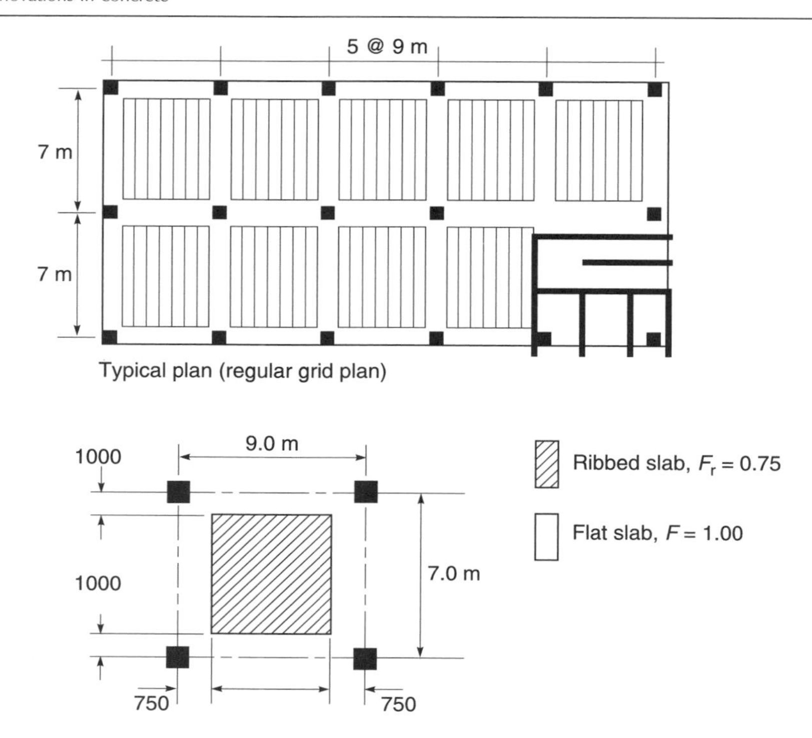

Fig. 3.14. Reduction factor for a composite floor

specified. There are two 'categories' of mesh panel types: *single-directional*, where the main reinforcement is in the long length; and *two-way* panels with reinforcement closely spaced in both directions.

BS meshes

Typical sizes of mesh are given in Table 3.3.

BS meshes with flying ends

Standard BS meshes can be supplied by reinforcement fabricators with flying ends to avoid lapping bar and to nest correctly with adjacent panels. The fixing times for these panels will be faster.

Designed mesh panels

Designed mesh can accommodate bar sizes up to 16 mm in both longitudinal and transverse directions, at a constant bar spacing. It is possible to increase the area of reinforcement in the principal direction

Table 3.3. Typical mesh sizes

Reference	Size	Weight: kg/m^2	Panel weight: kg
Square mesh			
A 393	10 mm @ 200, 4.8 m long 10mm @ 200, 2.4 m width	6.16	71
A 142	6 mm @ 200, 4.8 m long 6 mm @ 200, 2.4 m width	2.22	25
Structural mesh			
B1131	12mm @ 100, 4.8 m long 8 mm @ 200, 2.4 m width	10.90	125
B 385	7 mm @ 100, 4.8 m long 7 mm @ 200, 2.4 m width	4.53	52

by bundling pairs of bar, for example, 16 mm and 12 mm. The reinforcement supplier will advise on the maximum bar size and optimum spacing. The practical limits on spacing, bar diameter and panel size are governed by the economics of fabrication in relation to the transportation width and the weight that can be handled on site.

Designed mesh panels with flying ends are preferred. They can be made to suit grids of buildings in dimensional increments of 50 mm. For economic machine production it is best to have a minimum run of 30 panels to fabricate. The maximum panel size is governed by the maximum weight two men can lift, which is 130 kg, or the maximum width and length that can be placed on a lorry. Table 3.4 gives practical limits on sheet size.

Productivity for manhandled mesh panels

Productivity rates are taken from the reinforcement trials (see Bennett and MacDonald, 1992). Forty single-directional mats, each

Table 3.4. Summary of limits for mesh panels

	Minimum	Maximum
Bar diameter	6 mm	16 mm
Sheet length	2.0 m	12 m
Width	1.0 m	3.4 m

weighing between 75 and 100 kg, were laid in 90 min. These were made with flying ends. There were no lapping bars for continuity.

Assumptions:

- ☐ Allow an additional 2.0 min per mat for single-directional panels without flying ends, for fixing lapping bars.
- ☐ Allow an additional 5 min per mat for placing lapping bars for two-directional panels without flying ends.
- ☐ Increase the time for laying heavier panels by 0.5 min for 115 kg mats and 1.0 min for 130 kg mats.

Production rates for any size of panel (two-man gang, maximum weight lifted 130 kg) are given in Table 3.5.

Worked example. Production rates for manhandled BS standard meshes are given in Table 3.6.

BS standard meshes are restricted to panel sizes of 4.8 m by 2.4 m, which is not ideal for building grids based on 6 m, 7.5 m and 9 m modules.

Crane-handled production rates

Any panel can be handled by crane, but for efficiency it is usual to make the panel as heavy and as large as possible to minimise the number required to cover a given area. The time to lift a small or large panel will be the same.

The limiting factors on crane-handled panels are the maximum length and width that can be transported and the maximum panel

Table 3.5 Production rates for one-way and two-way mesh panels

	One-way panel		Two-way panel	
Weight: kg	With flying ends: min	Without: min	With flying ends: min	Without: min
80	2.00	4.00	2.00	5.00
90	2.25	4.00	2.25	5.00
100	2.50	4.00	2.50	5.50
120	2.75	4.25	2.75	5.50
130	3.00	4.50	3.00	6.00

Note
'Without' means that panels do not *overlap* with a full tension lap, so additional layers of single bars have to be fixed and this takes extra time. It is better to specify manhandled mesh panels with flying or lapping ends.

Table 3.6. Production rates for BS standard meshes with lapping ends

Type	Weight: kg/m^2	Total weight: kg	Time: min
A393	6.16	70.9	2.0
A252	3.95	45.3	2.0
A193	3.02	34.8	1.0
A142	2.22	25.6	2.0
B1131	10.90	25.6	3.0
B785	8.14	93.8	2.25
B503	5.93	68.3	1.0
B385	4.52	52.1	2.0

size that can be machined. This results in a maximum sheet length of 11 m and a width of 3.4 m.

The productivity times for crane-handled mats have been taken from the speed trials (Bennett and MacDonald, 1992). The crane placed seven reinforcement panels, all but two measuring 9 m by 3 m and weighing 300 kg each, in 30 min. Each panel was picked up from a storage point on the ground and hoisted to a height of 15 m before being landed in position 50 m away. This attempted to simulate the time on site for craning material from a lorry or storage place to the upper levels of a building.

We can assume conservatively that a panel can be positioned in less than 5 min, whether the panel weighs 500 kg or 100 kg. Any loose bar fixing needed to complete the assembly, for example, links or additional bars around column heads, must be added to the fixing time.

Total assembly time

In any reinforced concrete structure there will be scope to incorporate loose bar, standard mesh and designed mesh panels that can be man-handled or crane-handled into position. The speed of assembly will be dictated by the proportion of loose bar to prefabricated bar. The greater the prefabrication area, the faster the assembly time.

For a particular weight or area of reinforcement there will be an optimum time for assembly. To determine this optimum, rationalisation of reinforcement should be carried out at the conceptual design stage. Evaluation of the best compromise between speed and material cost must consider the overall construction cycle planned for the floors and structure. If the formwork cycle time is slow, there may be little point in accelerating the reinforcement fixing time. A balanced labour demand and continuity of work for all trades in each step of the construction process is an important consideration.

To determine the composite assembly speed for a particular reinforcement arrangement, it is necessary to consider the areas of loose bar and prefabricated bar and to adjust the productivity ratios to allow for the resulting work.

The following examples have been developed to show the effect of incorporating purpose-designed prefabricated mats into a conventional loose-bar design for a ribbed slab. The first example assumes a conventional loose bar design and is used as the benchmark against which to compare both prefabrication and the equivalent flat slab construction with crane-handled mats.

Example 1: Loose bar assembly

Ribbed slab: Building 15 (Fig. 3.11(b))
$a = 900\,\text{mm}$, $d = 200\,\text{mm}$ $F_r = 0.84$, composite $R_f = 0.87$.
Floor area per pour $= 1000\,\text{m}^2$.
Total weight of rebar ($230\,\text{kg/m}^2$) $= 30\,\text{t}$.
Outputs for flat and ribbed slabs are given in Table 3.7.

Table 3.7. Reinforcement assembly rates and times for loose bar assembly

(a) Flat slab output

Diameter: mm	8	10	12	16	20	25	32
Tonnes/day	0.35	0.50	1.00	1.80	2.50	2.50	2.50

(b) Ribbed slab output

Diameter: mm	Percentage	Tonnes	Rate: t/day	Time: days
10	8	2.40	0.50	4.8
12	5	1.50	0.85	1.7
16	10	6.0	2.12	4.0
20	20	6.0	2.12	2.8
25	47	14.1	2.12	6.6
Totals	**100**	**30.0**		**20.0**

Example 2: Designed panels plus loose bar

30% designed panels, 14 bays.

9 tonnes of top mat, hand laid in panels 4.5 m × 3 m, 6 per bay area, weight 120 kg.

70% loose bar = 21 t.

Loose bar production—see Table 3.8.

Designed mesh panel productivity. 100 kg panels with flying ends take 2.5 min to lay (see Table 3.5). Therefore, 84 panels in 1000 m^2 floor area will take (84 × 2.5/60) = 7.0 man-h = 1 man-day.

Total time to complete the area by a 2-man gang = (11.5 + 1.0) = 12.5/2 = 6.5 gang-days.

Example 3: Flat slab

Building 15 (see Fig. 3.10(b)), converted to flat slab with a 9 m by 7.2 m grid. The weight of rebar increases by, say, 10% over a ribbed slab and equals 32 t. Floor area = 1000 m^2.

Option A: Loose bar

See Table 3.9.

Option B: Manhandled design mesh panels. Assumptions:

80% rebar converted to panels = 28 t; loose bar 20% = 7 t. There are, say, 20 panels per bay, and a total of 14 bays.

Loose bar production figures are given in Table 3.10.

Panels weigh 100 kg each; fixing rate = 2.50 min per panel per gang.

Fixing time = (20 × 14) × 2.5 min/60 = 11.6 gang-h (say 1.5 days) = 3 man-days.

Total time to complete area by 2-man gang = (4.6 + 3)/2 = 3.8 gang-days.

Table 3.8. Loose bar assembly rates and times (ribbed slab)

Diameter: mm	Percentage	Tonnes	Rate: t/day	Time: days
10	8	0.6	0.50	1.2
12	2	0.6	0.85	0.7
16	5	1.5	1.53	1.0
20	25	5.2	2.12	2.4
25	36	7.5	2.12	3.5
Totals		**21.00**		**11.5**

Table 3.9. Loose bar production (Option A)

Diameter: mm	Percentage	Tonnes	Rate: t/day	Time: days
10	5	1.7	0.56	3.4
12	15	5.0	1.00	5.0
16	20	6.6	1.80	3.6
20	30	9.9	2.50	4.0
25	30	9.9	2.50	4.0
Totals	**100**	**33.0**		**20.0**

Option C: 90% crane-handled. Weight = 38 t, converted to prefab mats; loose bar is 10% = 4 t.

There are 7 panels per bay and a total of 14 bays = 98 panels.

Loose bar as for Option B: (4/7 × 1.7) = 3 man-days.

Crane-handled mats placed @ 5 min per panel: (98 × 5)/60 = 8 gang-h = 1 day for a 2-man gang.

Total time to complete an area by a 2-man gang = (3.0 + 2.0)/2 = 2.5 gang-days.

Summary

	Option A	Option B	Option C
Gang production	10 days	4 days	2.5 days

Minimum cost of manufacture

Once a design has been rationalised, with the weight and time of assembly optimised for the least direct and indirect cost, it may be possible to make an additional saving in the manufactured cost.

Table 3.10. Loose bar production (Option B)

Diameter: mm	Percentage	Tonnes	Rate: t/day	Time: days
10	2	0.7	0.5	1.4
12	2	0.7	1.0	0.7
16	2	0.8	1.8	0.4
20	4	1.4	2.5	0.6
25	10	3.4	2.5	1.4
Totals	**100**	**7.0**		**4.6**

In the course of manufacturing, when coiled wire and straight lengths of bar are being processed, certain bar diameters are cheaper to cut and bend than others. If bars are specified which are cheaper to process, then the specifier might expect the supplier to offer a reduction in price.

Distribution of cut and bent bar specified

Data collected on the average tonnage of bar specified in construction (1990), and on the percentage that is cut and bent, indicates that 16 mm bars are the most popular bar size in the market (Table 3.11).

Average labour time for cutting and bending

Typical outputs for labour time in a factory for cutting and bending bars are given in Table 3.12. Since there are more bars of smaller diameter than of larger diameter per tonne, they require more time to process.

Table 3.11. Percentages of bars specified

Diameter: mm	Total: %	Bent: %
8	2	1
10	9	5
12	17	10
16	24	14
20	18	11
25	15	9
32	11	7
40	4	2
Totals	**100**	**59**

Table 3.12. Typical bar bending and cutting outputs

Diameter: mm	Cutting: h/t*	Cut and bend: h/t
12 mm and less	1.87	3.56
16 mm to 20 mm	1.15	3.04
25 mm and above	0.87	2.41

* h/t: hours per tonne

Typical cost of manufacture (1990 prices)

This figure (Table 3.13) is based on an average labour rate of 2.28 h/t for all reinforcement types. The variable costs of manufacture include lighting, heating and power for the factory, consumables, repairs and labour.

Typical usage of BS fabric supplied to industry

See Table 3.14.

Conversion cost of single bar to BS fabric (1990 prices)

This figure (Table 3.15) is added to the cost of single bar to obtain the cost of manufacture of BS fabric.

If the current cost of single bar purchased from steel mills is, say, £220/tonne, then the cost of BS fabric of weight 125 kg is

$$(220 + 47) \times 0.125 = £33.38/\text{mat}$$

$$\text{Add profit of } 15\% = £5.00$$

$$\text{Allow for delivery } £15/\text{t} = 0.125 \times 15 = £1.88$$

$$\textbf{Total} = \textbf{£40.26 per mat}$$

Table 3.13. Typical reinforcement manufacturing cost

	£/t
Variable costs	22
Fixed costs	37
Total excluding profit and delivery	**59**

Table 3.14. Typical usage of BS fabric

Type	%	Type	%	Type	%
A393	20	B1131	3	C785	1
A252	20	B785	6	C636	1
A153	15	B503	4	C503	1
A142	14	B385	1	C385	4
A98	1	B383	2	C283	4
Totals	**70**		**18**		**11**

Table 3.15. Cost of converting single bar to BS fabric

	£/t
Labour	2
Other variable costs	12
Fixed costs	15
Total plus overhead (excluding profit)	**47**

Conversion cost of single bar to designed fabric (1990 prices)

This figure (Table 3.16) is calculated similarly to BS fabric except that the capital cost of machinery is greater and the set-up cost for different layouts requires two hours of labour. The minimum order should not be less than 15 sheets for economic working. The figure is added to the manufacturing cost of single bar to get the total cost of the mat.

The cost of time

A model for predicting the time-related cost of early or late completion of a project was developed by Reading University for the BCA in 1995. The model calculates the benefit of completion ahead of programme as a reduction in finance cost and early rental income, which can be offset against interest charges on a building loan.

Costs associated with construction preliminaries are usually taken as a single percentage, typically around 10% of the contract value. This is insensitive to particular resource savings and other cost savings during the project.

Table 3.16. Cost of converting single bar to designed fabric

	£/t
Additional labour	3
Variable costs	10
Fixed costs	15
BS fabric cost	47
Total	**75**

Using cash flow calculations for the construction cost reduces the errors based on lump sum percentages. Most cumulative cost on construction conform to an 'S'-type curve that can be divided into monthly or quarterly amounts. The closer refinement of monthly cost intervals is preferred for calculating the benefits of reinforcement assembly.

Financial calculations

A developer works on the assumption that short-term money is borrowed to finance a project. This includes the cost of land purchase, the cost of demolition, and of construction. The short-term loan is repaid at the end of the development period, from the proceeds of the sale or from raising a long-term loan against rental income. Money borrowed as a short-term loan is generally offered at 0.5%–6% above bank rate, depending on the size of the loan and the length of time the loan will be outstanding.

Loan interest periods

The development process can be divided into three periods: the acquisition period, which is the time from buying the land to the start of building work; the construction period; and the disposal period.

Total financial costs

Cash flow projections can model more accurately the complexity of cost and time variables on major building projects where, for example, a part of the scheme is let or sold before the building is finished. It allows for refinements such as inflation and taxation fluctuations during the period of borrowing as well as predicting the real saving that rationalisation can offer.

Issues related to reinforcement assembly can only affect the construction period and the construction finance cost.

The cost of construction finance C_{fc} in months is given by

$$C_{\mathrm{fc}} = C_{\mathrm{c}}(1 + i)^{(T/2)}$$

where C_{c} = construction cost, i = interest rate/month, and T = construction time (months). Then any saving in construction time of t months will show a saving S_{fc} in finance cost of

$$S_{\mathrm{fc}} = C_{\mathrm{c}}\ [(1 + i)^{(T/2)} - (1 + i)^{(T/2-t/2)}]$$

Case study

Background

The purpose of the case study is to use a model building and apply the formulae and factors developed earlier to determine the cost benefits of rationalising reinforcement. The seven-storey concrete-framed building (Fig. 3.15) used in this case study is situated on the M4 corridor (RCC, 1994). The analysis considers the following reinforcement assembly arrangements for a 300 mm deep flat slab:

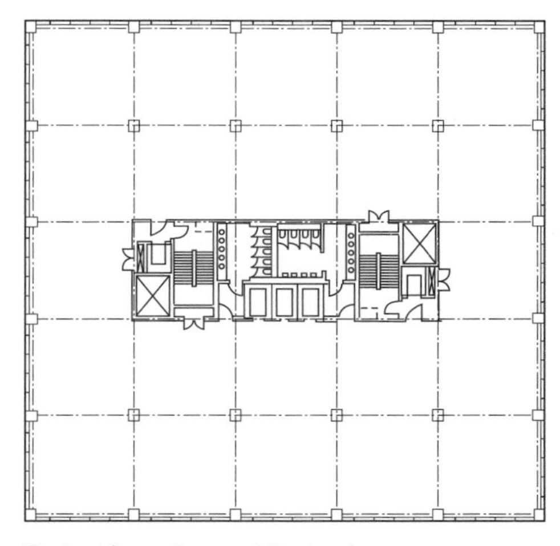

Typical floor plan: architectural

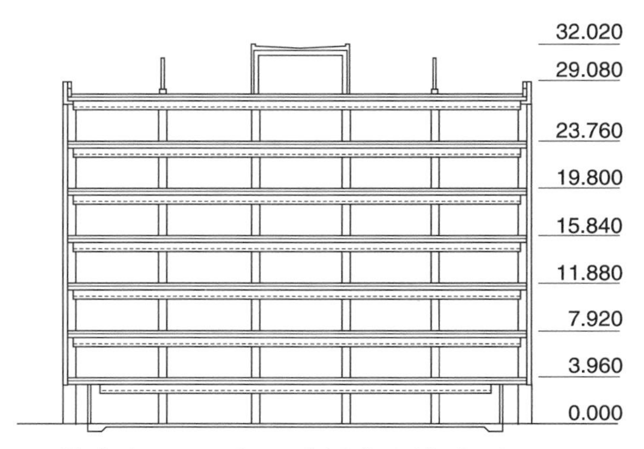

Typical cross-section and detail: architectural

Fig. 3.15. Seven-storey concrete-framed building

(1) traditional loose bar layout with shear links;
(2) rationalised loose bar assembly without shear links;
(3) single-directional (one-way) manhandled mesh panels;
(4) two-way crane-handled panels.

The alternative reinforcement layouts have been evaluated for cost of manufacture, cost of labour in assembly and saving in cost of finance for faster assembly.

The reinforcement arrangements for categories (1) and (2) have been prepared by an experienced detailer, whilst those for the factory-made prefabricated panels in (3) and (4) have been detailed by ROM Ltd and David Bennett Associates.

The flat slab fixing rate for loose bar and those for the manhandled and crane-handled panels have been taken from the tables given in the preceding sections. The cost of manufacture and supply of reinforcement has been based on estimated prices given by ROM Ltd.

(1) Traditional loose bar assembly (with shear links)

The weight of reinforcement required per m^2 for the floor slab (Fig. 3.16) is based on design guidance notes given in BS 8110 for flat slabs. The reinforcement detailing and layout arrangement for a typical section of the floor slab area is shown in Fig. 3.17. Quantities are shown in Table 3.17.

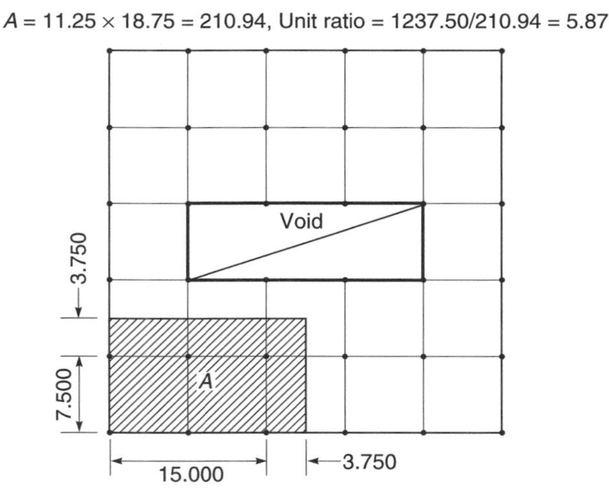

Fig. 3.16. Reinforcement area (unit ratio of area) detailed for traditional and rationalised loose bar

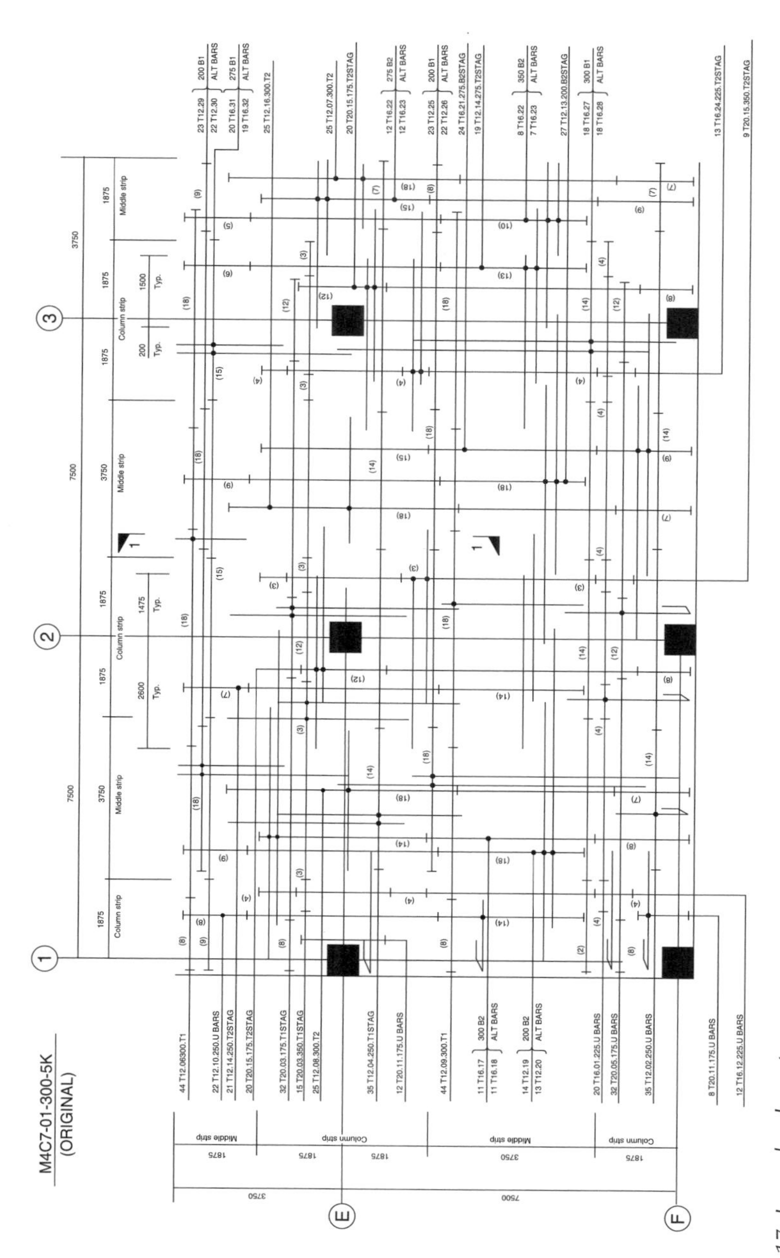

Fig. 3.17. Loose bar layout

Table 3.17. Reinforcement layout (loose bar) for alternative (1)

Bar mark	Type and size	Total no.	Length: mm	Shape code	A: mm	B: mm	C: mm
01	T16	20	3975	38	2925	250	
02	T12	35	2800	38	1800	250	
03	T20	47	4075	20			
04	T12	35	4075	20			
05	T20	32	3925	39	2925	250	
06	T12	44	1650	20			
07	T12	25	1650	20			
08	T12	25	3250	20			
09	T12	44	3250	20			
10	T12	22	2800	38	1800	210	
11	T20	20	3950	39	2925	210	
12	T16	12	3950	38	2925	210	
13	T12	27	6150	20			
14	T12	40	4075	20			
15	T20	49	4075	20			
16	T12	25	3250	20			
17	T16	11	6950	20			
18	T16	11	5950	20			
19	T12	14	6900	20			
20	T12	13	5950	20			
21	T16	24	6200	20			
22	T16	20	3950	20			
23	T16	19	2250	20			
24	T16	12	4075	20			
25	T12	23	6900	20			
26	T12	22	5950	20			
27	T16	18	6950	20			
28	T16	18	5950	20			
29	T12	23	3900	20			
30	T12	22	2250	20			
31	T16	20	3950	20			
32	T16	19	2250	20			
33	T8	102	425	85	100	250	65
34	T12	60	2400	20			

The production calculation for this alternative is given in Table 3.18. There are approximately 5.86 units of this section to cover the entire floor; therefore the assembly time for a total material weight of 26.47 t is **17 man-days.**

Table 3.18. Production calculation for alternative (1) (one man output/day)

Bar diameter: mm	Total weight: t	Rate of assembly: t/day	Time: days
8	0.018	0.35	0.05
12	1.597	1.00	1.60
16	1.488	1.80	0.82
20	1.414	2.50	0.57

(2) Rationalised loose bar assembly (without shear links)

A minimum quantity of reinforcement is aimed for. Refer to Fig. 3.18 and the schedule (Table 3.19) for the reinforcement details.

Rationalisation is interpreted in this study as the replacement of smaller diameter bars with larger diameter bars at wider spacings, without increasing the weight of reinforcement provided. This is possible provided the cover to the reinforcement is not reduced and the bar spacing does not exceed the permissible distance given in BS 8110.

Rationalisation will speed up the assembly time of single bars because fewer bars are actually handled.

The production calculation for this alternative is given in Table 3.20. There are approximately 5.86 units of this floor section to cover the entire floor; therefore the assembly time for a total material weight of 27.14 t = **13 man-days.** Table 3.21 gives a breakdown of bar diameter and weight.

(3) Single-directional panels (without shear links)

Refer to Figs. 3.19 and 3.20 for the panel layout and the schedule (Tables 3.22 and 3.23) for the individual panel reinforcement.

On inspection of the loose bar details, it was evident that lightweight prefabricated panels with both one-way and two-directional reinforcement could be of benefit to minimise the weight of reinforcement placed. Bar spacing at irregular intervals is possible with prefabricated panels, and in this exercise spacings of 112 mm and 214 mm have been used to keep to the minimum area of reinforcement over a given location. A small amount of loose bar was required at the slab edges and elsewhere and has been called off separately.

We have calculated the productivity times for the prefabricated panels and loose bar separately (Table 3.24) before finding the cumulative time for the composite assembly.

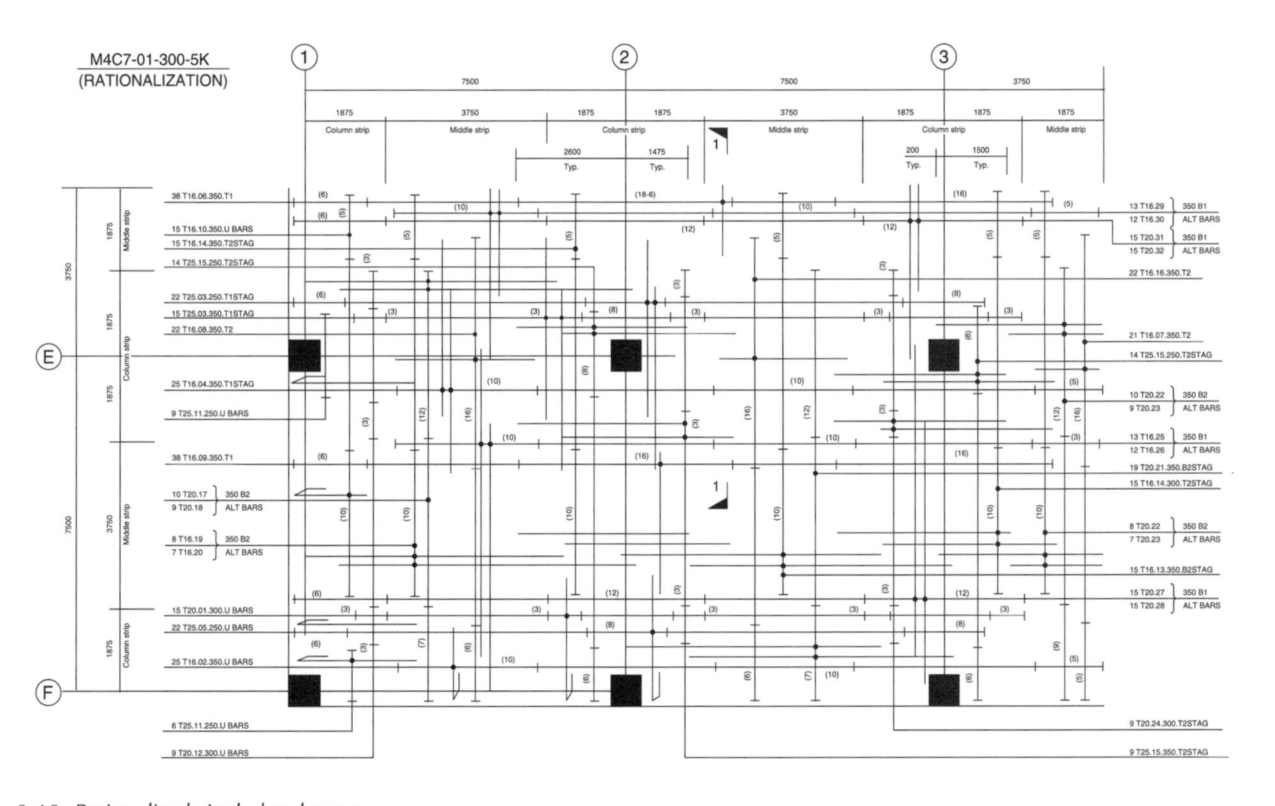

Fig. 3.18. Rationalised single bar layout

Table 3.19. Reinforcement layout (rationalised loose bar) for alternative (2)

Bar mark	Type and size	Total no.	Length: mm	Shape code	A: mm	B: mm	C: mm
01	T20	15	4125	38	2925	250	
02	T16	25	2950	38	1800	250	
03	T25	37	4275	20			
04	T16	25	4075	20			
05	T25	22	4125	39	2925	250	
06	T16	38	1800	20			
07	T16	21	1800	20			
08	T16	22	3250	20			
09	T16	38	3400	20			
10	T16	15	2800	38	1800	210	
11	T25	15	4150	39	2925	200	
12	T20	9	4100	38	2925	200	
13	T16	15	6200	20			
14	T16	30	4225	20			
15	T25	37	4275	20			
16	T16	22	3400	20			
17	T20	10	6950	20			
18	T20	9	5950	20			
19	T16	8	6900	20			
20	T16	7	5950	20			
21	T20	19	6350	20			
22	T20	18	4100	20			
23	T20	16	2400	20			
24	T20	9	4075	20			
25	T16	13	6900	20			
26	T16	12	5950	20			
27	T20	15	6950	20			
28	T20	15	5950	20			
29	T16	13	4050	20			
30	T16	12	2400	20			
31	T20	15	4100	20			
32	T20	15	2400	20			
33	T8	102	425	85	100	250	65
34	T12	60	2400	20			

The crane may take five lifts to deposit all the bundles of panels on the floor area. We should allow an additional time for awaiting panels. Allow, say, for two lift delays of 5 min per panel, totalling 10 min of time to add to the assembly time.

Table 3.20. Production calculation for alternative (2) (one man output/day)

Bar diameter: mm	Total weight: t	Rate of assembly: t/day	Time: days
8	0.018	0.35	0.05
12	0.128	1.00	0.13
16	1.525	1.80	0.85
20	1.569	2.50	0.63
25	1.342	2.50	0.56
Totals	**4.631**		**2.22**

There are approximately 11 units of this floor section to cover the entire floor; therefore the assembly time for a total panel material weight of 28.99 t = **4 man-days.**

If we now include the loose bar, there are approximately 11 units of this floor section covering the entire floor; therefore the assembly time for a total loose bar weight of 1.86 t = **1.25 man-days.**

Cumulatively, allowing 4 man-days for prefabricated panels and 1.25 man-days for loose bars, the total assembly time is **5.25 man-days**.

(4) Two-way crane-handled panels (without shear links)

On inspection of the reinforcement required over the column supports and the middle strips, the large variation in the reinforcement

Table 3.21. Slab rationalisation for time study (original totals shown in brackets)

(a) Reading building

Dia.	T8	T12	T16	T20	T25	Total	Total: kg/m^2
Weight: kg	18	128	1828	1929	2586	6489	29.3
	(18)	(1597)	(1488)	(1414)	(0)	(4517)	(20.4)
No. of bars	102	60	316	165	111	754	% saving: 19.10%
	(102)	(479)	(204)	(148)	(0)	(933)	

(b) Rochdale building

Dia.	T8	T12	T16	T20	T25	Total	Total: kg/m^2
Weight: kg	27	107	1568	1645	229	3576	20.5
	(27)	(1399)	(1115)	(1060)	(0)	(3601)	(20.6)
No. of bars	162	50	288	153	60	713	% saving: 21.70%
	(162)	(474)	(166)	(109)	(0)	(911)	

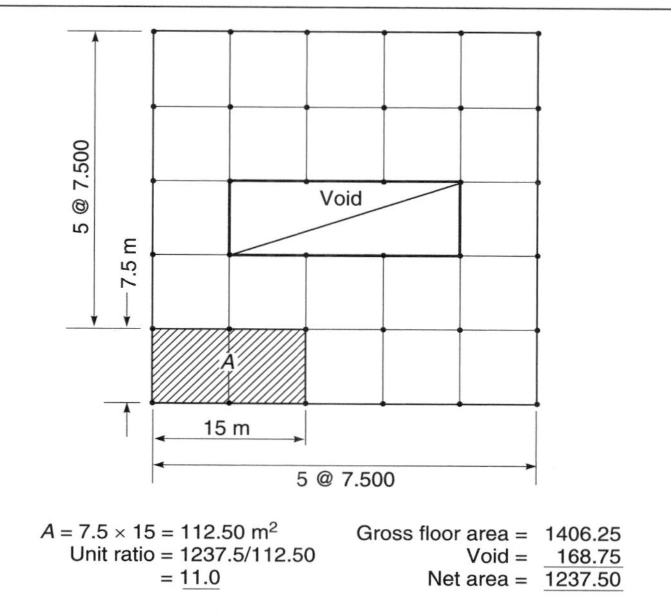

Fig. 3.19. Reinforcement area (unit ratio of area) detailed for one-way and two-way panels

quantities has made it difficult to design the two-way panels to be efficient in the use of material. If the design of the flat slab were analysed to smooth out the high peaks in bending moments over the column supports and to match them to midspan moments using, say, a finite element mesh or yield line theory, then an economical panel arrangement would be possible.

To minimise the number of panels handled by the crane it is necessary to standardise the reinforcement in both directions. The worst case was fixed as the reinforcement required for the entire panel. Consequently, extra reinforcement has been provided where it is not required and this will result in a higher tonnage of steel. *This exercise is for illustration purposes only.* The panel layout shown in Fig. 3.21 may be impractical to install as the flying ends of both bottom and top panels lap over the column positions.

There are four bottom panels and seven top panels to complete the sectional area. It is important to make each panel as big and as heavy as possible to reduce crane handling time. Loose bars in the form of U bars are required for the edges of the slab.

Crane-handled panels production

Bottom panels (7.5 m × 3.5 m) 4 No.

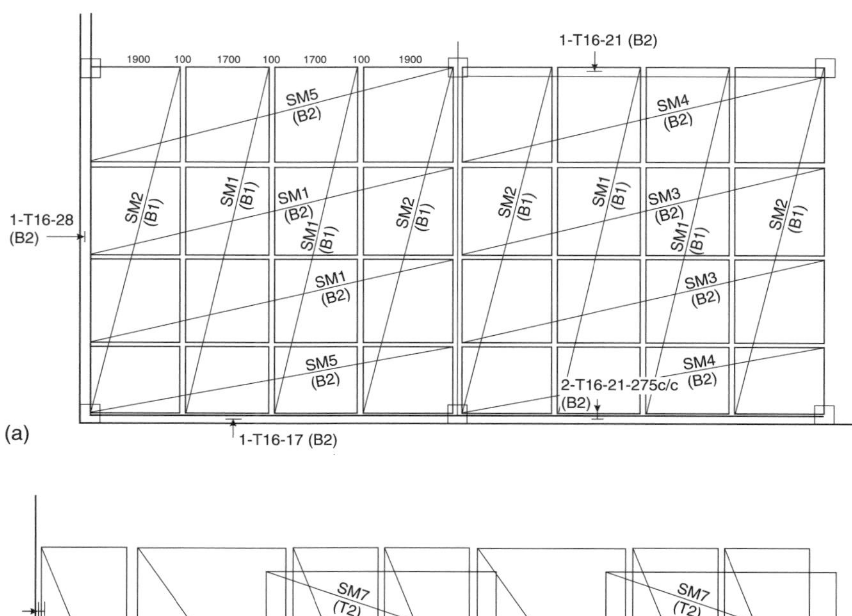

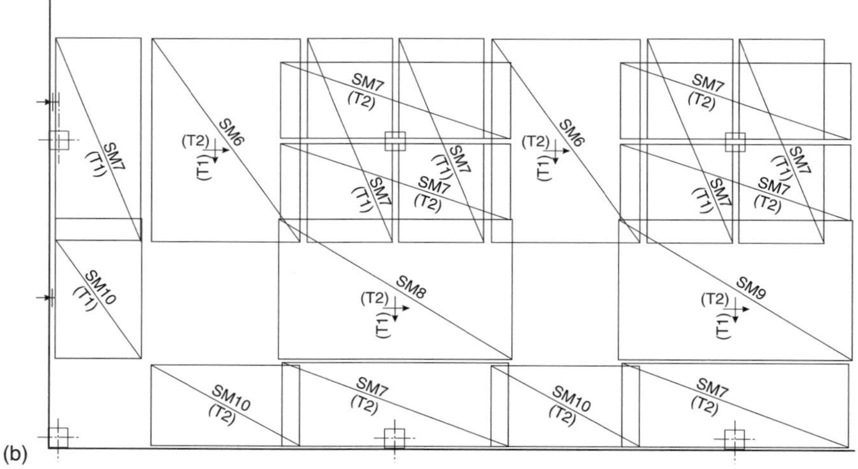

Fig. 3.20. (a) Single-directional reinforcement: bottom layer mats; SM1–SM5 refer to the mark numbers on Table 3.22. (b) Single-directional reinforcement: top panel layout; SM6–SM10 refer to the mark numbers on Table 3.23

Top panels (7.5 m × 3.75 m) 5.5 No.

Total panel tonnage = 2.800 + 0.928 = 3.728 t

There are 11 units of this sectional area to cover the entire floor. Using the crane and one extra man results in an assembly time of 55 min per panel for a total weight of 3.728 t.

Table 3.22. Bar schedules for alternative (3): single-directional (one-way) panels, bottom layer

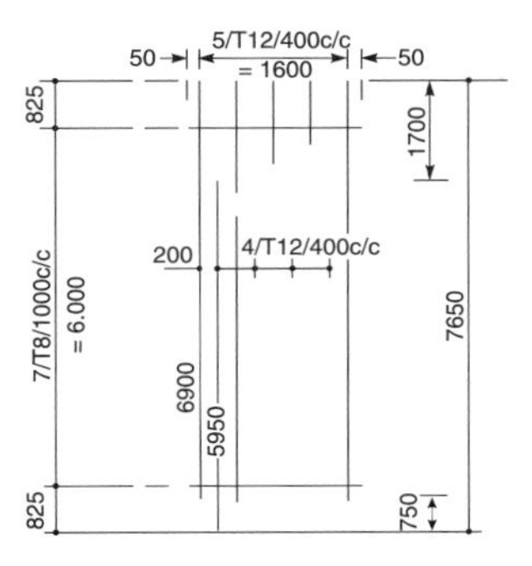

(a) Panel SM1

Wires	No. per sheet	Size: mm	Length: m	Weight: kg/m	Weight of sheet: kg
Longitudinal	5	T12	6.900	0.888	30.636
	4	T12	5.950	0.888	21.134
Cross	7	T8	1.700	0.395	4.700
Fabric to BS 4483. Bending to BS 4466					Total: 56.470

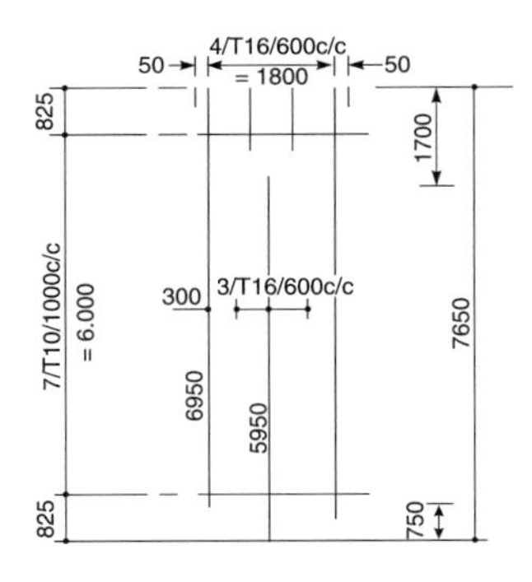

(b) Panel SM2

Wires	No. per sheet	Size: mm	Length: m	Weight: kg/m	Weight of sheet: kg
Longitudinal	4	T16	6.950	1.579	43.896
	3	T16	5.950	1.579	28.185
Cross	7	T10	1.900	0.616	8.193
Fabric to BS 4483. Bending to BS 4466					Total: 80.274

(continued)

Table 3.22 (continued)

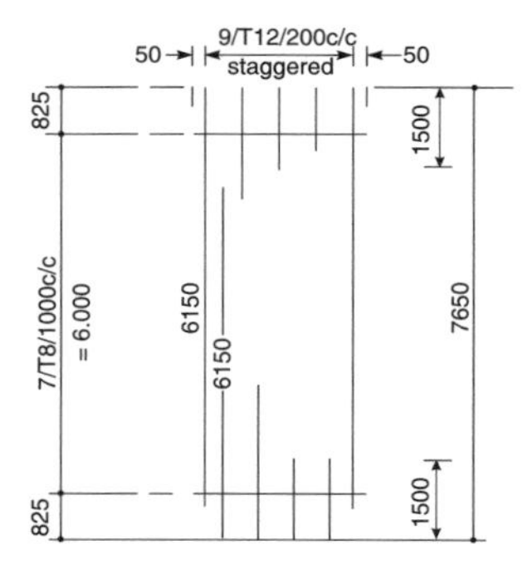

(c) Panel SM3

Wires	No. per sheet	Size: mm	Length: m	Weight: kg/m	Weight of sheet: kg
Longitudinal	9	T12	6.150	0.888	49.151
Cross	7	T8	1.700	0.395	4.700
Fabric to BS 4483. Bending to BS 4466					Total: 53.851

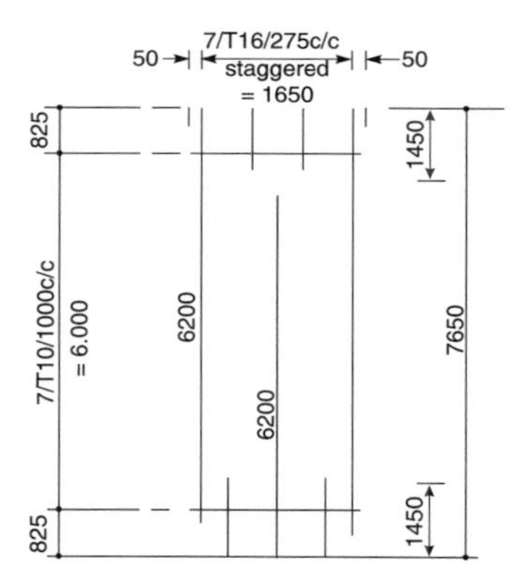

(d) Panel SM4

Wires	No. per sheet	Size: mm	Length: m	Weight: kg/m	Weight of sheet: kg
Longitudinal	7	T16	6.200	1.579	68.529
Cross	7	T10	1.750	0.616	7.546
Fabric to BS 4483. Bending to BS 4466					Total: 76.075

Table 3.22 (continued)

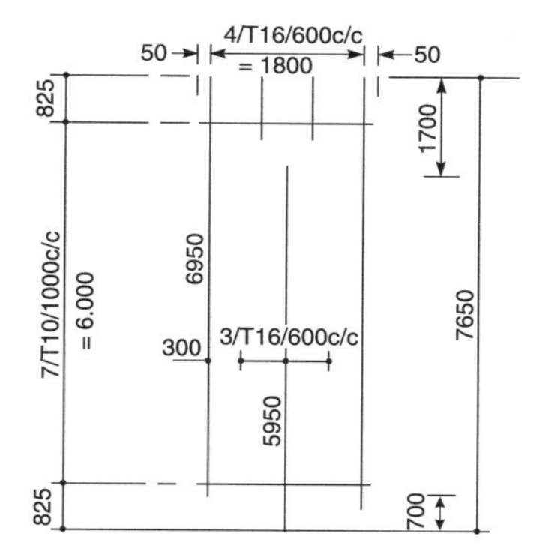

(e) Panel SM5

Wires	No. per sheet	Size: mm	Length: m	Weight: kg/m	Weight of sheet: kg
Longitudinal	4	T16	6.950	1.579	43.896
	3	T16	5.950	1.579	28.185
Cross	7	T10	1.900	0.616	8.193
Fabric to BS 4483. Bending to BS 4466					Total: 80.574

Table 3.23. Bar schedules for alternative (3): single-directional (one-way) panels, top layer

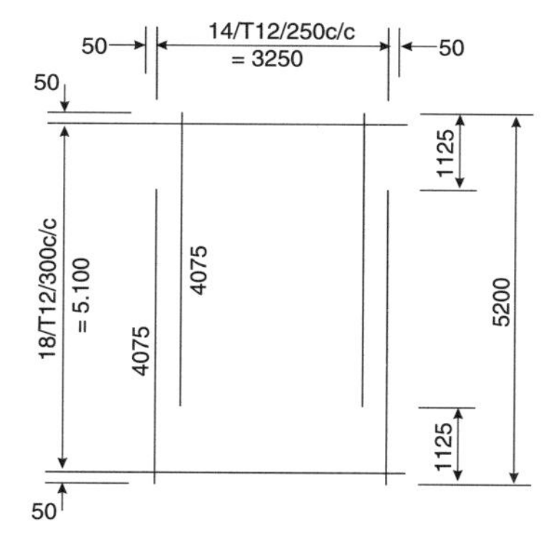

(a) Panel SM6

Wires	No. per sheet	Size: mm	Length: m	Weight: kg/m	Weight of sheet: kg
Longitudinal	14	T12	4.075	0.888	50.660
Cross	18	T12	3.350	0.888	53.546
Fabric to BS 4483. Bending to BS 4466					Total: 104.206

(continued)

Table 3.23 (continued)

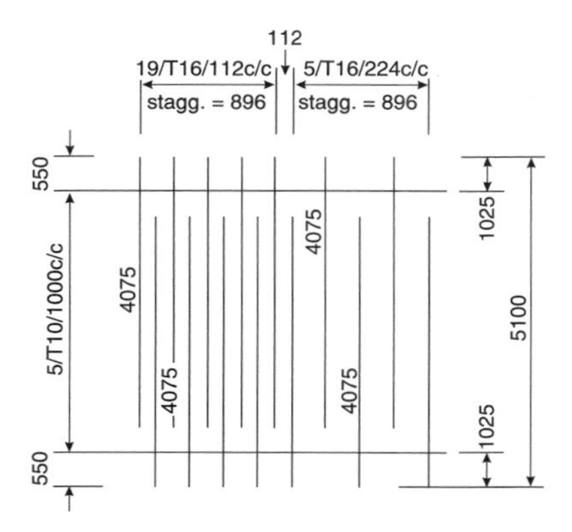

(b) Panel SM7

Wires	No. per sheet	Size: mm	Length: m	Weight: kg/m	Weight of sheet: kg
Longitudinal	14	T16	4.075	1.579	90.082
Cross	5	T10	2.000	0.616	6.160
Fabric to BS 4483. Bending to BS 4466					Total: 96.242

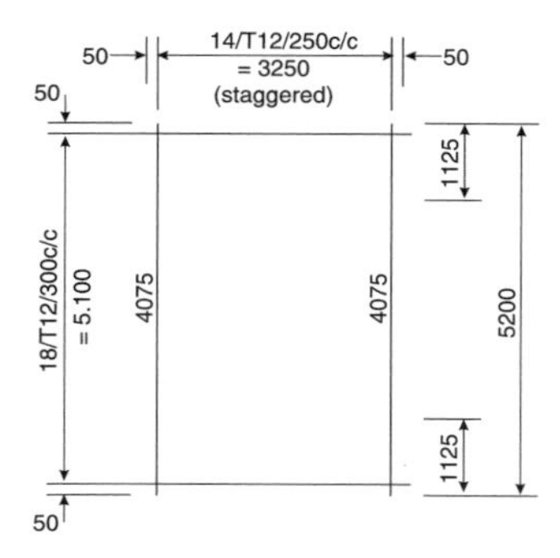

(c) Panel SM8

Wires	No. per sheet	Size: mm	Length: m	Weight: kg/m	Weight of sheet: kg
Longitudinal	14	T12	4.075	0.888	50.660
Cross	18	T12	3.350	0.888	53.546
Fabric to BS 4483. Bending to BS 4466					Total: 104.206

Table 3.23 (continued)

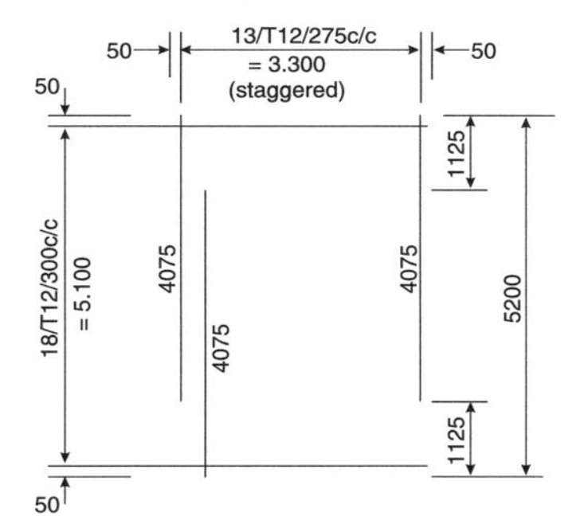

(d) Panel SM9

Wires	No. per sheet	Size: mm	Length: m	Weight: kg/m	Weight of sheet: kg
Longitudinal	13	T12	5.200	0.888	60.029
Cross	18	T12	3.400	0.888	54.346
Fabric to BS 4483. Bending to BS 4466					Total: 114.375

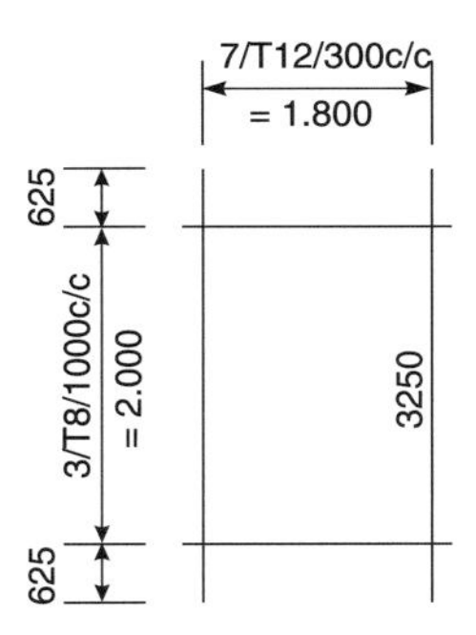

(e) Panel SM10

Wires	No. per sheet	Size: mm	Length: m	Weight: kg/m	Weight of sheet: kg
Longitudinal	7	T12	3.250	0.888	20.202
Cross	3	T8	1.900	0.395	2.252
Fabric to BS 4483. Bending to BS 4466					Total: 22.454

Table 3.24. Production times for alternative (3)

(a) Prefabricated one-way panels (with flying ends)

Panel ref.	Size: m	Weight: kg	Number	Total weight: kg	Time: min
sm1	6.9 × 1.70	56.470	6	338.82	12.0
sm2	6.95 × 1.90	80.274	4	321.09	8.0
sm3	6.15 × 1.70	53.851	2	7.70	4.0
sm4	6.20 × 1.75	76.546	2	153.09	4.0
sm5	6.95 × 1.9	80.574	2	161.15	4.0
sm6	4.75 × 3.35	104.106	2	208.41	5.0
sm7	4.75 × 2.00	96.242	11	1058.66	27.5
sm8	4.75 × 3.35	104.206	1	104.21	2.5
sm9	5.20 × 3.40	114.375	1	114.38	2.75
sm10	3.25 × 1.90	22.454	3	67.36	6.0
Totals			**34**	**2634.87**	**75.75**

(b) Additional loose bar production: bar schedule and rates

Diameter: mm	Mark	Length: m	Number	Unit weight	Total: kg
16	21	6.20	3	1.579	29.37
	17	6.95	1	1.579	10.97
	18	5.95	1	1.579	9.40
20	03	4.075	2	2.467	20.11
12	09	3.250	1	0.888	2.89
	U bars	1.20	90	0.888	95.90

(c) Production calculation (one man output/day)

Bar diameter: mm	Total weight: t	Rate of assembly: t/day	Time: days
12	0.0988	1.00	0.098
16	0.0497	1.80	0.028
20	0.0201	2.50	0.018
Totals	**0.1686**		**0.134**

Loose bar production. The production calculation (one man output/day) shows that, for 12 mm U bars with a total weight of 0.095 t and an assembly rate of 1.00 t/day, the assembly time is 0.094 days.

Cumulative time. As shown in Table 3.25, this is 18.25 h = 2.28 working days.

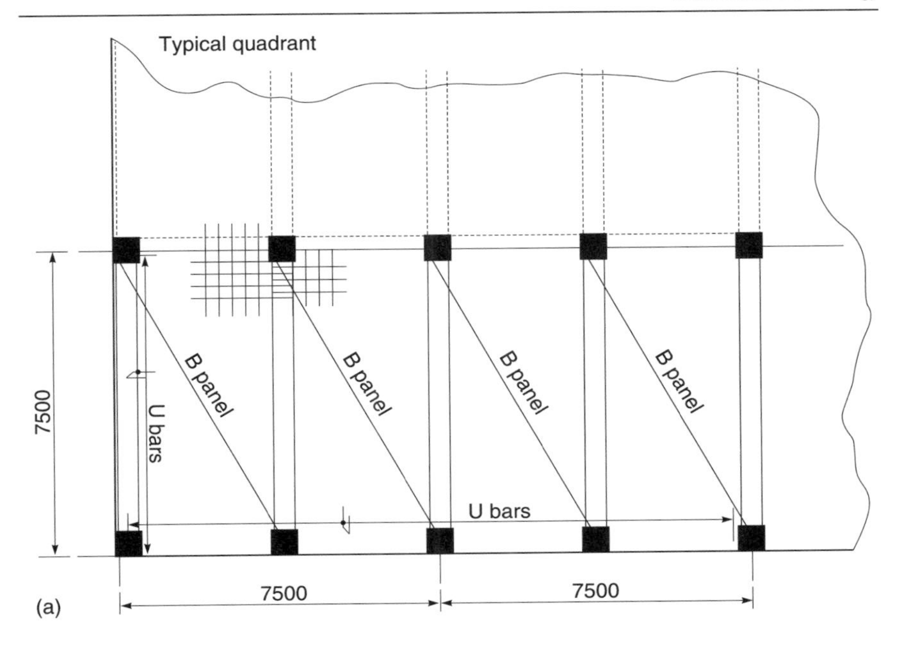

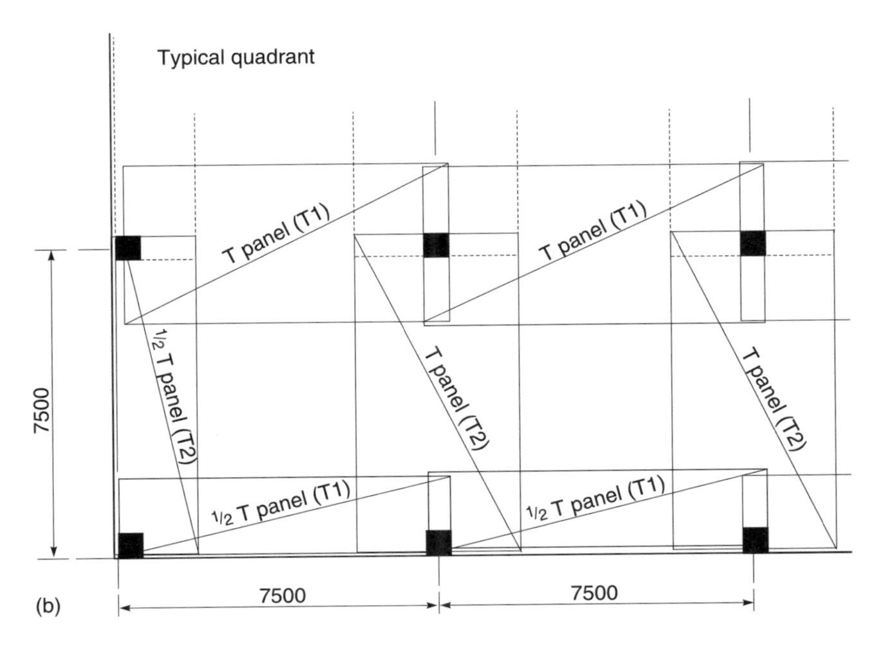

Fig. 3.21. Two-way crane-handled mat layout: (a) bottom layer; (b) top layer

Table 3.25. Cumulative time for alternative (4)

	Assembly time: h	Total weight: t
Crane-handled	10.00	41
Loose bar	8.25	1
Totals	**18.25**	**42**

Summary of material cost

Material quantities for the seven floors—each floor is assumed identical—are summarised in Table 3.26 for each category of reinforcement and the material costs calculated. (Guide prices for reinforcement are supplied by ROM Ltd.)

Labour cost

In current practice the cost of labour for fixing reinforcement is based on a built-up rate per tonne. This figure is on average £150/t for an all-in rate, irrespective of bar diameter, and is frequently used to price the labour cost of fixing loose bar. For a more detailed breakdown, the labour cost will vary from £70/t for large diameter bars to £200/t for the smallest diameter bar.

In order to make a fair comparison of the labour cost of fixing any type of reinforcement, the loose bar labour rate has been used as a benchmark and adjusted for the comparative speed of assembly. For example, if it takes less time to assemble prefabricated manhandled panels of reinforcement, then the loose bar labour cost must be adjusted to account for the saving in time.

Table 3.26. Material cost summary for the four alternatives

Alternative	(1)	(2)	(3)	(4)
Loose bars	185.3 t	190 t	13 t	7 t
Cost/tonne	£275	£275	£275	£275
One-way panels			202.9 t	
Cost/tonne			£350	
Two-way panels				294 t
Cost/tonne				£375
Total	**£50 958**	**£52 250**	**£74 596**	**£112 175**

It is important to recognise that the unfactored loose bar rates must not be applied to any alternative option. In many cases the tonnage of reinforcement handled may be greater, whilst the time for installation will be less. In the absence of qualitative data on the actual cost of installing any type of reinforcement, the best compromise is to use the loose bar rate and to adjust this, which is what has been done here. At present these rates are not used in the construction industry, as this is new ground.

The productivity times for a two-man reinforcement gang have been used to compare the rate of assembly against loose bar and then to estimate the cost of labour for the four categories of reinforcement assembly (Table 3.27).

Savings in construction finance

Reductions in overall construction time will make a saving in the cost of construction finance. This is a relative value which compares the savings in time against a traditional loose bar assembly. In this study only the savings in construction finance cost have been calculated. Savings from reductions in preliminaries, which can account for 10% of construction cost, and early rental income from faster assembly, have been ignored.

It has been assumed that the reinforcement is on the critical path of frame construction for the duration of the reinforcement assembly time. We also assume that the formwork and concrete placing cycles closely match the speed of reinforcement assembly. We have compared the productivity gains in construction times, assuming that a four-man gang of steel fixers is deployed on each scheme (Table 3.28).

The seven-storey building has a construction cost of £7.397 million, taking 62 weeks or 15.5 months to complete.

Cost of construction finance

$$C_{\mathrm{fc}} = C_{\mathrm{c}}(I + i)^{\{(T_2/2)/12\}}$$

Table 3.27. Factored labour rates

Alternative	(1)	(2)	(3)	(4)
Time	8.5/8.5	6.5/8.5	2.5/8.5	1.75/8.5
Loose bar ratio	1	0.76	0.29	0.21
Factored labour rate	£150/t	£114/t	£43.5/t	£30.9/t
Total tonnage	183	190	15.9	350
Cost	**£27 795**	**£21 124**	**£7960**	**£5700**

Table 3.28. Production output for the whole building, based on a four-man gang and assuming a 5.5-day week

Alternative	(1)	(2)	(3)	(4)
Days per floor	4	3.5	1.25	1.75
7 floors	28	22.75	8.75	12.25
Relative saving (days)	nil	5.25	19.25	15.75
Savings	**nil**	**1 week**	**3.5 weeks**	**2.75 weeks**

where C_c = construction value, i = interest rate/annum, T_2 = original programme time (months).

Savings in construction finance

$$S_{cf} = C_c[(I + i)^{\{(T_2/2)/12\}} - [(1 + i)^{((T_2 - t)/2)/12}]$$

where t = savings in months due to speed of rebar assembly.

Case: manhandled panels. $t = 3.5$ weeks $= 0.875$ months; $i = 10\%$.
Construction cost $C_c = £7\,397\,120$.

$$\text{Then savings} = £7\,397\,120[(1 + 0.10)^{(15.5/2)} - (1 + 0.10)^{((15.5-0.875)/2)/12}]$$

$$= C_c[1.063\,49 - 1.059\,799] = £27\,300$$

Case: crane-handled panels. $t = 2.75$ weeks $= 0.688$ months; $i = 10\%$.
Then savings $= C_c[1.063\,49 - 1.060\,587] = £21\,476$

Case: rationalised loose bar. $t = 1$ week $= 0.25$ months; $i = 10\%$.
$C_c[1.063\,49 - 1.062\,433] = £7818$.

Summary of finance cost savings

Alternative	(1)	(2)	(3)	(4)
S_{fc}	**nil**	**7818**	**27 300**	**21 476**

Evaluation of overall cost benefits

As can be seen from Table 3.29, based on just the material costs, traditional cut and bent steel (1) is the most cost-effective solution. If the analysis includes labour cost, then option (2), rationalised loose bar designed for minimum weight, becomes the cheapest option. Only a significant change in fabric reinforcement prices will produce a large overall saving because of the reduced labour cost.

Savings in finance cost are real if it can be shown that pre-fabricated reinforcement assembly, option (3), will reduce the overall

Table 3.29. Overall costs for the four alternatives: (1) loose bar; (2) rationalised loose bar; (3) one-way panel; (4) two-way panel

Alternative	(1)	(2)	(3)	(4)
Materials	£50 958	£52 150	£74 596	£112 175
Labour	£27 795	£21 124	£7 960	£5 700
Sub-total	£78 753	£73 374	£82 556	£117 875
Finance	nil	−£7 818	−£27 300	−£21 476
Totals	**£78 753**	**£65 556**	**£55 256**	**£96 400**

construction time by that saving in time. Savings in overall construction will also reduce preliminary costs and give the clients or the developer the opportunity to recover early rental income. However, this is based on the premise that the whole of the reinforcement installation is on the critical path. This depends upon a variety of factors. If the bay sizes are relatively small, i.e. less than one-third of the floor area, then formwork would be the critical task. If the bay sizes increase, then the reinforcement placing becomes more critical. Rapid reinforcement placing must be matched by rapid forming cycles for savings to be sustainable.

Conclusions

The issues associated with rationalisation were first identified, and then a survey of experts in the industry was undertaken to confirm that the issues were correct. They were required to supply quantification of the input values of reinforcement assembly time. These were developed into a series of calculations to value each step in the chain. A case study was used to develop an integrated view of the value chain for various rationalisation options. This showed that, in broad terms, at the current input cost levels there was very little cost difference between the forms of rationalisation that had been chosen. The conclusion is that there is little economic justification for rationalisation unless the design, detailing, purchase, supply and installation are vertically integrated such that all of the components in the value chain contribute to the whole of the potential saving.

Minimum steel weight

Both rationalistion alternatives increased the amount of steel with the objective of an overall reduction in cost. In option (2) the increase was 24% and in option (3) the increase was 46% with respect to option (1) which is the current standard for industry. The material proportion of the total cost was 70% and 90% respectively. With a design industry that both designs for minimum steel weight and prices everything in cost per tonne there is a natural emphasis on final weight of reinforcement. Both of these solutions would look expensive if the decision were made at concept design stage.

Design for faster assembly

The labour cost reduces due to the improved installation buildability. It can be reduced by as much as 80%, as can be seen in options (3) and (4). However, the reduction in labour cost is offset by the higher price for prefabrication and the increased tonnage of steel. The status quo may change in the future as labour rates increase and machine prefabrication prices normalise. Another issue that must be considered is the availability of the correct skills. A boom in demand will bring automatic pressures to reduce the quantity of skilled resources that are needed.

Overall speed of construction through the use of prefabrication will bring benefits in reduced finance cost, as indicated in options (3), (4) and (2), but this can only be realised in terms of total construction output, which includes forming cycle, cladding, services and fitting out.

Recommendations for future action

Rationalisation of reinforcement, as defined in this study, leads to an increase in the weight of steel and therefore, because of single aspect pricing (price per tonne of steel), the design must be rationalised to achieve greater efficiency on site and in manufacturing operations. The decision to rationalise and prefabricate reinforcement must always include an evaluation and justification of the following for their contribution to offsetting the volume of steel:

- ☐ speed of total assembly;
- ☐ labour efficiency;
- ☐ deskilling and risk avoidance;
- ☐ saving in construction finance cost.

As has been demonstrated in this study, these are not simple calculations. To attempt to assign values to individual details, say column sections or beam details, without knowing their context would be unwise until the industry has much more knowledge about the time costs associated with handling different bar sizes on site.

Shearheads and shear reinforcement

In structures where large uninterrupted floor areas and minimum depth of construction are the prime requirements, a flat slab, in situ floor is often the preferred solution. However, to resist punching shear induced around columns, it has been the practice to thicken the concrete slab locally as a drop panel or to widen the column with a column capital, as this may be easier than detailing a large quantity of shear links and reinforcement for a flat slab.

On the other hand, to maintain a flat soffit without any drop panels or column capitals, as in Fig. 3.22, will speed up the forming of the slab, encourage the use of large prefabricated formwork systems and make it easier to install mechanical services. A number of shear reinforcement systems have been developed for flat slabs. Whatever option is preferred, it is important that the shear reinforcement is properly anchored in the concrete compression zone and to the main reinforcement. All stirrup reinforcement should bind around all the bottom reinforcement.

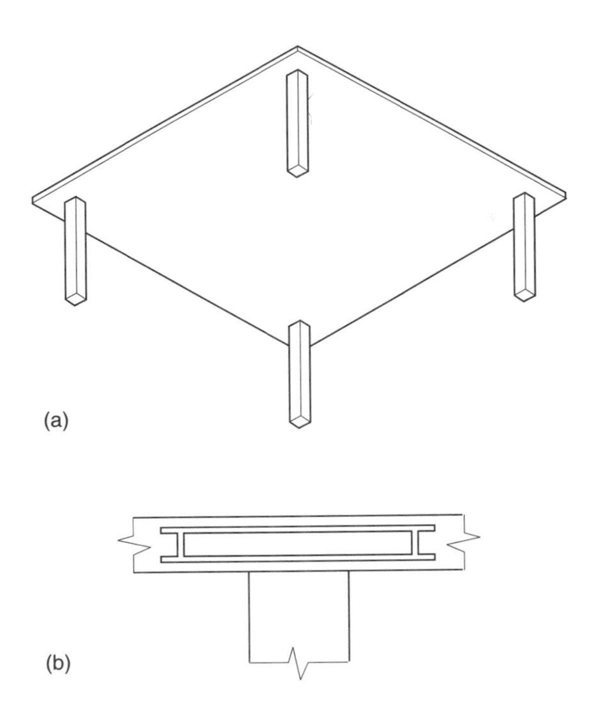

Fig. 3.22. Using a shearhead in a solid flat slab produces a flat soffit

Shear combs, studrails and dowel strips

Shear combs, studrails and dowel strips are comb-like reinforcing elements, the teeth of which consist of studs with truncated cone heads which are welded onto flat steel strips (Fig. 3.23). Compared with other forms of shear reinforcement, the combs minimise congestion and ease fixing of normal reinforcement. They are also capable of positively positioning shear steel and securing an anchorage even in thin slabs.

There has been considerable test evidence on the efficiency of shear combs in Germany, where they have been in use for many years. Professor Paul Regan (Westminster University) carried out a collaborative test of shear combs to verify their design according to CP 110 and BS 8110. He confirms that the Code limits the use of shear reinforcement to slabs with a depth greater than 200 mm, whereas the tests on shear combs show no such limitation is necessary.

Performance

Tests conducted by Sieble showed that conventional stirrup shear reinforcement for a slab should encompass both layers of bottom reinforcement. Even when this is done the full capacity of the slab in bending is not reached, because of slippage and shear failure between the slab and column edge. Failure occurs between the column–slab junction zone and the compression struts, with a steep separating crack which causes the slab to slip.

Understanding this failure mode led to the development of the shear comb or dowel strip reinforcement. When the dowel strips are arranged radially around the column, they create dowel action with the concrete at the column–slab support. The dowel strip binds the junction zone across the compression strut and allows a beam-type rotation, rather than a rupture of the compression strut (Fig. 3.24).

The design method is as follows:

(a) Calculate the shear loading and check whether shear reinforcement is necessary. If it is, check that the maximum load for a slab with shear reinforcement is not exceeded.
(b) Determine the first perimeter at which the concrete alone can resist the loading. This defines the extent of the area requiring shear reinforcement.
(c) Calculate the shear to be carried by the shear reinforcement by subtracting the resistance due to the concrete from shear loading.

(a)

(b)

Fig. 3.23. (a) Studrails and (b) shear ladders in place in a flat slab

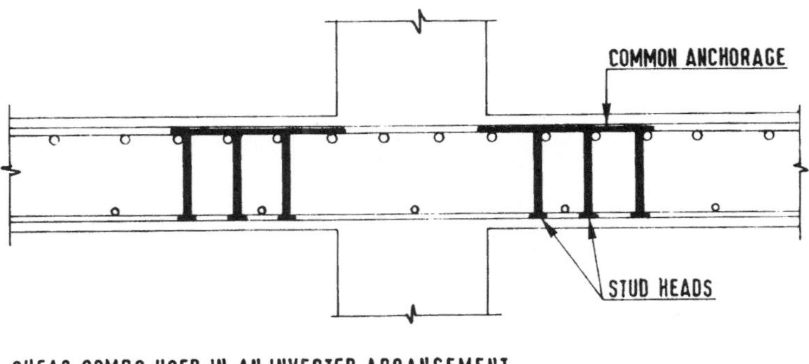

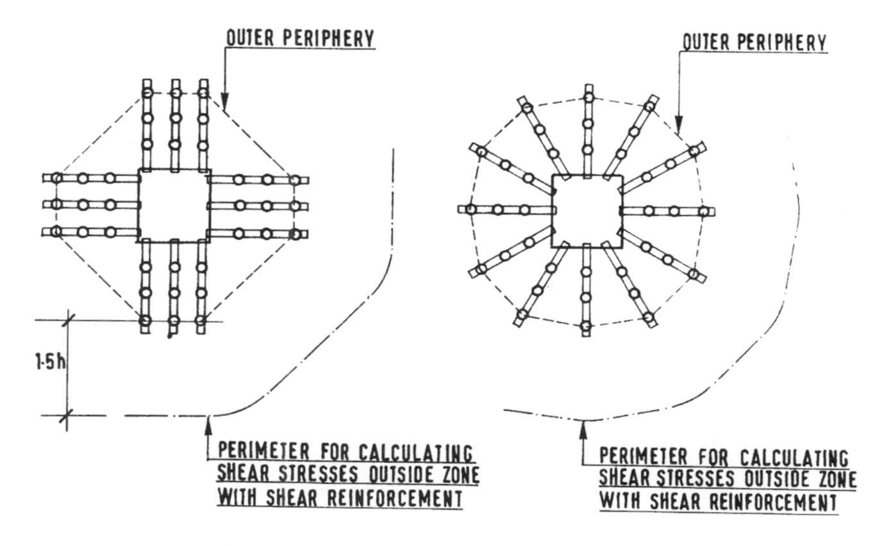

ALTERNATIVE PLAN ARRANGEMENTS OF SHEAR COMBS

Fig. 3.24. (a) Use of shear combs or studrails, showing shear perimeter calculation

Layout of the shear combs should be fully detailed by the designer and should show whether they are laid out radially or circumferentially to the column. The flat bar should be fixed to the formwork with studs pointing upwards.

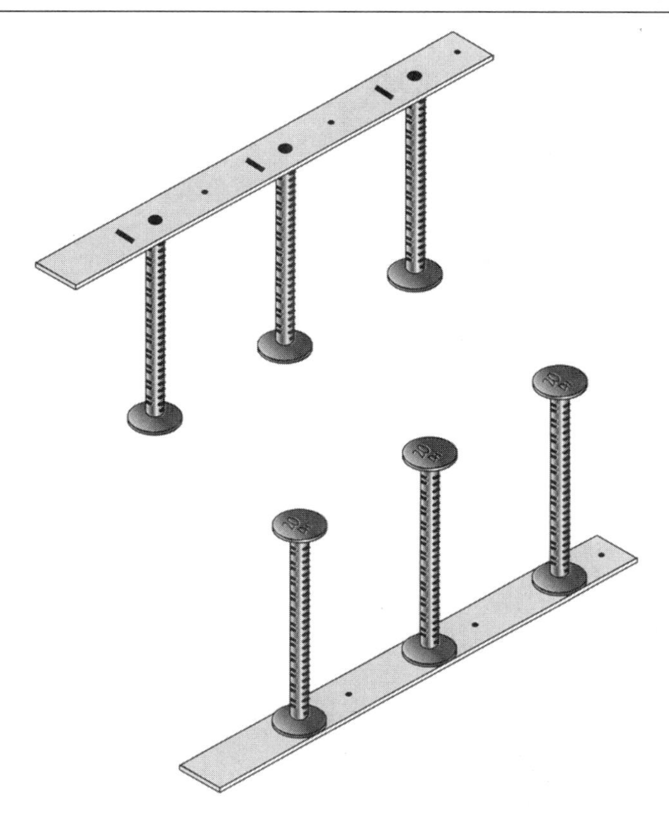

Fig. 3.24 (continued). (b) Studrails can be used either way up

Open-type shearheads

This type of shear reinforcement has been used for hydraulically lifting floors and sleeving them through support columns that are already in position. They have been used extensively with lift slab construction (Fig. 3.25).

Shearheads consist of four steel I or channel sections, welded together in pairs at right angles to one another and continuous through the column, extending the full width of shear perimeter on either side of the supporting column. The ends of the shearheads may be cut at an angle, provided that the plastic moment capacity of the remaining section is adequate to resist the shear carried by the arm of the shearhead (Fig. 3.26).

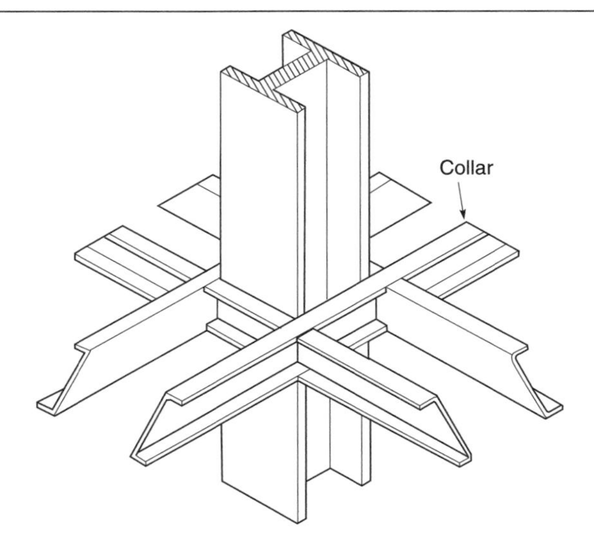

Fig. 3.25. Lift slab column assembly

Design guidance (ACI Building Code)

The full plastic moment of resistance M_p required for each arm of the shearhead (Fig. 3.27) is given by

$$M_p = \frac{V_u}{8p[h + K(L_s - C/2)]}$$

Fig. 3.26. Open-type steel shearhead in lift slab construction

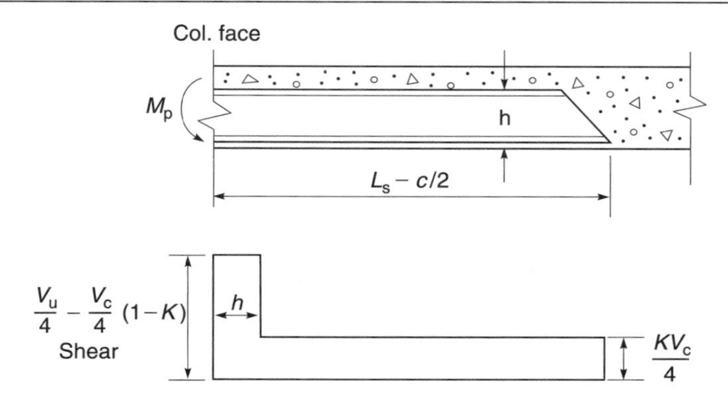

Fig. 3.27. Idealised shear distribution at ultimate load

where p = coefficient of flexure (reduction factor), h = depth of shearhead section, L_s = minimum length of the shearhead arm from the column centre, K = ratio of the stiffness of the shearhead to that of the composite section made up of a cracked section of slab with a width C equal to that of the column plus the slab effective depth and the shearhead ($K \geq 0.15$), V_u = total ultimate shear, and C = one-fourth the perimeter of the square column or of the shearhead reinforcement.

The critical section for shear in the slab, to be used as a measure of diagonal tension, should be perpendicular to the slab. The section should cross each shearhead arm at points three-quarters of the distance $L_s - C/2$ from the column face to the end of the shearhead. However, the critical section need not approach closer than $d/2$ to the periphery of the column, where d is the depth of the slab.

The ultimate shear stress V_u must not exceed $4p\sqrt{f_c}$ on the critical section, where p is the coefficient of shear. Generally, the ultimate shear strength of a slab with shearhead reinforcement should not exceed 1.75 times that of the slab without the shearhead.

The shearhead can also assume to contribute a resisting moment M_s to each column strip of the slab, given by

$$M_s = \frac{pKV_u}{8(L_s - C/2)}$$

where p = coefficient of flexure, and L_s = the length of the shearhead arm. M_s, however, must not exceed 30% of the total moment resistance required in the column strip, nor the change in column strip moment over the length L_s, nor the value M_p.

A series of exhaustive tests has been carried out by the PCA in Skokie, USA, using different open shearhead arrangements, some with double sections which were designated as 'over-reinforced', some with single sections designated as 'under-reinforced', and these were compared with slabs with just normal shear reinforcement and no shearhead. It is clear from the test results that the double-section shearhead failed with widest shear perimeter and followed the perimeter of the shearhead. It was reported to have a 75% increase in shear capacity over a normally reinforced slab of the same depth. When the slab contained a single 'under-reinforced' shearhead, the failure perimeter fell inside the ends of the shearhead. The test confirmed that when the flexural capacity of the shearhead at the column face was not exceeded, the failure surface generally followed the perimeter of the shearhead. This type of failure only occurred with an 'over-reinforced' shearhead. When the flexural capacity at the column face was exceeded, the failure perimeter fell inside the ends of the shearhead and this was defined as an 'under-reinforced' shearhead. The double shearhead arrangement was the most effective for shear reinforcement in flat slabs.

It was from these and other corroborative tests that the design rules for open shearheads were developed and adopted by the ACI Committee for use in the ACI Building Code. Three important and basic design criteria for open shearheads became evident from the test results:

- ☐ Minimum flexural capacity must be provided to ensure that the required shear capacity of the slab is reached before the flexural capacity of the shearhead is reached.
- ☐ The nominal shear stress in the slab at the end of the shearhead must be limited.
- ☐ If the first two conditions are satisfied, the negative slab reinforcement in the slab can be reduced in proportion to the moment contribution of the shearhead.

The closed 'Geilinger' steel shearhead

A significant advantage of the closed 'Geilinger' shearhead (Fig. 3.28) is the ability to provide holes through the slab close to the column face, to permit service pipework and conduits to be taken through the floor at the most preferred location (Fig. 3.29).

Closed shearheads are square grillages of channel or I section. The section size and length can be determined by the open shearhead design rules, but enhanced by the greater rigidity of the braced member and the confinement of the concrete within the shearhead.

Safety against punching shear in the case of an internal column support must be verified in the slab section at a distance of half the effective depth beyond the shearhead perimeter. With edge and corner

Fig. 3.28. Geilinger shearhead

Fig. 3.29. Shearhead allows for duct holes

column locations the permissible shear stress outside the shearhead perimeter must be limited by a reduction factor.

The shear force of the slab is transferred to the column supports exclusively through the webs of the shearhead sections. All weld joints are to be designed for this condition, especially when connecting the shearhead to a steel column support. The difference in moment between the collar edge and the support axis is to be carried by the shearhead.

Openings in the floor through the shearhead are limited and must result in a large concrete cross-section in order to take the bending moments. For this reason the sum of the openings in each direction must not exceed one-third of the length of the shearhead perimeter.

As regards fireproofing, Geilinger steel shearheads meet the requirements of fire resistance class F 30, provided the concrete cover on the steel sections is at least 10 mm.

Shearheads can be supplied integrally with storey-high steel or spun concrete columns for flat slab construction (Fig. 3.30).

Fig. 3.30. Geilinger integral column and shearhead used in flat slab construction

Shearhoop reinforcement

A shearhoop consists of a prefabricated assembly of specially bent links and hoop bars that provide shear reinforcement. These hoops come in various sizes, and are placed around the supporting columns to form a complete system of hoop reinforcement that extends out to the shear perimeter and to the point beyond which the shear resistance of the concrete slab is adequate. The shearhoop greatly simplifies the detailing and positioning of shear reinforcement in flat slabs between the bottom reinforcement and is fully anchored to the top reinforcement.

Component sizes

There are three types of shearhoop, a type SS used for all internal, edge and corner columns and two others, types RE and RN, which are used in conjunction with type SS for internal columns when opposite sides of the shearhoop would be greater than $0.5d$ from the opposite column faces (d is defined as the effective depth of the slab).

The depth or height of the shearhoops can vary from 130 mm to 410 mm, in increments of 5 mm, to suit slab depths between 200 mm and 450 mm, and they are available in square shapes with sides that can vary in length between 300 mm and 2900 mm, increasing in 200 mm increments. Shearhoops are manufactured from Grade 460 type 2 deformed bars that are weldable and conform to BS 4449:1988.

Placement

Shearhoops are supplied to the site ready to assemble. After the bottom mat of reinforcement is fixed, the innermost shearhoop is positioned around the column starter bars, followed by successively larger shearhoops until the appropriate shear perimeter is reached. The first layer of top reinforcement is then positioned and located under the top hooks of the shearhoops. The top layer of reinforcement is then placed. However, shearhoops are slow to install (Fig. 3.31).

Anchorage

Shearhoops can be considered fully anchored in the compression zone provided they are placed between the bottom slab reinforcement, and have the same cover as the bottom layer of reinforcement.

In the top of the slab, shearhoops are fully anchored if the upper legs or bobs have the same cover as the top layer of reinforcement T1, and the top slab reinforcement T2 is fixed under the ends of the bobs of the shearhoops.

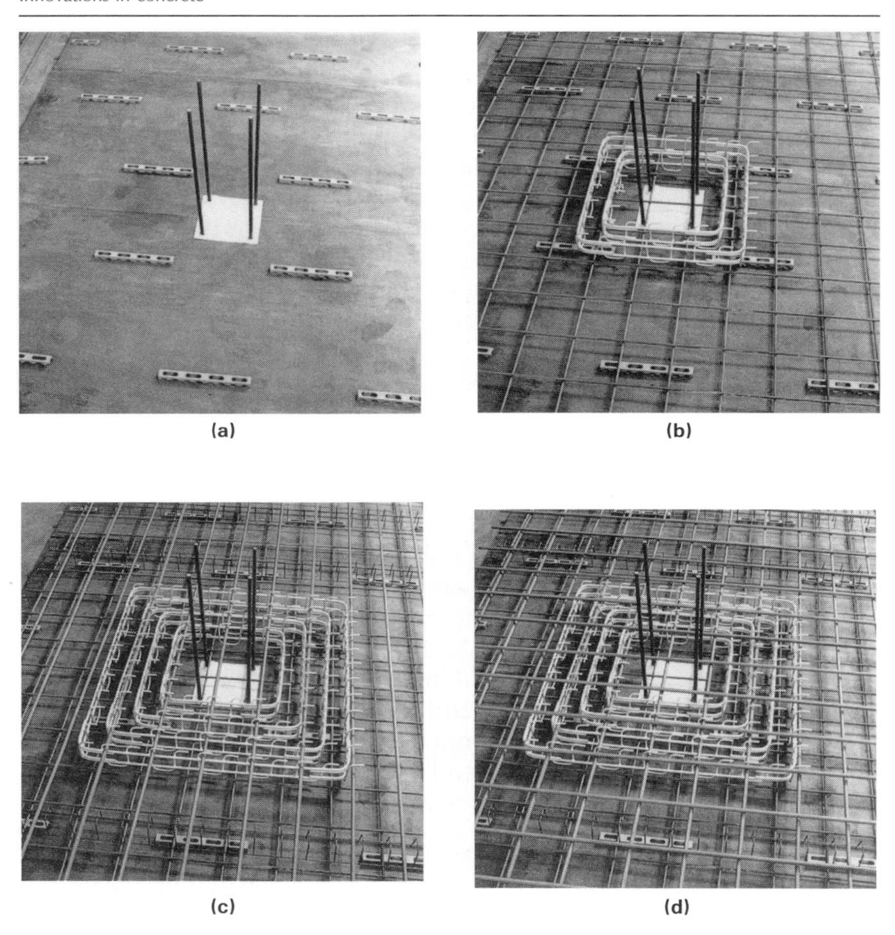

Fig. 3.31. Shearhoop construction sequence: (a) place and fix bottom reinforcement on spacers; (b) place shearhoops on spacers that provide the same cover as for B1 reinforcement; (c) place top reinforcement T2 on chairs and locate fully under the ends of the shearlegs, or fix additional facing bars; (d) finally, place and fix top reinforcement T1

Critical distances

The perimeter of the innermost shearhoop should not be further than 0.5d from any column face, where d is the effective depth of the slab.

Holes through the slab

Shearhoops provide very little scope for service openings close to columns. With the first shearhoop positioned 0.5d from the column, a very small service hole can be inserted whose width must not be greater than one-quarter of the width of the column, provided that cover is maintained to the shearhoop reinforcement.

Post-tensioned concrete

Introduction

Post-tensioned construction has been used throughout the world since the 1950s. The number of completed post-tensioned slabs for office buildings is in the thousands in the USA, several hundreds in Australia, Europe and the Far East. They speak volumes for the reliability and proven technical merits of this technology (Fig. 3.32).

Yet in the UK there are less than 100 buildings using this form of construction. Why has this economical form of construction not proven more popular? One reason may be that the potential advantages of post-tensioning have not been properly evaluated by design and construction professionals. Another may be because sufficiently practical and detailed guidance for the UK building industry has not been available. How does prestressing a slab improve the strength and behaviour of a concrete floor? The fundamental advantages are the reduction in deflection of the slab, the shallower depths of construction that result (Fig. 3.33) and the longer spans that are economical for a given floor loading.

Fig. 3.32. Post-tensioned floor slab under construction

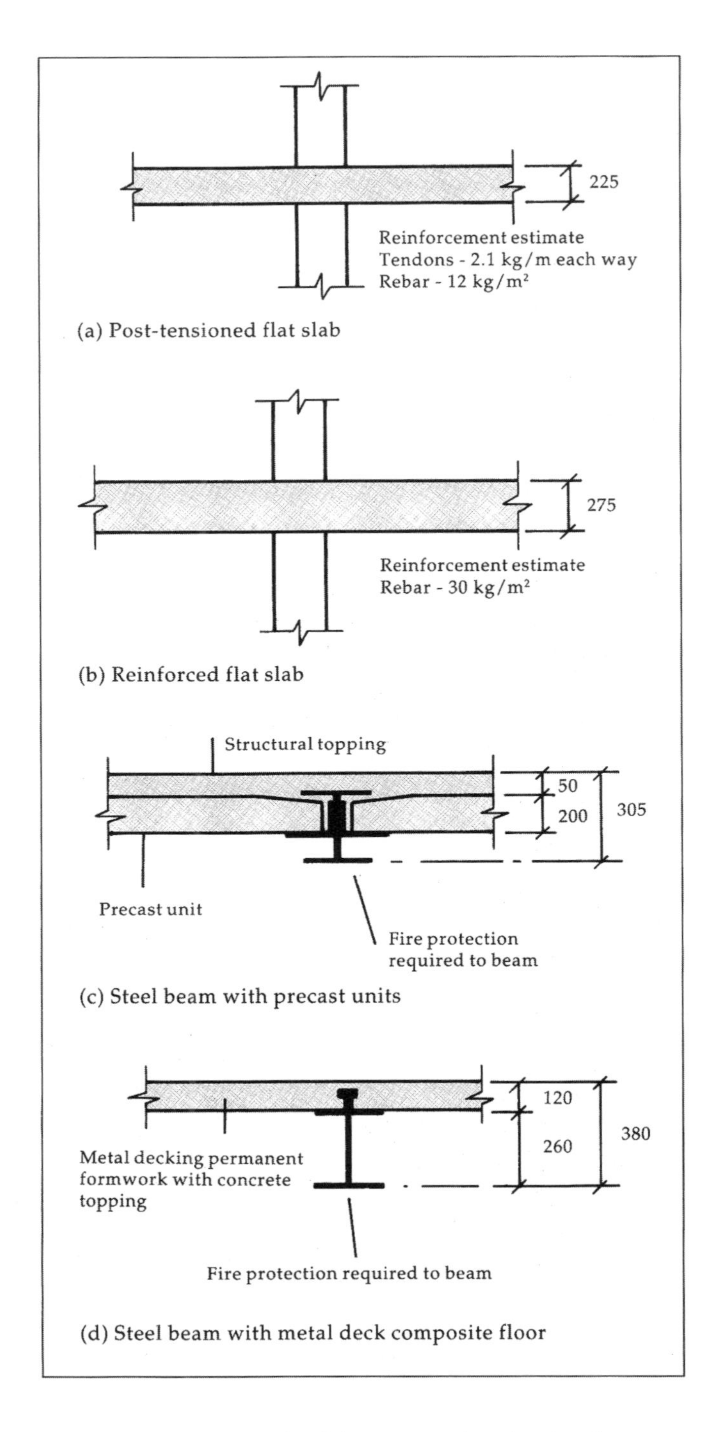

Fig. 3.33. Construction depths of flat slabs compared with other floor types

In a reinforced concrete slab, tensile cracking of the concrete occurs after load is applied in order to bring the reinforcement into action and to resist further tensile stresses building up. In a post-tensioned slab the act of prestressing puts the concrete into compression before gravity loads are applied. This significantly reduces the tensile stresses in the concrete and can even be enough to prevent any tensile cracking of the concrete from occurring. Besides increasing the moment and the shear capacity of the slab in this way, the prestressing enhances the stiffness of the slab section because the concrete remains uncracked, and consequently the slab thickness can be reduced. The tendon profile or drape of the prestressing strand follows the shape of the bending moment diagram. This is usually parabolic for uniformly distributed load distributions. The eccentricity of the tendon relative to the centre of gravity of the slab induces a deflection in the opposite direction to that resulting from self-weight and the applied loads. By varying the amount of prestress in the tendons and the position of the tendon within the slab depth, any proportion of the slab self-weight and the applied load deflections can be 'balanced out' (Fig. 3.34).

Naturally, the most economic solutions are those where a high amount of prestress can be applied to balance out the imposed loads and self-weight of the slab. Thus prestressing works best for relatively long spans with low ratios of live to dead load. Solutions with shorter spans and large live loads can prove economic, but the large difference between the maximum and minimum load conditions necessitates a limitation on the amount of prestress that can be applied. Thus, for longer spans that are lightly loaded, large amounts of prestress and small amounts of reinforcement are used, whilst for shorter spans that are heavily loaded, more reinforcement than prestress is applied. Average prestress levels usually range from 0.7 to 2.5 N/mm^2 for solid slabs and occasionally up to 6 N/mm^2 for ribbed or waffle slabs.

Methods of prestressing

There are two methods of prestressing in situ concrete:

- ☐ pretensioning;
- ☐ post-tensioning.

This note deals essentially with post-tensioned construction, but before we move on to discuss it in more detail it is useful to touch on the difference between the two methods.

Pretensioning is used mainly in precast concrete and bridge design. The steel tendons, in the form of wires or strands, are tensioned between rigidly held end-blocks. The concrete is then poured to encase

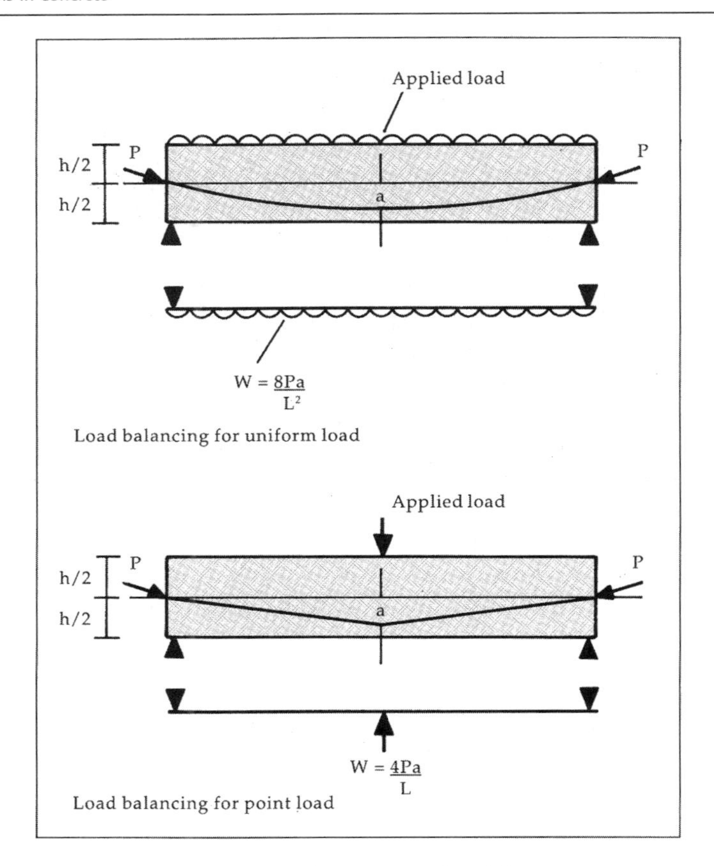

Fig. 3.34. Load balancing of unbonded tendons in flat slab

the prestressing strands and to tie in the end-blocks. Once the concrete has hardened sufficiently, the anchors in the end-blocks are released and the prestress force is transferred to the concrete through the bond that develops between the steel strand and the concrete.

Post-tensioning is common in building construction for suspended and ground-supported slabs. Here the prestress force is applied after the concrete has been cast and has achieved a minimum compressive strength. Post-tensioned flat slabs can be constructed using bonded or unbonded tendons. Each method has its particular advantages, which can be summarised as:

- *bonded* (Fig. 3.35)—develops higher ultimate flexural strength, does not depend upon the anchorages after grouting, localises the effect of damage;

Fig. 3.35. Post-tensioned floor slab using bonded tendons

□ *unbonded* (Fig. 3.36)—gives greater economy, reduces friction losses, simplifies prefabrication of the tendons, is faster to install and provides a larger available lever arm.

Fig. 3.36. Floor slab using unbonded tendons

Bonded tendons are either sleeved into metal ducts that are 75 mm wide by 20 mm deep before the concrete is poured, or pulled through the ducts after the concrete is cast. The ducts are grouted up after the tendons are prestressed. Unbonded tendons come prefabricated in a PVC sleeve impregnated with a rust-inhibiting grease which can also act as a lubricant to reduce friction loss. In unbonded design no bond exists between the concrete and the tendon; thus the integrity of the process is dictated by the effectiveness of the end anchorages. The allowable net force in the tendon over time is greater for a bonded system because there is relatively less prestressing loss.

Generally, tendons consist of seven-wire strands, each 12 mm or 15 mm in diameter, with a characteristic strength of 1850 N/mm^2 (reinforcement has a characteristic strength of 460 N/mm^2). The stress is applied to the tendons by portable hydraulic jacks situated at the live end anchor. Special jaws of the jack fit over the multiple strands and uniformly extend them with a given force. Live end anchors are fitted with wedges which draw in as the force in the tendon is transferred from the jack to the anchor.

Construction details

Restraint and infill strips

In order to achieve maximum efficiency in post-tensioned construction, the slab pour should be configured in a rectangle with the maximum number of tendons laid out in the longitudinal direction. For prestressing to be effective, the concrete must be free to undergo elastic shortening due to the prestress force. If the slab is restrained by solid core walls or shear walls, for example, much of the precompression in the slab will be absorbed by the restraints and the benefit of the prestressing will be lost.

It is possible to overcome this by allowing the slab to move independently of the restraint, by incorporating a reinforced concrete infill joint (Fig. 3.37). The tendons do not continue through the strip and are anchored on each face of the adjacent slab. Thus it is important early in the design process to rationalise the position of construction joints and position any movement strips away from wall or other restraints (Fig. 3.38).

Extent of pour size

The economical limit on the length of a tendon, whether bonded or unbonded, is usually 30 m when stressed at one end and 60 m when stressed from both ends. The longer slab and beam lengths are achieved with the use of intermediate anchors, and without interruption of ten-

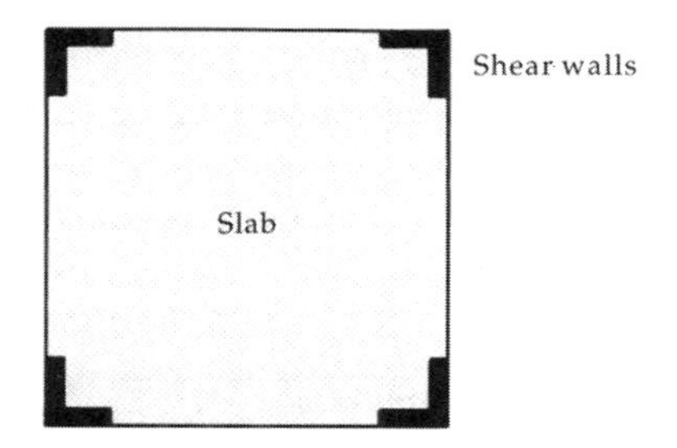

(a) Plan of slab with four corner shear walls

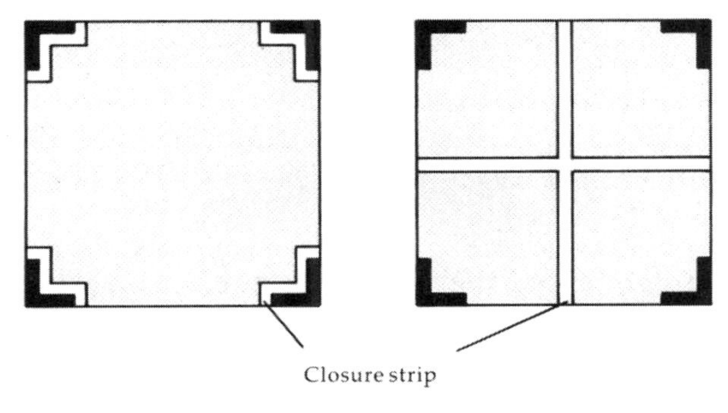

(b) Alternatives for arrangement of closure strips

Fig. 3.37. Controlling restraint cracking

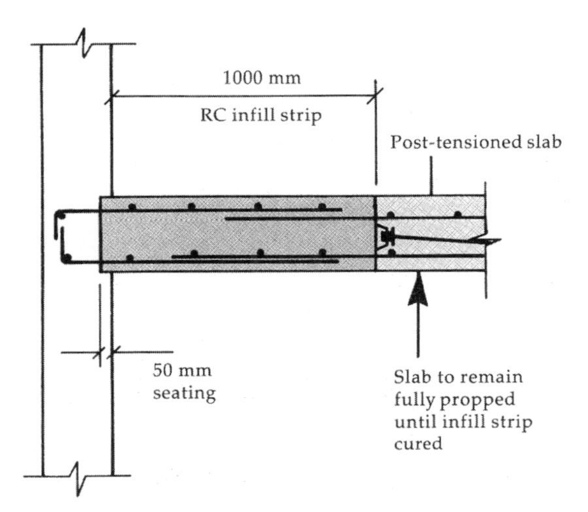

Fig. 3.38. Typical infill strip at supporting wall

don lengths. This is usual when detailing tendon layout and anchorages for successive concrete pours.

Joints

Generally, construction joints in floors should be as few as possible and located at points of low stress, i.e. points of contraflexure. This is usually at a distance between one-quarter and one-third of the span from the supports.

Tendon layout

Various tendon layouts have been used for post-tensioned flat slabs (Fig. 3.39). The accepted layout of tendons is a 'beam–slab' configuration, which concentrates tendons in one direction along beam lines and uniformly distributes them in the transverse direction for the slab span (Fig. 3.39(a)). The beam–slab arrangement is preferred because it makes tendon installation easy and eliminates interweaving of tendons. The tendons concentrated along the beam strip should be placed within a half slab depth from the column face to obtain maximum shear contribution. Generally, for ribbed and waffle slabs, the distribution of the tendons will be dictated by the rib spacing.

The following guidelines for tendon spacing should be noted:

- ☐ maximum tendon spacing: 6 × slab thickness when unbonded, 8 × slab thickness when bonded;
- ☐ grouped or bundled tendons should not exceed four;
- ☐ minimum tendon horizontal spacing: 2 × tendon diameter or 75 mm, whichever is greater;
- ☐ vertical spacing: tendon diameter.

It is usual for the post-tension specialist to prepare working drawings indicating the tendon layout and spacings for a particular project. As the tendon is normally stressed from one end, live end access is required beyond the live end anchor for applying the prestress. Usually a 1 m wide platform is required. This is an important point in preparing falsework plans.

Stressing sequence

The stressing sequence should be specified by the post-tension specialist and agreed with the structural engineer. It is usual to stress the tendons for the beam strips first, before the transverse or distribution tendons, to avoid overloading the falsework. Some specialists prefer to stress alternate tendons, working from one end of the pour. It is believed that this minimises the prestress loss due to elastic shortening.

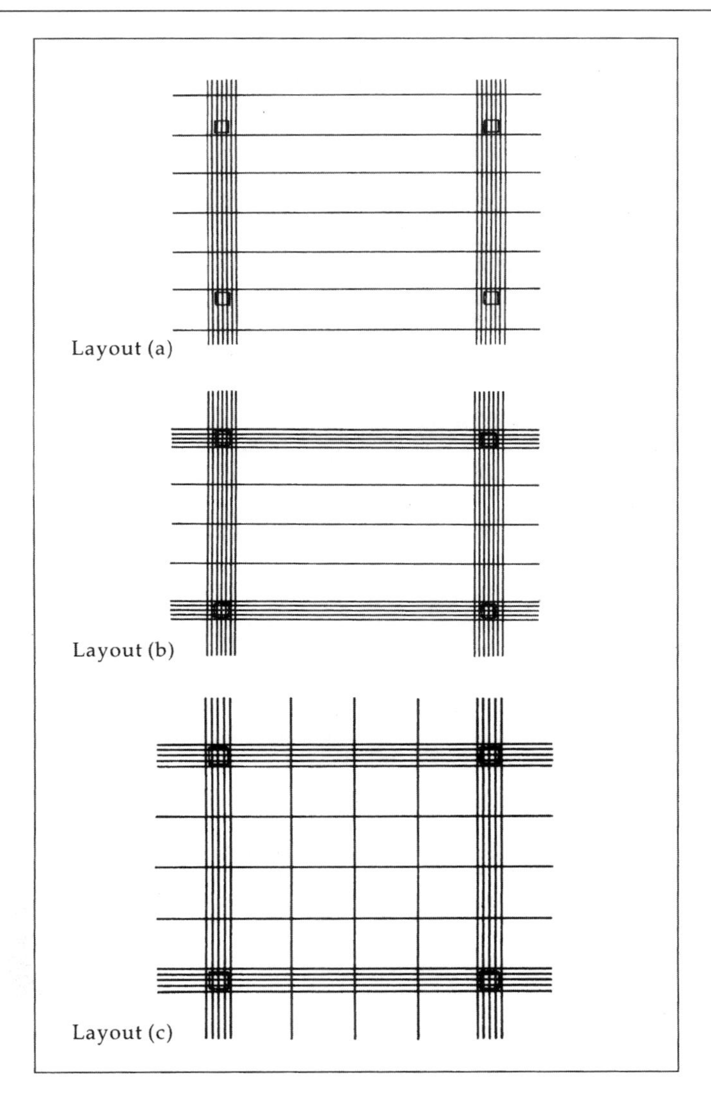

Fig. 3.39. Common layouts of unbonded tendons in a flat slab

No common guidelines on this point are generally agreed by designers because prestressing forces are fairly low. Therefore it is largely a matter of judgement and past experience which decides the sequence of stressing. For efficiency of construction it is preferable to stress the tendons in sequence, working from one end to the other. This is often adopted in practice.

(a)

(b)

Fig. 3.40. Stressing of tendons: (a) access is important; (b) positioning of stop-ends to divide the slab into bays

Stressing (Fig. 3.40) is applied in two stages. The initial 50% stress is applied as soon as the concrete has achieved an agreed minimum compressive strength, usually 15 MPa. The earlier the stress is applied the better, because it minimises the risk of concrete developing cracks due to thermal contraction. The final stress is applied after the 28-day strength is achieved, which is usually 25 MPa. The final stress can be left for up to two to three weeks after the slab is cast, as the initial 50% prestress will allow the floor to support its own weight and the construction loads.

(c)

(d)

Fig. 3.40 (continued). (c) Stressing anchorage assembled with plastic recess form; (d) recessed anchorages, showing wedges

Stressing records are submitted to the engineer for approval before tendons are cut. The extension of the tendons is recorded at each stressing stage. After the tendons have been cropped the end anchors are sealed with a grease cap and the recesses filled with a sand–cement mortar. Calibration certificates for the jack friction and pressure gauges are required before stressing operations begin.

Formwork and cropping

The formwork requirements and construction details are provided by the post-tension specialist and show details for anchorages, bolt hole dimensions, tendon positions and box-outs, etc. Formwork procedure should follow the principles of good formwork practice, with additional recommendations for ease of construction; these are:

- ☐ edge forms should be predrilled for end anchors and fixed in position before tendons are installed;
- ☐ the edge form should be robust enough to withstand the initial pull on the forms as the tendons are positioned.

Props for post-tensioned slabs can be removed once the slab has received its initial post-tensioning, as mentioned before. However, two levels of back propping are usually necessary to transfer wet concrete load and high construction loads from the upper floors. This must be fully considered in the design stages and agreed with the contractor.

Concrete and construction checks

Concrete grades between 30 and 40 MPa are usually specified for post-tensioned construction. Prior to placing of the concrete, tendon profiles should be checked by measuring the profile dimension from the formwork to the centre of the tendon diameter (Fig. 3.41). Tendon sheathing, particularly at the anchorages and grout vent pipes for bonded systems, should be carefully checked to ensure that it has not been dislodged or damaged.

Special attention must be given to vibration of concrete at tendon anchorages to ensure good compaction. Proper walkboards should be used as tendons must not be stepped on during concreting operations. It is prudent to ensure that tendon profiles are checked and maintained at the correct dimensions just ahead of concrete placement.

Cubes taken for minimum strength requirements for first-stage prestressing should be cured alongside the slab to simulate the slab curing regime. Early strength is an essential requirement in post-tensioned construction in order that the prestress can be applied as soon as possible. This will minimise any thermal and plastic cracking that may occur.

Once the initial 50% prestress is applied the formwork can be removed. The construction sequence for post-tensioned construction is slightly different from conventional reinforced concrete procedure. A typical sequence is: fix formwork; place rebar; fix tendons and end anchors; pour concrete slab; apply 50% prestress; remove formwork;

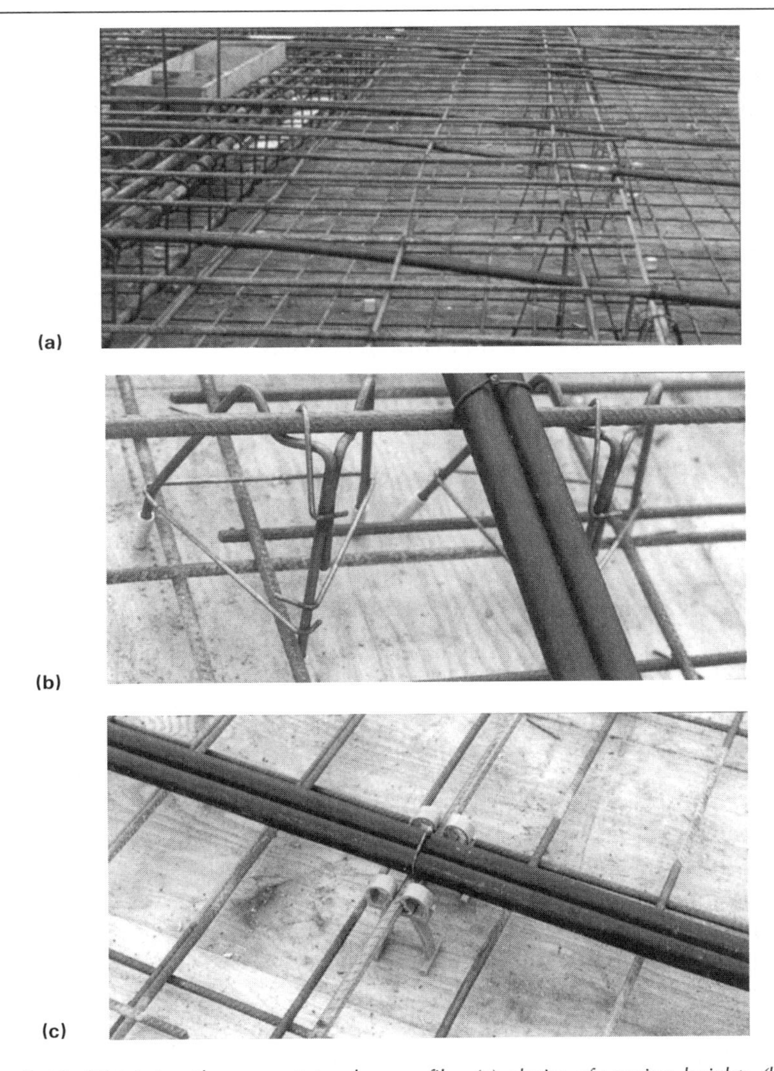

Fig. 3.41. Obtaining the correct tendon profile: (a) chairs of varying height; (b) wire chairs; (c) smaller plastic chairs for cover

reprop; apply final-stage stress; make good pockets; grout up bonded ducts; and cut off excess strand.

Speed and relative costs

Post-tensioned construction requires less material to be placed and concrete to be poured than traditional reinforced concrete, making savings in material cost and labour. Generally, designs using unbonded tendons are quicker to install than bonded construction because the unbonded tendons are factory assembled and delivered to site ready to use. Bonded tendons require preparation on site to bundle and sheath the strands into ducts before they can be installed; consequently they take much longer to fix. To inject speed into post-tensioned concrete construction, the designs must review the number of sequential activities and the extra access needed for stressing operations outside the slab. Single-stage stressing instead of two-stage stressing may help, along with designs that utilise the benefits of prefabricated reinforcement and unbonded tendons to increase speed of fixing. Additionally, high early strength concrete can achieve 20 MPa after 24 h, and can be incorporated to allow early removal of formwork to achieve a three- or four-day floor cycle.

There is a small cost difference between a reinforced and a post-tensioned floor carrying the same imposed loads. However, whilst the cost of tendons may be 30% more than that of rebar, the reduction in material content of about 25% and the reduction in overall slab thicknesses and storey height probably more than balance out this cost difference. As a general rule of thumb, if a reinforced concrete slab costs around £100/m^2 a post-tensioned slab will cost 5–10% more. However, when evaluating the overall benefits of post-tensioned construction for a building such as the Exchange Tower (Figs. 3.42 and 3.43), the additional benefits weigh heavily in its favour:

- ☐ reduced dead load, reduced foundation load;
- ☐ thinner floor slab, reduced overall building height;
- ☐ reduced building volume reduces overall cost of heat and light, therefore lowering running costs;
- ☐ longer-span floors reduce the number of columns and increase the lettable floor area.

On the downside, post-tensioned construction is not as flexible as reinforced concrete for accommodating service penetrations and holes after the floor has been cast. Repairs and any remedial works have to be carried out by specialists and therefore may cost more. Developers and funding institutions have reservations on the suitability of this form of construction for commercial projects. Most of these reservations arise mainly out of ignorance on the flexibility of this form of construction for cutting holes, i.e. staircases and openings after the floor has been cast, and for the safe demolition of such structures.

Notes on these have been compiled to provide an understanding of the limitations in this area.

Fig. 3.42. Exchange Tower, Docklands, London

Cutting holes in slabs

When forming large post-construction openings in suspended floors, their position has to be selected with due regard to the original design concept. Post-tensioned slabs are no different in this respect.

When considering a flat slab design, it will be necessary to locate openings away from areas of bonded tendons in order to minimise any reinstatement which may be required (Fig. 3.44). This does not unduly restrict internal planning, despite tendons covering two-thirds of the floor, as they usually run in one direction and are evenly spaced apart.

Exposure of tendons

Prior to any breaking, it is necessary to locate the tendon positions. This can be done easily, using a covermeter. The surrounding concrete

FLOOR DESIGN OPTIONS

	Option (9 × 7.5 m grid)	Floor depth (mm)	Total depth 17 floors (m)	Cost index (frame only)
STEEL	9 m primary beams 7.5 m secondary beams 130 mm metal deck	410	6.97	1.35
	7.5 m primary beams 9 m secondary beams 130 mm metal deck	460	7.82	1.25
CONCRETE	Post-tensioned 7.5 m spine beam 9 m ribbed slab	250	4.25	1.0
	Post-tensioned solid slab (with column drops)	250	4.25	1.05
	Reinforced 7.5 m spine beam 9 m ribbed slab	425	7.22	1.0

Fig. 3.43. Exchange Tower: floor design options

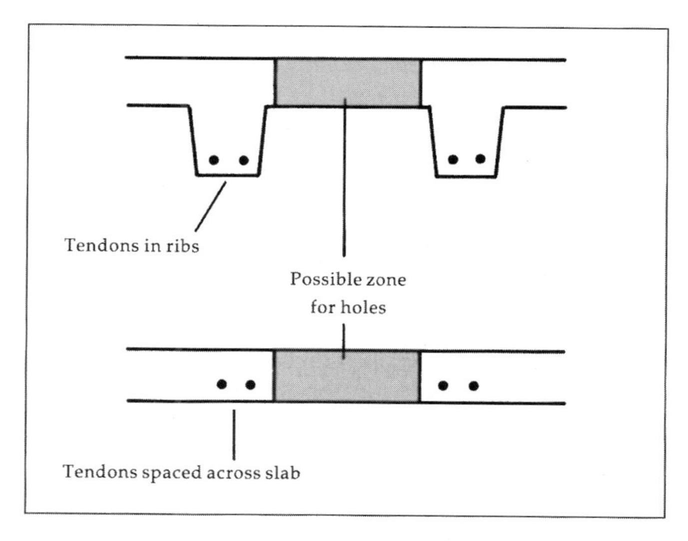

Fig. 3.44. Zones where holes can be readily formed in post-tensioned slabs

can then be broken away, exposing the tendons along their entire length (Fig. 3.45). It is usual to have an engineer present during this stage of the operation to ensure that no premature damage occurs to the tendons.

De-tensioning

Once the tendons are uncovered (Fig. 3.45(a)), they can then be de-tensioned one by one. The procedure is to slip a bearing plate over the tendon and against the concrete face. Special 'Gemi' jacks can then be positioned against the bearing plates and used to take up the load in the tendon (Fig. 3.45(b)). This has the effect of relieving the load in the

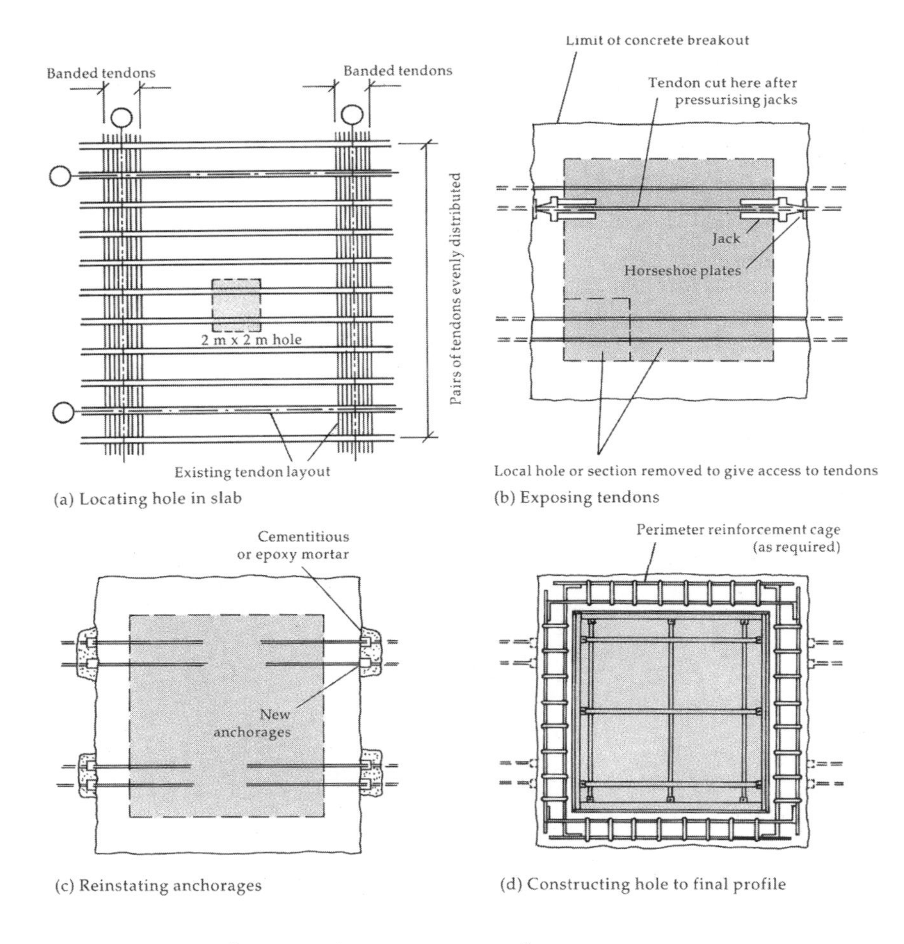

Fig. 3.45. Forming large openings post-construction

tendon length between the jacks. The tendon can then be cut, using either a disc or a flame cutter. The pressure is then released from the jacks and the tendon length across the opening can be cut. Having de-tensioned one pair of tendons, it is necessary to reinstate and stress the anchors to both before continuing to de-tension the second pair (Fig. 3.45(c)).

Once the tendon sheathing is stripped back, a standard anchorage can is fitted to the remaining tendon. This is then set into the parent concrete, using a fast-setting epoxy mortar. Once the mortar has set the tendon will be stressed. When stressing is completed satisfactorily, the tendon is cut off and sealed. Once the anchors are reinstated, any further work is mainly cosmetic. Any desired edge profile can be formed, e.g. a nib or upstand, using a small cage of links around the perimeter (Fig. 3.45(d)). Final pouring of the concrete into the edge strip completes the operation.

Forming a stairwell opening

Procedures have been adopted by Genstar Structures Ltd to cut large stair openings in post-tensioned slabs to interconnect floors at ten separate levels (Figs. 3.46 and 3.47). Stair openings of 4.3 m × 4.3 m were cut into the floor slab without the use of extensive shoring or additional support beams. Interior refit designers were asked to avoid beam locations and end spans when locating the stair opening positions. After removal of the concrete in the opening location, the tendons were cut with a torch while being restrained at the edges of the opening by heavy metal clamps. The metal clamps were used to provide a slow release of the tendon force after the tendons were cut. This process permitted de-

Fig. 3.46. Example of hole formed post-construction

Fig. 3.47. Atrium formed post-construction

tensioning of all tendons through the stairwell openings with no damage to the glazing which had already been installed around the perimeter of the buildings. Tendon anchors were then reset and, after the concrete front of the anchor had reached full strength, the tendons were restressed at the perimeter of the opening. Finally, forms were installed, and concrete was placed around the forms to finish the openings.

Demolition work

Prestressing locks energy into the tendons and this energy must be carefully released before demolition. The level of prestress in a building is low compared to that in civil engineering structures such as bridges. The energy released is too low to cause any dramatic structural effect. A prestressed building is demolished in the same way as a reinforced concrete building, but with additional measures to contain the anchorages.

Tendons are exposed at a single point in sequence and destressed by heating the tendon or by cutting it. As areas of floors are destressed, they can be broken out readily. Because there is less reinforcement and tendons in a post-tensioned slab, it should be quicker to demolish than a conventional reinforced concrete slab.

Case Study: Sainsbury Centre, Cannock

(with reference to a report written by David Ramsay, formerly of DHV, now with White Young Green)

Introduction

Of all the variety of engineering activities, foundation design is perhaps the one area where the greatest amount of engineering judgement has to be regularly exercised. All site investigations are, by their very nature, mere snapshots of the specific area sampled. The real performance of the ground may differ significantly from the theoretical figures produced in the laboratory. Any site investigation is a compromise between accuracy, cost and time available. Moreover, the degree of investigation undertaken may vary considerably from one project to another depending on the severity of the site and the sensitivity of the structure to be designed.

The Cannock site, located east of Wolverhampton, has had a very varied past. In the late 19th century it contained a flight of locks which carried the Birmingham to Stafford canal. Between 1955 and 1965 the entire canal and most of the hill was removed when opencast coal mining took place, The normal method of working the mines was to use draglines to remove the overburden and deposit it in spoil heaps behind the working face. Other excavators would be employed to load the exposed coal into trucks for removing to the coal plant. When mining finished, what was left was quarry discard material, loosely compacted and interspersed with numerous large boulders. During the 1970s and 1980s the area was landscaped and became local leisure space, until it was sold for development in 1988.

Project details

The site was purchased for a superstore development in late 1989. In the six months prior to this a detailed site investigation took place to allow a design to be developed. The superstore was to be a single-storey structure with a floor area of approximately 5000 m^2, with a back-up store, administration and food preparation area on the same level. The client's brief was for a brick-clad, pitched-roof building in keeping with local house style but incorporating colours and detailing to match local features. Externally, there were to be extensive paved areas and substantial car parking. The development had to be completed by January 1991.

The general topography of the site was gently sloping northeast to southwest at a gradient of approximately 1:30. The site was partly encroached by the ends of two earth bunds which had been formed at the time of the leisure scheme. On the southeast boundary there was a small potholed road which became known as Not Washbrook Lane. The road was bounded by ditches which were known to run full in times of rain. The provisional location of the store moved around during the development of the design.

Site investigation

A series of ground investigations, consisting of 24 boreholes and three trial pits, was carried out between August and October 1989. The results of the investigation proved the depth of the made ground and provided data from which bearing pressures and settlements could be calculated. The material was tested for contaminants and found to be non-toxic. Results from the early investigations in August allowed preliminary foundation options to be determined. However, the preferred location of the store necessitated a further series of boreholes being sunk in October.

Design problems

The boreholes proved that sound strata were at a significant depth, in excess of 30 m over most of the site. Subsurface contour plans produced from the site investigation data showed that the proposed store straddled a deep pit. Equally significant was the fact that the edge of the pit was coincident with the southern site boundary, creating large differences in the depth of unconsolidated material. Due to the variability of the fill material little reliance could be placed on testing samples; consequently the site investigation included extensive in situ penetration tests to probe the strata. The results from these tests indicated a very wide scatter of results at all depths, showing that severe differential settlements could be expected with shallow foundations.

Foundation options

The nature of the ground was such that no foundation design recommendations were made in the original geotechnical report. The solution that was chosen was developed after a series of detailed discussions between the geotechnical engineers and their structural counterparts. There were three main options:

(1) rigid raft;
(2) piles;
(3) flexible raft.

Option 2 was quickly ruled out after discussions with piling contractors, since there were a number of potential obstructions in the depth to sound strata.

Option 3 was ruled out after preliminary calculations indicated that settlements could be up to 500 mm, with differentials of up to 200 mm.

This left Option 1. An estimate of the costs showed that a flexible raft was economically viable. Detailed costings of the scheme were carried out, based on a deep reinforced concrete raft under the sales floor and a separate raft under the back-up store. Although this still fell within the original rough budget, it proved in the end that an economic-depth raft was reliant upon the integrity of the soil strata, which was too great a risk for the client and could not be sustained. The design team were instructed to stop work.

Sudden loss of work is a great innovating force. Although Option 3, the flexible raft, had been ruled out due to excessive deflection it was still technically feasible. Some lateral thinking was required. Foundation design in buildings is generally based on satisfying two criteria. The first is that the ground shall safely carry the imposed loads without shear failure. The second is that it shall do so without excessive settlement. In this case it was the second criterion that was presenting the problem. What constitutes excessive settlement is a very subjective issue. In the context of a superstore the key performance requirements are that finishes shall remain in good condition and that trolleys will not run away from customers! In addition, some of the refrigeration plant can be adversely affected if abnormally shallow falls are found in the condensate drains. After further calculation and study of the ground conditions, it was agreed with the client that the maximum floor gradient could be 1:300, provided any settlement would not adversely affect the finishes and cladding. This latter point was critical, since internally the store which was to be tiled with terrazzo and externally with brick paviours, both of which are sensitive to settlement.

Developing the design

Within two weeks of having the project stopped it was moving again, but this time with a unique design never before seen in the UK. There were two separate problems to solve in terms of foundation design. One was to accommodate the depth of settlement and the other was to carry the concentrated loads near the surface. To provide a hard 'crust' near the surface to take the principal loads, it was decided to carry out dynamic compaction. This consists of placing a stone blanket over the site and then dynamically compacting it by pounding it with a heavy weight. It is important to avoid excessive build-up of water in the fill material below, so at the time of placing the stone blanket cut-off drains were installed, above and below the blanket, to keep water levels as low as possible. The dynamic compaction (DC) operation took place at the end of October, less than a month after the project received the go-ahead and with the design still not fully resolved.

Almost immediately after the DC was complete, the weather broke. With the stone blanket down, a good working platform now existed and the land drains were now proving their value. Plate bearing tests and additional penetration tests were carried out on the compacted fill material. With post-compaction testing complete, soil design parameters could be finalised and design progressed.

Settlement analysis

A simplified model was established for the purposes of the settlement analysis. The most significant contribution to settlement was from bulk earthmoving operations. The fill over the store area was essentially wedge shaped, resulting in greater potential settlement towards the southern end of the site. The zone of influence of any loading on the fill was quite large because it was unconsolidated. The planned removal of the earth bunds on the north and south edges of the store site would have a significant effect on negative settlement, but as it was not known when this operation would occur we had to make provision for this occurring during or after the permanent construction.

The scheme that was proposed consisted of perimeter edge beams supporting brick cladding and the perimeter columns, with ground-bearing internal slabs and internal columns supported on isolated pad foundations. Due to concern about differential settlement at joints, and to provide a structure with inherent resistance to cracking, a fully post-tensioned solution was adopted. Fully post-tensioned ground beams of this size have never been used in the UK. The surface of the floor slab was kept flat across the site, except where special finishes were required.

Settlement analysis was carried out by DHV's Geotechnical Division and summarised in two sketches. Localised differential settlements under concentrated point loads were estimated to be a maximum of 16 mm, while the overall settlement due to the consolidation of material was a maximum of 120 mm. Long-term creep settlement was estimated to be of the same order again and was generally beneficial in reducing differential movement.

Construction

Site operations began in January 1990. The major portion of the site was infilled in order to raise the prominence of the building on the sloping site.

Good quality type 2 imported fill was used under the store, whilst the car park predominantly utilised site materials won from bund levelling operations which were going on at the same time. During this period the north bund remained in place. Its subsequent removal was to have some unforeseen results.

The perimeter ground beams were constructed on the dynamically compacted stone blanket to give uniform support. As the surface was not level, even after reinstatment following the DC works, the depths of the beams varied. The sub-base below the beams was sand blinded and covered with a polythene slip-membrane to allow the beam to contract when post-tensioning was applied. Ordinary reinforcement in the beam was nominal except at locations where it was designed to spread high loads. Nominal prestressing force was applied to the beams. Anchor groups were positioned centrally on the beam so as not to induce bending, except under heavy loads at the column points. The principal benefit of the prestressing was to prevent and to limit the size of any cracks in the event of excessive differential settlement. Due to long-term creep, which shortens the beam slightly, the brickwork receives a gentle compression which greatly enhances its crack resistance.

Prestressing

The type of tendons used comprised unbonded 'Dyform monostrands'. These tendons consist of 15.2 mm diameter strands surrounded by a grease-filled plastic sheath. The covering allows the strand to be stressed after the concrete is cast and also gives it long-term corrosion protection. The strands are delivered in 2 m diameter coils on pallets. Each of the strands is pre-cut to the correct length and colour coded to correspond to those shown on the sub-contractor's drawings. The stressing anchor is installed on the end shutter and set at the correct distance to provide end cover. The static anchors are fixed clear of the shutter face at the opposite end. Careful detailing of anchor zones is essential to avoid congestion.

The anchors are fixed at the centroid of the beam. The number of tendons used varies according to the beam cross-section. The 90 m long edge beams were cast in short sections and stressed when all the sections had been cast. Stressing was carried out using a small portable jack powered by a portable electric pump. Since unbonded tendons were used, there was no grouting to carry out as there is with conventional tendons.

As the beams were completed, the filling of the store area was progressed. At the same time, erection of the steel frame began. Consideration was given to ensuring that differential settlement of column bases would not cause excessive stress at the connections. In critical positions articulated joints were provided. The inherent flexibility of the steelwork was adequate to cope with the anticipated movements.

Bracing to the steelwork was accomplished predominantly through the diagonal steel flats built into the walls. However, at the loading dock area, solid walls were used as bracing and to give impact resistance. The next important stage of construction was that of the floor slab, which started as soon as an area of roof was watertight.

As with the beams, the post-tensioning in the floor slab gives it inherent resistance to cracking and the ability to avoid joints that would be susceptible to differential settlement. To allow the prestress to be taken up in the slab it is essential to have a good slip membrane on smooth and level blinding. Lack of attention to the uniformity and smoothness of the blinding will result in cracks.

The slab dimensions were approximately 90 m × 70 m. Post-tensioning consisted of pairs of tendons running the whole length of the 200 mm deep slab. The slab was initially detailed to be cast in 4.5 m strips, on 'a hit and miss' sequence; however, to maintain the programme the contractor requested a change to 'wide bay' pours. Where the tendons continue through the stop-end, the remainder of the coils are stacked ready to be carried on into the next pour. Whereas narrow-bay pours only require one-way stressing initially, with transverse stressing being carried out when all pours are complete, wide bays require to be stressed in both directions soon after pouring to avoid shrinkage cracking. Intermediate anchors are placed at the stop-end to allow this to be done. Where steps in the slab level occurred, a gradual transition was detailed; sudden steps in the blinding would have resulted in cracks. Pours took place on consecutive days to allow the areas to be stressed as one.

All brickwork joints were made over-size and joint spacings kept to 7 m to accommodate curvature developing in the supporting beams or any angular movement in the steel columns. Flexible couplings were detailed where services left the building. External drain runs were constructed with half-length pipes to accommodate movement. Because the settlement was predominantly due to settlement of the fill material, the external finishes were not expected to suffer large differential movements with respect to the building.

Monitoring

Because of the nature of the design it was decided to instigate a regular settlement monitoring programme. This consisted of monitoring points, which were established on the perimeter beams as soon as possible and subsequently onto the brickwork when this was constructed. Although not carried out to a high degree of accuracy, this has given us valuable information on the performance of the building. The first monitoring points were established in April 1990. It is likely that 50–60 mm of settlement had taken place before this time. By the time slab construction began in July a further 45 mm had taken place at the most severe points. It should be noted that up to this time all settlement was being built out by the contractor. By the time the store opened a further 10 mm of settlement had taken place. The general movement noted was in line with predictions. However, the rise that began to occur at the northern edge was not expected. After some consideration it was realised that this correlated with the removal of the northern bund by the adjacent developer. Although this was off our site, the effects were quite marked, causing a 15 mm rise of that edge of the building. This effect was short lived because, after three months, the movement of this edge virtually ceased.

The 12-month defects inspection took place in January 1992. We took the opportunity to take a further set of levels. These showed that during the year a further 15 mm of settlement had taken place. One significant crack was noted in the back-up area. This had previously been identified during construction and was not related to settlement. In the sensitive area of the terrazzo floor in the main floor there were no significant faults and it was considered by the architect to be one of the best floors he had seen. Externally, the facing brickwork was in perfect condition with no sign of settlement cracks anywhere.

Long term

The store is intended to have a minimum design life of fifteen years. There is no doubt that settlement will continue. Settlement to date from the time of start of construction has been 120 mm. Differentials are currently around 60 mm. It is likely that further settlement in the order of 100 mm will take place, with possibly a further 50 mm of differential. The client has had the courage to accept a structure with a far from standard performance and in return has gained the benefit of a 10% reduction in construction costs.

Steel-fibre-reinforced ground-supported floors

(based on technical notes written by the author for Bekhaert)

Materials

Fibres have been used to reinforce brittle building materials since the cradle of civilisation. The ancient Sumerians used straw to reinforce sun-baked bricks to build ziggurats, the forerunner of the great pyramids of Egypt built several hundred years later. In the 16th century the Elizabethans used horse hair to reinforce plaster and mortar in construction. Asbestos fibre, first milled from asbestos rock in Italy at the turn of the 20th century, was used extensively in the manufacture of concrete drainage and roofing materials.

In recent times various natural and synthetic fibres have been incorporated in cement-based building products. They range from nylon, polypropylene, glass and cellulose to carbon, aramid and steel. Interest in steel-fibre-reinforced concrete burgeoned in the 1980s, the special properties of steel fibre concrete offering particular structural and economic advantages over conventional reinforced concrete. Nowhere has this been more noticeable than in the application of fibres for ground-supported slabs (Fig. 3.48).

For Europe alone, over 25 million m^2 of floor area have been specified with steel fibre reinforcement. During 1991, steel fibres specified in

Fig. 3.48. Dispensing Dramix fibres into a readymix truck

Belgium and the Netherlands accounted for over 40% of the industrial flooring market. A number of fibre products have been introduced into the market, all claiming to have properties that reinforce concrete and reduce cost. Apart from the obvious advantages of no reinforcement fixing costs and faster production, the structural advantages need to be carefully considered.

Generally, industrial fibres can be divided into two categories:

- fibres for primary reinforcement — flexure and crack control;
- fibres for secondary reinforcement — early-age crack control.

Those in the first category have the mechanical properties of conventional steel reinforcement. Those in the second category have a low modulus of elasticity and high creep and cannot be substituted for conventional reinforcement, but may be used for early-age crack control.

Generally nylon, polypropylene and polyethylene fibres will fall into the secondary reinforcement category. Steel, carbon and aramid fibres fall into the primary category. The effectiveness of fibre reinforcement depends on the shape, length, diameter and mechanical properties of the material. Steel fibres are primarily classified according to their method of manufacture and then further characterised by their mechanical properties and aspect ratio. According to the American ASTM A820-90, the four basic categories are:

- type 1 — drawn wire;
- type 2 — slit sheet;
- type 3 — melt extract;
- type 4 — other.

The steel fibres most often used in ground-supported slabs are short lengths of type 2 fibre, typically 20–25 mm long with a wavy profile, or longer lengths of type 1 fibre, typically 50–70 mm long with hook ends. Type 1 drawn wire fibres have the highest tensile strength and aspect ratio of length/diameter, with a fibre count of around 2000–4000 per kg and a reinforcing dosage rate of between 20 and 40 kg per cubic metre of concrete. The short type 2 fibres have a lower tensile strength and a higher ductility, requiring a very high dosage rate in concrete to be as effective as type 1 fibres. A typical fibre count is in the range of 15 000–40 000 per kg. However, type 2 fibres are considerably cheaper than type 1. The principal performance difference between the two fibres is that the longer type 1 has greater post-crack toughness and residual strength, while the shorter gives a finer distribution of reinforcement and tends to be more effective at controlling the propagation of a crack (Fig. 3.49).

STEEL	SHAPE		L/D	K/f	FIBRE FACTOR
SLIT STEEL	straight		35	0.50	18
	deformed		60	0.63	38
DRAWN WIRE	hooked ends		100	0.75	75
			75	0.75	56
			60	0.75	45
DRAWN WIRE	ribbed		40	0.75	30
			52	0.75	39
MILL CUT	straight		38	0.50	25

FIg. 3.49. Typical fibre effectiveness factors

Type 1 fibre characteristics

The mechanical properties of conventional reinforcing bars and steel wire are worth comparing to get some idea of the quality and integrity of steel fibre. In processing reinforcing bars from single-sized mild steel coils, cold drawing can reduce a bar diameter from 20 mm to 12 mm. In processing the reinforcement the tensile strength and flexural toughness or ductility are enhanced, the characteristic tensile strength increasing from 250 N/mm^2 (mild steel) to 460 N/mm^2 (high yield). In processing steel fibres for concrete the mild steel coil diameter is reduced further by cold drawing, down to 1 mm diameter or less. The stretching of steel during the drawing and reduction process further toughens the material, enhancing its tensile strength to 1100 N/mm^2, about 2.5 times that of high yield steel. The drawing process aligns the molecules to create a more densely packed crystalline structure, reducing also the percentage of internal flaws from the steel casting process. The fracture toughness and fatigue resistance of the drawn wire are also improved

It is useful to compare the mechanical properties of fibres used in industrial floors, e.g. polypropylene and glass fibre, with bar and steel wire fabric (Table 3.30).

Table 3.30. Mechanical properties of fibres

Material	Tensile strength: MPa	Elastic modulus: GPa	Specific weight: g/cm^2
High yield steel	460	210	7.85
Polypropylene	400	8	1.00
Glass fibre	1700	50	2.00
Drawn wire	1100	210	7.85

Design of ground-supported floors

The merits of the use of plain concrete, steel fabric and steel-fibre-reinforced concrete for industrial ground floors have been the subject of considerable debate. The Concrete Society's *Technical Report No. 34* refers to the increasing use of steel fibre in ground floors on the Continent, in airport and highway pavements in Scandinavia and North America, and acknowledges their suitability for reinforcement in ground floor slabs.

Most design methods for ground slabs are based on Westegaard's theory that the sub-base and soil act as an elastic medium. The soil resistance is proportional to the slab deformation under load. This displacement gives an upward force proportional to the vertical deformation and leads to bending stresses in the slab. The slab must be capable of distributing the applied loads to the sub-base without exceeding the subgrade bearing capacity or the flexural strength of the slab. Thus, in the design of floors, two elements need to be calculated:

- the depth of the slab to resist the bending or flexural stresses created by the applied loading and sub-base reaction;
- the joint spacing of the slab to resist subgrade friction activated by concrete movement from drying shrinkage and thermal contraction.

In designing a ground-supported slab it is generally assumed that:

- the use of steel fabric reinforcement does not increase the flexural capacity of the slab, but has the primary function of crack control;
- the slab depth is related to the flexural strength of the plain concrete;
- if steel fibres are introduced which exhibit a superior post-cracking behaviour to plain concrete, then an enhanced design value for flexural strength can be assumed, based on test evidence.

Calculation for flexural strength

Design of steel fibre concrete floors follows the guidelines laid down in the Concrete Society *Technical Report 34* and differs only by the value of the tensile strength used in calculating slab depth. The slab is assumed a homogeneous material whose characteristic flexural strength must be greater than the imposed flexural stress caused by the imposed loads.

The imposed stresses will depend on the following factors.

Value of the subgrade reaction k. As the design is not sensitive to the modulus of subgrade reaction, in the absence of plate loading tests the subgrade reaction can be determined with safety from a knowledge and classification of the soil type (Table 3.31). For most industrial ground slabs a granular sub-base between 150 and 250 mm is compacted over the subgrade and sand blinded to form a uniform and level surface. This helps to reduce skin friction and maintain the integrity of the subgrade during construction.

Type and severity of the applied load P. The most frequent loading categories are uniformly distributed loads, wheel loads and point loads caused by storage racks. Since a totally uniform load on a uniform subgrade has no resultant bending moments, the slab design is based upon the most critical combination of actual loading. More often this is taken as alternate floor strips, usually an aisle width wide, with and without load applied. The maximum moment is calculated using the rigorous method shown in C&CA TR 550 or the simplified value given in *Technical Report 34*. Where the loading distribution is known, the characteristic load is multiplied by a load factor of 1.5. Where the distribution is unknown, to cater for future processes and change of use, a load factor of 2 is used.

The thickness of the floor slab h. The slab thickness chosen will ensure that the imposed flexural stresses from the various loading categories do not exceed the permissible flexural strength of the concrete.

Table 3.31. Typical values of k

Classification	Dry weight: kg/m^3	CBR	*k*
Organic clay and silt	1280–1680	3–5	14–27
Fine-grained silt and clay	1440–1760	4–8	27–54
Coarse-grained sandy soil	1760–2000	20–40	54–82
Coarse-grained gravelly soil	2000–2240	60–80	82–150
Well compacted crushed stone		80–100	150–250

Generally the Westergaard or Heteneyi formula is used, modified to allow for spread of load from contact area, wheel base, leg load separation etc. A typical equation for an idealised point load and leg load with spreader plate is shown in Fig. 3.50 to illustrate the relationship between the subgrade reaction k, slab depth h and the resulting flexural stress f, arising from the imposed load P.

It is useful to prepare a design schedule setting out all the data assumed, the properties of the concrete, modulus of subgrade reaction and loading category. The flexural stresses calculated from the different loading regimes are then checked against the flexural strength of the steel fibre concrete. The slab depth h that satisfies this limit state is chosen.

From the results of full-scale tests and correlation with actual slab behaviour, it is evident that slabs reinforced with type 1 fibres with a high aspect ratio and hook ends have a large reserve in load-supporting capacity and strain deformation. This is explained by the post-crack performance of slabs, with plastic yielding helping to redistribute moments induced by the imposed loading. To allow for this when using Westegaard's or similar elastic design theory, an enlarged spread of the imposed load is assumed, together with a modification for E of 200 GPa based on a partially cracked section.

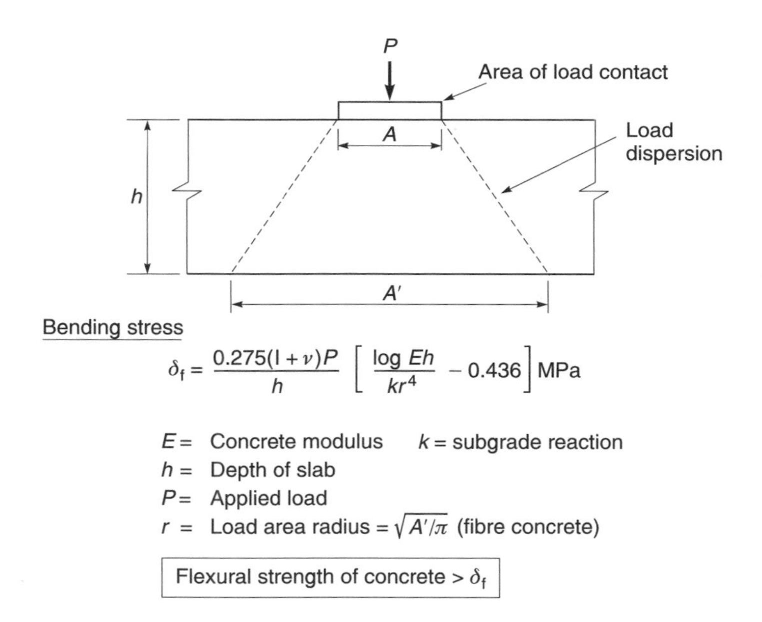

Fig. 3.50. Typical leg load: applied stress

Determining the flexural strength of steel fibre concrete

For a plain concrete section, tensile forces cannot be transmitted across a crack (Fig. 3.51(a)). Once the tensile strength of the concrete is exceeded, either in bending or in direct tension, micro-cracks will extend rapidly through the concrete, resulting in brittle failure.

Introducing a steel fibre that has good anchorage properties and high tensile strength, the flexural capacity of the slab can be enhanced by the ability of the fibre to hold the cracked section until ductile failure occurs (Fig. 3.51(b)). The load-carrying capacity of the ground slab is thus enhanced because of superior post-cracking performance.

Typical load–deflection curves for plain and fibre concrete are shown in Fig. 3.52. For plain concrete the difference between the first and peak load is negligible, whereas for steel-fibre-reinforced concrete the difference is substantial. Moreover, the load–deflection curve for fibre concrete has a significant plateau, coincident with plastic yielding. The design flexural strength of steel fibre concrete is thus based on this yield strength, with an appropriate safety factor.

For many years independent researchers investigated the optimum parameters for the shape, length and density of fibre for reinforcing concrete. Typical strength characteristics of different fibre types have

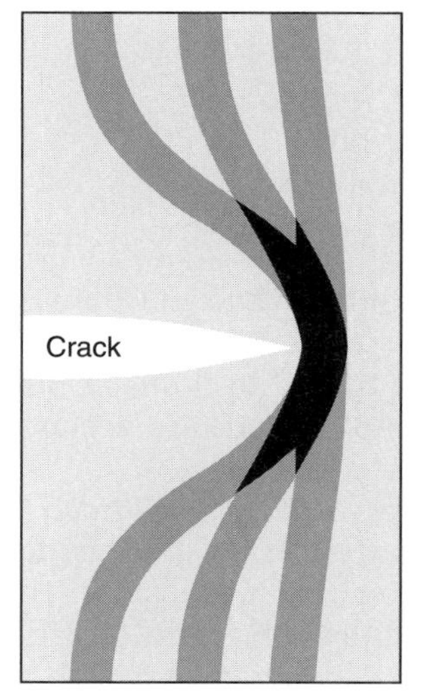

Plain concrete: no tension can be transmitted across the crack

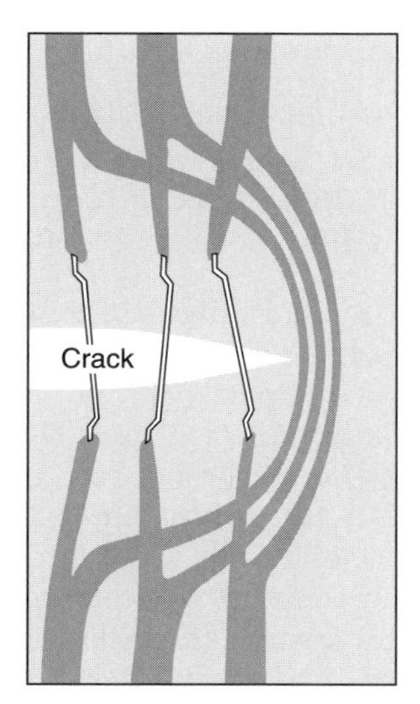

Fibre-reinforced concrete: some of the tension is transmitted by the fibres across the crack

Fig. 3.51. How fibre reinforcement controls cracking

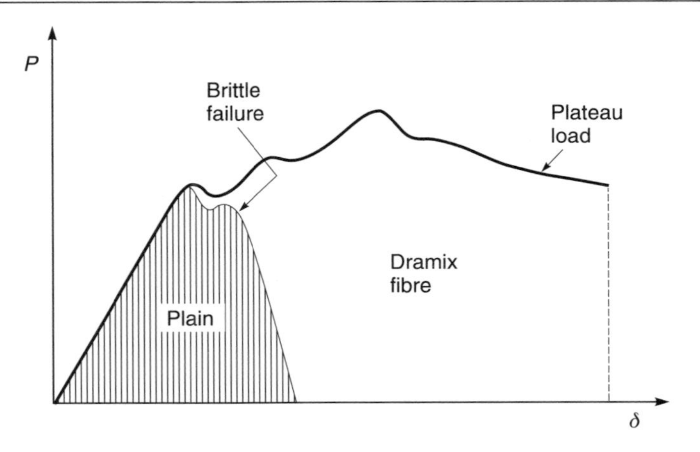

Fig. 3.52. Load–deflection curves for plain and fibre-reinforced concrete

been independently reported. The research work of Thames Polytechnic, Pennsylvania State University and the CUR Foundation in Holland illustrate the relationship between fibre shape, length, diameter and dosage in concrete.

The highest working stresses were achieved with steel wire having the following characteristics:

- cold drawn wire, tensile strength of 1000 N/mm^2;
- aspect ratio (length/diameter) 60 to 100;
- deformed fibres (hooked ends, twisted or ribbed);
- economic fibre density: 20–50 kg/m^3.

It is fairly evident, from these test results, why steel fibres with the characteristic hooked ends were preferred. The flexural strength for any steel fibre concrete can be determined by the Dutch CUR Recommendation 10 test, or ASTM C 1018; both are accepted international tests for fibre-reinforced concrete.

In the CUR test, a specified number of fibre-reinforced concrete prisms — all made with the same quantity of fibre and grade of concrete — are cured for 28 days before undergoing load testing. The load is applied to the prism at a constant rate of deflection until the deflection reaches 3 mm. The area under the load–deflection curve, D_b, is calculated (Fig. 3.53(a)). This value is then factored by 3 mm to give the equivalent flexural load, and then by the section modulus to give the equivalent flexural strength of the fibre concrete (Fig. 3.53(b)). The average equivalent strength values from 12 prism tests is factored by the standard deviation to give the design flexural strength. Thus a

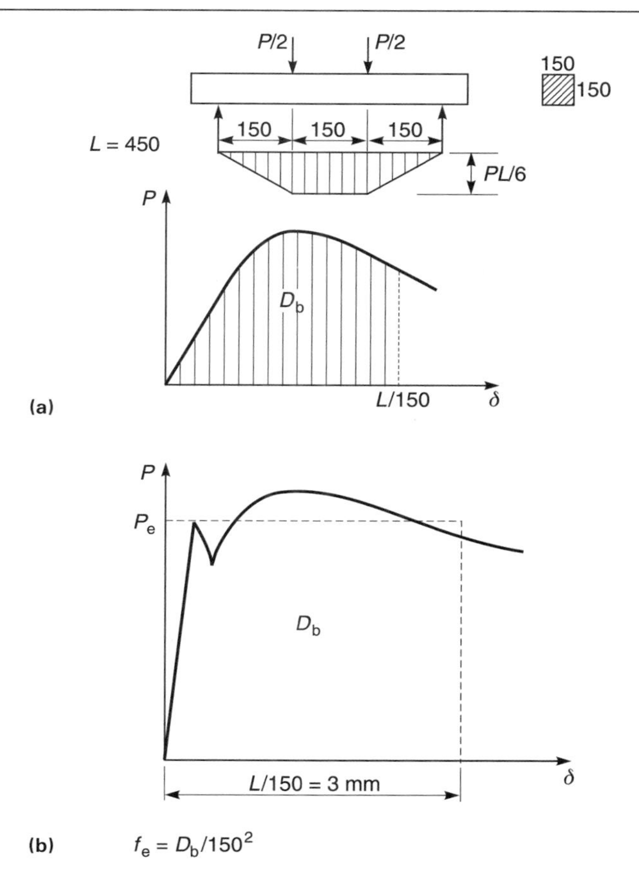

Fig. 3.53. Load–deflection curves for test specimens: (a) CUR test; (b) equivalent flexural strength

whole suite of tests can be done to determine design values for different fibre aspect ratios and dosages.

For the ASTM test the same standard prism is load tested, and the deflection d and load at first crack are noted as the load is increased until the beam deflects to 15.5 times the first-crack deflection (i.e., $15.5d$). Typical design flexural strength for 30 MPa concrete is shown in Table 3.32. The flexural strength is checked against the imposed flexural stresses and the depth of slab h, chosen to keep the stresses within this limit.

Type 1 steel fibres are available in a range of sizes and wire diameters for ground-supported slabs, offered in high tensile steel with or without zinc coating, in stainless steel or high carbon steel for very

Table 3.32. Design flexural strength f_f of 30 MPa concrete incorporating different dosages of steel fibre

Dramix fibre	Dosage: kg/m^3				
ZC 60/1.00	20	25	30	35	40
ZC 60/0.80		20	25	30	35
f_f: MPa	2.4	2.8	3.2	3.5	3.8

high tensile strength. The choice will depend on the nature of the application, the floor specification, the imposed floor loading and durability.

Design approach for Type 1 and Type 2 steel fibres

Steel fibre type, shape and dosage have a direct effect on the tensile strength of concrete. As a result, steel-fibre-reinforced slabs may be significantly thinner than conventionally designed ground-supported slabs with mesh reinforcement. Two design methods may be adopted. The implications of each method should be considered with reference to the subgrade, potential settlement, slab deformation and long-term behaviour.

Elastic method. This follows the basic principles outlined in the Concrete Society's *Technical Report 34* and C&CA *Technical Report 550*, based on the work of Westegaard. At low dosage rates the modulus of rupture is similar to that of plain concrete. If the critical load applied to the ground floor slab is static, there will be little difference between the slab depths. If the critical load is dynamic, the greater fatigue resistance of the steel fibre concrete may result in a reduction in slab thickness. At high dosage rates, the first crack modulus of rupture may increase and this may be adopted if substantiated by test results.

Plastic limit state. Suitable design methods, such as those proposed by Meyerhof and Losberg, may be adopted. The steel fibre must show, by test results, that the concrete will have an increase in flexural strength. Prism tests based on CUR 10 or similar must show that the steel-fibre-reinforced concrete has an Re3 value (ratio of residual flexural strength at 3 mm deflection to that at first crack) greater than 30. This is likely to be achieved with drawn wire fibres with a high aspect ratio. The plastic design method will show a significant reduction in slab thickness in comparison with a conventionally designed floor slab. The thickness must be sufficiently robust to resist surface tolerance changes caused by slab deflection. Basic design guidance can be found in Concrete Society *Technical Report 34*, Edition 2, under Appendix F.

Construction specification

The construction procedures for steel fibre concrete are the same as for conventional ground slab construction. The main points and special procedures are summarised below.

Handling. Steel fibre with an aspect ratio less than 50 (thick and short fibres), i.e. type 2 fibres, can be added to concrete like coarse aggregates, without balling up. If the aspect ratio exceeds 50 — most type 1 fibres will be closer to 75 — the fibre can easily ball up during handling and mixing. Long thin fibres can be blown into the concrete mix using compressed air, or added very slowly, while monitoring the mixing process. Both options require careful control, extra labour and can be a potential problem during batching operations.

The other solution is to fuse the fibres together with a water-soluble glue to form thick plates. These plates or bundles are squat, having an aspect ratio of nearly 1. Such a fibre is excellent for handling and batching with concrete. The fibre bundles can be added to the truck mixer on site or at the batching plant by conveyor, or added to the concrete mix with the aggregates. After the fibre bundles have dispersed into the mix, the glue breaks down, allowing the individual fibres to spread evenly through the mix (Fig. 3.54).

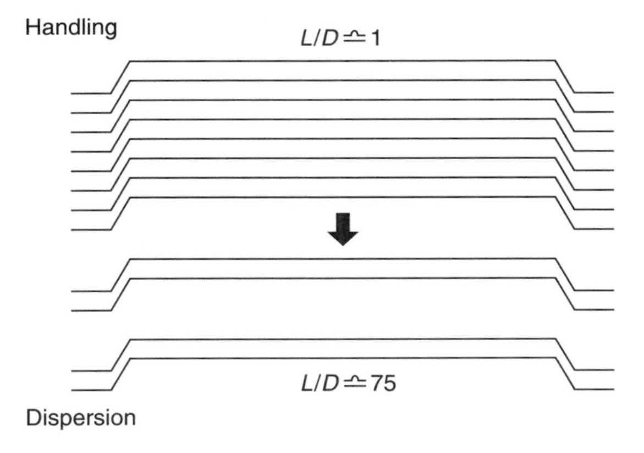

Fig. 3.54. Bundling of fibres improves fibre dispersion

Subgrade preparation. The quality of the floor is closely related to the quality of the subgrade. A sub-base of granular material, carefully compacted and sand blinded, is recommended to reduce friction restraint and to provide a working surface for construction traffic. Undulations in the sub-base should be within a tolerance of ±10 mm before concreting. The modulus of subgrade reaction can be found by a plate loading test or by using the characteristic values given in tables in *Technical Report 34*.

Slip membrane. A polythene layer should be placed between the slab and the sub-base to reduce frictional restraint.

Concrete quality. Minimum concrete grade is 25 MPa, with a cement content of 320 kg/m^3 and a water/cement ratio not exceeding 0.55. Sand content should fall between 750 and 850 kg/m^3. Well graded aggregates should be used with maximum aggregate size of 28 mm for gravel and 32 mm for crushed stone. The fraction retained on the 14 mm sieve should be between 15% and 20% of the coarse aggregate. A pumped concrete mix will require a higher fine aggregate content to avoid segregation during pumping.

Addition of steel fibres. Steel fibres can be added at the batching plant or on site. On site the concrete must have a 125 mm target slump before the fibres can be added. The fibres are dispensed into the truck mixer by conveyor (Fig. 3.55). The truck mixer drum must rotate at maximum speed. After all the fibre is loaded, the truck mixer will continue to rotate for a further 5 min or until all the fibres are uniformly dispersed. Fibre distribution is visually inspected during pouring (Fig. 3.56).

Quality control. Collect a random sample of 10 litres of fibre concrete and extract the fibres using a magnet. Wash, dry and weigh the fibres accurately. The fibre content must be within 20% of the dosage specified to be within specification.

Finishing and curing. All traditional methods of finishing the concrete surface are acceptable, particularly laser screeding, razor back and vibration beam. An abrasion-resistant topping of quartz and cement can be applied, followed by power floating (Fig. 3.57) and trowelling.

Good curing of the surface is important to avoid rapid drying out and curling of the slab. It is critical when the slab depth is 150 mm or less. For thin slabs (125–150 mm) all openings into the building must be closed with the exception of entrances for construction plant and ventilation points. A double layer of curing compound is usually applied to ensure an unbroken surface-curing film.

(a)

(b)

Fig. 3.55. Steel fibre concrete: (a) conveying fibres to the mix; (b) placing the concrete

Joints. Joints are defined by their function and method of construction. Floor joints in steel fibre concrete should provide as smooth a passage as possible for trafficking across the joint. The differential vertical displacement at the joint must also be limited. Where joints

Fig. 3.56. Laser screed laying a steel fibre concrete floor

Fig. 3.57. Power floating the concrete floor

are provided, the need for high standards of workmanship in their formation cannot be over-emphasised. Particular attention must be paid to the alignment of the joint, the fixing of dowels and the compaction of concrete at pre-formed joints.

There is no doubting the superior quality of a post-formed saw-cut joint to a pre-formed crack control joint. All induced crack joints are saw-cut in steel fibre floors, for this reason. Satisfactory crack control is achieved by limiting the bay size between induced crack joints to:

Typical type 1 fibre dosage	20 kg/m^3	30 kg/m^3
Bay size	8 m × 8 m	12 m × 12 m

Isolation joints around columns, manholes, perimeter of floor area, position of construction joints and crack control joints must be agreed before concreting begins.

Saw-cut joints. Induced crack joints are saw-cut 3 mm wide to one-third of the slab thickness. Saw cuts must be carried out as soon as possible after the concrete has hardened. In the summer this is normally 24 h after casting, extended to 48 h in winter months.

Pre-formed tied construction joints. The construction joints require special attention to ensure that the arisses formed are square. Mesh reinforcement should be placed at mid-depth, with a full bond length, to ensure the joint is tied. Mesh type will vary according to bay size and slab thickness. Saw-cut induced joints should be made, each side of the construction joint, at half the nominal bay size, to limit movement at the tied joint.

Jointless floors

Steel-fibre-reinforced concrete floors, typically up to 50 m × 50 m in plan area, can be cast using a laser screed, with no intermediate joints. Steel fibre dosages will be upward of 35–40 kg/m^3. The free movement joints on the perimeter of these floors will be subject to large movement and will require special detailing if they are to be bridged by vehicles. This type of floor is best suited to large open-plan buildings with minimal intrusions. Some cracking may be expected, but this is generally not greater than 0.5 mm in width. A jointless floor will perform better for small-wheeled vehicles than a floor with induced joints and will reduce the potential for curling.

Steel fibre durability

The addition of steel fibres in concrete does not affect the permeability, durability or density. It has no influence on the rate of carbonation, chloride ion or sulphate attack. Uncoated steel fibres have generally good corrosion resistance; the small diameter and large surface area of the wires are more effectively shielded by the lime-rich cement paste than conventional bar reinforcement. Moreover, the corrosion expansion of the steel fibres at the surface is so small that they are insufficient to spall the concrete. However, steel fibres close to the exposed concrete surface can show signs of rust staining in the long term. For a blemish-free concrete finish, zinc-coated or stainless steel fibres should be specified.

Alternative materials for reinforcement and prestress

Plastic reinforcement

(edited from a paper written by John Clarke, BCA)

The concrete industry has traditionally tackled the problem of steel reinforcement corrosion in aggressive environments by improving the quality of the concrete itself, often with limited results. Recently, other approaches have been considered, such as using stainless steel or bars with an epoxy coating. In extreme cases, cathodic protection is installed, though generally as part of a repair system rather than for new structures.

A significant recent development is to replace the steel with fibre-reinforced plastic (FRP), with glass, carbon or aramid fibres set in a suitable resin to form a rod or grid. With the correct choice of resin and fibre, the materials, which are well accepted by the aerospace and automotive industries, should be more durable. Yet the construction industry is being slow to adopt them. One reason is that their properties are significantly different from those of steel and so the behaviour of elements reinforced with them will be different. Another is that there are no agreed standards for the materials or for design. These, and other aspects, are considered below.

Bar sizes and shapes

As yet there are no standard bar sizes or surface profiles. Each supplier has its own method of producing a bar with adequate bond in a range of diameters. Most bars are circular in cross-section, though square or rectangular profiles are available. In general, the maximum diameter is about 25 mm, although 32 mm bars are being produced.

The most common method of manufacturing FRP reinforcing rods is protrusion — essentially pulling the uncured material through a heated die — with a secondary process to produce an appropriate surface to ensure bond. A limitation is that thermosetting resins are used which, once fully cured, do not allow the rods to be bent into the range of shapes typical of steel. Thus 'specials' are required. Some Japanese manufacturers produce spiral reinforcement and two- or three-dimensional grids. Other manufacturing techniques are being developed.

Material properties

The mechanical properties of FRPs are chiefly determined by the amount and type of fibre, the durability being a function of both the resin and the fibre. Table 3.33 gives typical properties. FRP thermoset materials have a linear response to failure, with no plasticity.

Overview of research

Over the past 15 years or so, much research has been carried out, initially in Japan and North America, mostly rather uncoordinated. Much of the experimental work has been on simple beams and slabs, though some columns have been tested. The work has led to a wide range of products and has been reported at conferences. In Europe the first coordinated development programme on FRP reinforcement was Eurocrete, a collaborative project partly funded under the Eureka scheme, with partners from the UK, France, the Netherlands and Norway. It was probably the first project of its kind in the world to bring together the disciplines involved with FRP, including materials suppliers, processors, research organisations and designers.

Demonstration structures

One of the best ways of demonstrating the potential for FRP reinforcement, and building confidence in the material, is to construct demonstration structures; a significant number have been built worldwide. Initially, because of concerns about the lower stiffness of FRP compared with steel, most were prestressed, with conventional steel for the sec-

Table 3.33. Mechanical properties of glass- and carbon-fibre-reinforced plastic reinforcement

Material	Tensile strength: N/mm^2	Stiffness: kN/mm^2
Glass fibre reinforcement	800	45
Carbon fibre reinforcement	1500	150

ondary reinforcement. Some footbridges and highway bridges have been built, chiefly in Japan and North America.

The first significant structure in Europe was a highway bridge with glass FRP prestressing cables, built in Düsseldorf in 1986. More recently, demonstration structures formed an important part of Eurocrete. Two footbridges were built, one at Chalgrove near Oxford and the other in Oslo. In addition, part of a berthing facility in Qatar was reconstructed with FRP reinforcement. A further application was as reinforcement for the fencing around a test facility for sensitive electrical equipment, where conventional steel would have caused interference. Other applications, which were tested in the laboratory and may soon move into practice, included retaining wall units and cladding panels. Other programmes are looking towards larger structures fully reinforced with FRP, such as an 80 m long footbridge being designed and built in Denmark.

Design

Standards for the design of reinforced concrete structures have developed over the past 100 years or so from simple prescriptive rules to a mixture of methods based on sound scientific principles and rules of thumb. Some aspects of behaviour, such as shear, are still not well understood and empirical approaches are used. When introducing a new type of reinforcement with very different properties, the existing approach can be adapted or completely new rules formulated. The latter is obviously more technically correct but is costly and time consuming. So it is generally considered that modifying the existing approaches is the only feasible option.

The Japanese Ministry of Construction has published draft guidelines for design, the Canadian Bridge Code will shortly have a chapter dealing with FRP, and guidance is being prepared by the American Concrete Institute. Proposed modifications to British Standards were developed within Eurocrete and are now being validated by the Institution of Structural Engineers. The intention is that they will provide an approved document for use by design engineers in the absence of a formal code of practice. The design approaches will lead to safe structures but probably not to the most economic use of FRP materials. Examples of differences between FRP-reinforced and steel-reinforced concrete structures include:

- ☐ Failure is likely to be by compression of the concrete, not rupture of the reinforcement.
- ☐ Partial safety factors will be higher than for steel reinforcement to take account of the variability of the materials and uncertainty about their long-term performance.

- ☐ The bond stresses, and hence anchorage lengths and lap lengths, will depend on the bar being used.
- ☐ Crack widths need to be limited only from considerations of aesthetics, and possibly water-tightness, and not for durability reasons.
- ☐ Deflections are likely to be higher than for equivalent steel-reinforced units.
- ☐ FRP rods have low compressive strengths compared to their tensile capacities, so the traditional design approaches for columns are no longer valid.
- ☐ Fire will be a significant design consideration for some structures but is an area that has received little attention. With embedded reinforcement, the prime concern will be to limit the temperature rise at the surface of the bar so that it stays below the glass transition temperature for the resin. Once this is exceeded, the fibres will cease to act compositely.

Costs

FRP costs more than normal high-yield steel bar. This higher initial cost can be justified only on the basis of whole-life costing for structures in aggressive environments. So potential users of the materials must be encouraged to consider the true total costs, not just the material costs. Further savings will be made as design approaches are developed to take account of the properties of the new materials.

Even in the absence of fully developed appropriate design methods, design by testing can lead to FRP-reinforced products that are effective in first-cost terms. For example, a bearing plate for soil nails was redesigned, reducing the weight from 190 kg to 50 kg, much of the saving coming from the significantly reduced cover requirements. The redesigned unit could he manhandled on site without a crane — a significant cost saving. It is likely that similar savings could he made on other precast concrete units.

Durability

Probably the single most important aspect of the use of FRP as reinforcement is its durability in concrete, as glass fibres and some resins are degraded by the high alkalinity. Work has concentrated on the development of alkali-resistant glass (AR glass) or the use of carbons or aramids, but little attention has been paid to the resin. Manufacturers are, for commercial reasons, reluctant to disclose details of their materials and considerable work is needed to develop acceptance criteria for durability. A major programme on durability carried out in Eurocrete included work on the materials themselves and on the composites embedded in representative concretes. Samples were stored in the laboratory under various conditions and on exposure sites in Europe and the Middle East. The results that are applicable to the resin and fibre combinations showed that the composite rods resisted the alkaline environment well.

Suppliers

It is unlikely that FRP reinforcement will ever replace steel for most structures. But the experimental work and demonstration projects have shown that FRP reinforcement is a practicable and cost-effective alternative to steel in some circumstances, such as highly aggressive environments. Several companies are therefore producing FRP reinforcing bars commercially, including Fibreforce in the UK and Hughes Brothers and Marshall Industries in the USA.

The future

The UK construction industry is extremely conservative, probably justifiably in view of the long structural lives expected. The most likely development route will be the use of the new materials in non-structural applications or where the consequences of failure are not too severe. With increasing confidence, more highly loaded and critical applications may be considered. But before FRP reinforcement becomes widely accepted for concrete structures, several significant aspects have to be addressed, including:

- the ability to produce suitable reinforcement shapes;
- the ability to produce large quantities of materials of a consistent quality;
- the development of materials with a standard range of properties;
- better understanding of the behaviour of structures in fire; and
- the development of design approaches that are more appropriate to the materials.

All are essential if the materials are to move from their present peripheral applications to those where their true potential can be realised.

Glass fibre prestressing strand

Post-tensioned and prestressed concrete are the bread 'n' butter designs in most long-span bridge structures today. If prestressed technology had not been developed, we would be seeing relatively squat bridges dominated, perhaps, by arched structures strengthened with steel reinforcement only — not the sleek, aesthetic lines of prestressed bridges designed by Maillart or Freysinnet, or the Pont du Normandie or the minimalist outline of the Kirchheim motorway overbridge.

The forces bridge structures have to resist are far larger in magnitude than those of a suspended floor in a building. Perhaps it is the apparent lack of adaptability of prestressed concrete to accept post-formed holes or openings that have led to its unpopularity in buildings. In time, awareness of its obvious advantages and competitiveness will lead to changes in the building industry. That process is one of a broader first-degree education, promotion of the technology and greater curiosity on the part of consulting engineers.

Assuming the reader has a certain familiarity with the technology, there are a few nagging long-term concerns for the designer of prestressed concrete structures: the integrity of the end anchors, the grouting operations in a bonded system and the prestress losses. The first two are largely problems of workmanship and adequate protection. There have been recent cases where grout has not properly filled the tendon ducts, leaving voids where water and air can mix to cause corrosion of end anchors and strand in the duct. The latter concern of prestress loss is related to the strain relief of high-tensile wire over long periods, and to the drying shrinkage of concrete which shortens the concrete slab. These losses can be accommodated in the design, and are a function of the material properties.

It would be an advantage if these losses were negligible, so that more of the applied stress could be used in the design calculations. Also, if the material properties of a strand were such that it did not corrode, as a steel strand does, then we would have a more durable prestress system. This is what SICOM in Germany was thinking about when it embarked on a research programme to develop a glass fibre strand that could replace steel strand, that is also alkali resistant and stable in concrete (Fig. 3.58).

Why glass fibre?

Glass fibre is strong enough to take high tensile loads, it has a low modulus of elasticity, it is light in weight and resistant to chlorides and alkalis present in concrete. Because it has such a low modulus of elasticity, the glass strand has to be extended (stretched) almost four times more than a steel strand to induce the required prestress into the strand. The prestress losses caused by concrete creep and shrinkage are, conversely, reduced by a quarter, as this movement represents a negligible percentage of the extension length of the strand.

Concrete structures are prestressed in order to improve serviceability, i.e. to control deflection and reduce or eliminate cracking. In conventional prestressed structures an initial compressive force is imposed

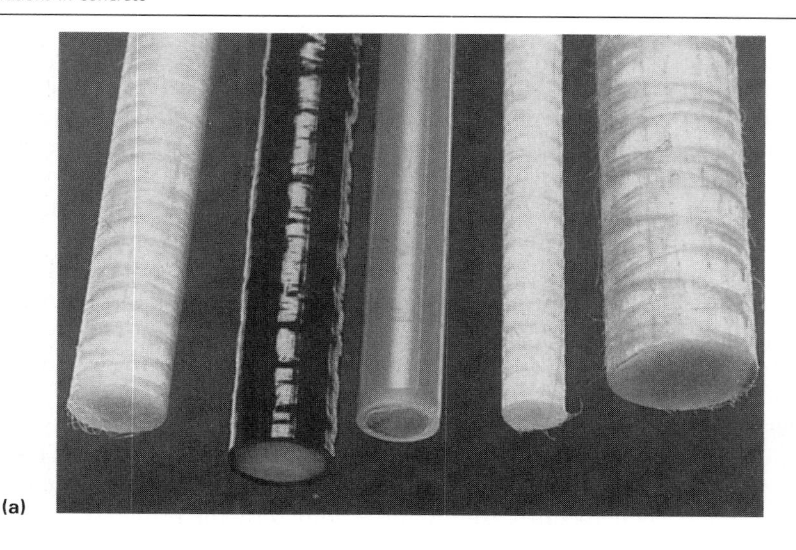

(a)

(b)

Fig. 3.58. (a) A variety of glassfibre composite bars. (b) GRP reinforcing bar placed in floor slab

and sustained by highly tensioned steel strand reacting to the concrete. Non-metallic materials have been developed for use as prestressing tendons as alternatives to high-tensile steel; these include high-tensile material made from carbon fibres, polyaramid fibres and glass fibres (Table 3.34).

(c)

Fig. 3.58 (continued). (c) Bridge at Oppegard Golf Course, Oslo, showing use of GRP reinforcement

One word of caution, however, must be expressed over the long-term creep characteristics of non-metallic tendons, particularly glass fibre strand. Gilbert and Gowripalan showed that the small loss of prestress of non-metallic strand having a low elastic modulus was negated by the higher creep loss of the material. In general terms it was found that the loss of prestress, and other time-dependent deformations of concrete, were similar in magnitude to those obtained with conventional high-tensile steel tendons.

Therefore, the principal advantage offered by these new materials, compared with traditional steel strand, is their corrosion resistance. Since the non-metallic strands do not corrode, it is possible to design a concrete structure with very little cover, even in the most aggressive conditions.

Polystal

Glass fibres and synthetic resin serve as the two basic components of Polystal. The glass fibres are saturated with synthetic resin, then heat treated and formed into a round bar. The bar has a diameter of 7.5 mm and consists of about 60 000 individual glass fibres. A polyamid sheath 0.5 mm thick is fused around glass fibre composite to protect the Polystal bar against chemical attack. The orientation of the glass fibre strands results in high longitudinal tensile stress, similar to that of high-grade prestressing steel, but a low modulus of elasticity of 51 kN/mm^2. When tensioning Polystal, this results in a fourfold increase in the

Table 3.34. Fibre material characteristics and comparison

	Reinforcing steel BSt 500	Prestressing steel St 1470/1670	Polystal (68% glassfibre)	Arapree (aramid fibre)	Carbon fibre composite material
Tensile strength: N/mm^2	> 550	> 1670	1670	1610	1700
Yield strength: N/mm^2	> 500	> 1470	–	–	–
Ultimate strain: %	10	6	3.3	2.5	1.1
Modulus of elasticity: kN/mm^2	210	205	51	64	146
Specific weight: g/cm^3	7.85	7.85	2.0	1.3	1.5
Fields of application	reinforced concrete structures	prestressed structures			stay cables, bracings

extension of the tendon for a given stress, compared to steel (Fig. 3.59). The losses due to concrete shrinkage and long-term creep, on the other hand, are reduced by the same magnitude. However, this may be negated by the long-term creep of the glass fibre.

Anchoring Polystal is more difficult than anchoring high-tensile steel and results in greater anchorage lengths. A profiled metal tube is sleeved over the Polystal tendon and resin-bonded to the strand (Fig. 3.60). The bond value between anchor and Polystal is only 10% of the tensile strength. This has to be done in the factory owing to Polystal's lack of workability at ambient temperatures. Experience gained to date from monitoring full-scale field trials on actual bridges has confirmed that this type of anchorage gives adequate fatigue resistance.

Applications

Ulenbergstrasse road bridge. This bridge was built in 1986 in Düsseldorf and is the world's first prestressed glass-reinforced structure. It has two spans of 21.3 m and 25.0 m, comprising a continuous beam supporting the 15 m wide road deck. The 1.44 m deep beam has been prestressed with 59 Polystal tendons, each carrying 600 kN. Measurements were taken when the prestress was first applied and some months later when the road had been in use. The deflections and performance of the bridge showed good compliance with the design values.

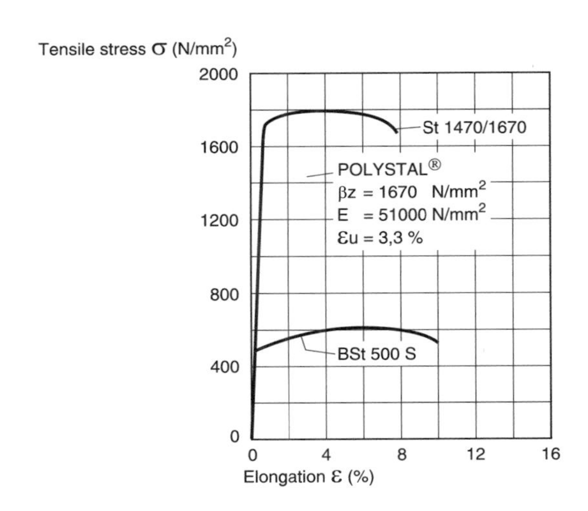

Fig. 3.59. Stress–strain diagram of Polystal bars in comparison with prestressing steel (St 1470/1670) and reinforcement (BSt 500 S)

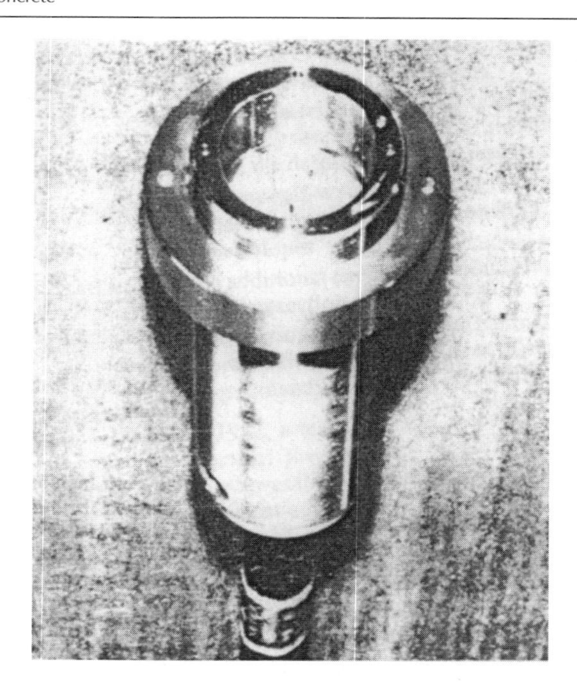

Fig. 3.60. Anchor head of a glassfibre composite prestressing tendon

Marienfelde footbridge, Berlin. This was built in 1988 and was externally prestressed with Polystal tendons. The deck structure comprises a double-T beam, 1.10 m deep, and a deck slab, with spans of 22.93 m and 27.61 m. Seven Polystal strands, each one stressed to carry 600 kN, were draped under the deck structure over the midspan section, and curved to run on the top of the deck slab over the central pier support (Fig. 3.61). The strands were anchored back to the bridge deck at each end of the span.

Design considerations

Design is carried out at ultimate limit state to ensure that the structure has adequate strength, and at serviceability limit state to ensure that its behaviour is satisfactory during the intended design life. The significant design aspects include bending, shear and torsion, deflections and cracking, and cover to reinforcement.

Bending. Failure of a reinforced or prestressed concrete member is due to yield or rupture of the reinforcement or crushing of the concrete. Knowing the stress–strain behaviour of the alternative material, the bending response can easily be determined from first principles.

Fig. 3.61. Underside view of the Berlin–Marienfelde bridge, showing glassfibre composite prestressing tendons

However, there are significant characteristics that must be allowed for. The first is that the material has a straight-line elastic stress–strain response and shows no significant yield before failure. The second is that, under sustained load, the alternative material will creep to failure, known as stress rupture. It would appear that this will lead to sudden failure of a concrete element without warning. However, in composite behaviour the cracking of concrete will give considerable pseudo-ductility to the structure and thus the design is likely to be governed by cracking. It would be prudent to apply a higher partial safety factor in the design.

Shear and torsion. Shear failure in a beam, or punching shear in a slab, usually occurs in conventional reinforced concrete design with no warning. Current design methods are based on empirical rules which are based on the strength of concrete and the amount of tensile reinforcement in the section. Extensive testing will be required to determine similar empirical design approaches for alternative reinforcement.

Deflection and cracking. For an uncracked section the deflection will be a function of the gross cross-section and not influenced by the type of reinforcement. But the deflection will depend on the scale and distribution of cracking, which is influenced by the bond between the concrete and the reinforcement. It is likely that deflection rather than strength will be the governing criterion for structures reinforced with non-ferrous materials. Crack widths have traditionally been limited because the cracks are assumed to accelerate the corrosion of reinforcement. It is now accepted that crack width may have no significant influence on the long-term durability of concrete. The main reason for limiting and controlling crack width in building structures is aesthetic. Notwithstanding such research findings, it would seem prudent to continue to design concrete structures on the assumption that cracks lead to greater risk of early corrosion of steel reinforcement.

Cover. The concrete cover to reinforcement serves three functions: corrosion protection, fire resistance and bond strength. Cover for durability of a non-ferrous material is not a requirement, so cover is only essential for developing an adequate bond between the reinforcement and the concrete, and for fire protection. Limited test data is available on fire resistance of alternative reinforcement, although some of the materials may satisfy current BS standards. The fire resistance of glass/resin composite and other materials is problematic; unless test data is available, they should not be specified as replacement reinforcement bar. They may be acceptable as prestressing strand provided the anchors are well protected.

Sensors for long-term monitoring systems

Prestressed glass fibre cable is made up from many individual fibre strands that are bundled together and then fused with a special resin. It is possible to incorporate an optical glass fibre strand within this bundle to act as a sensor, continuously monitoring stress changes and macro-deformations along the length of the structure (Fig. 3.62).

The sensors react to every longitudinal variation by a proportional change in the light permeability through the optical fibre, which is then measured and recorded (Fig. 3.63). The frequency bands of the electromagnetic waves that fall on the sensitivity area of our eyes are known as light and range between wavelengths of 0.35 μm and 0.78 μm.

Light in the infrared band is often used for measuring distances, e.g. in laser levels and distomats, and these wavelengths can often be just

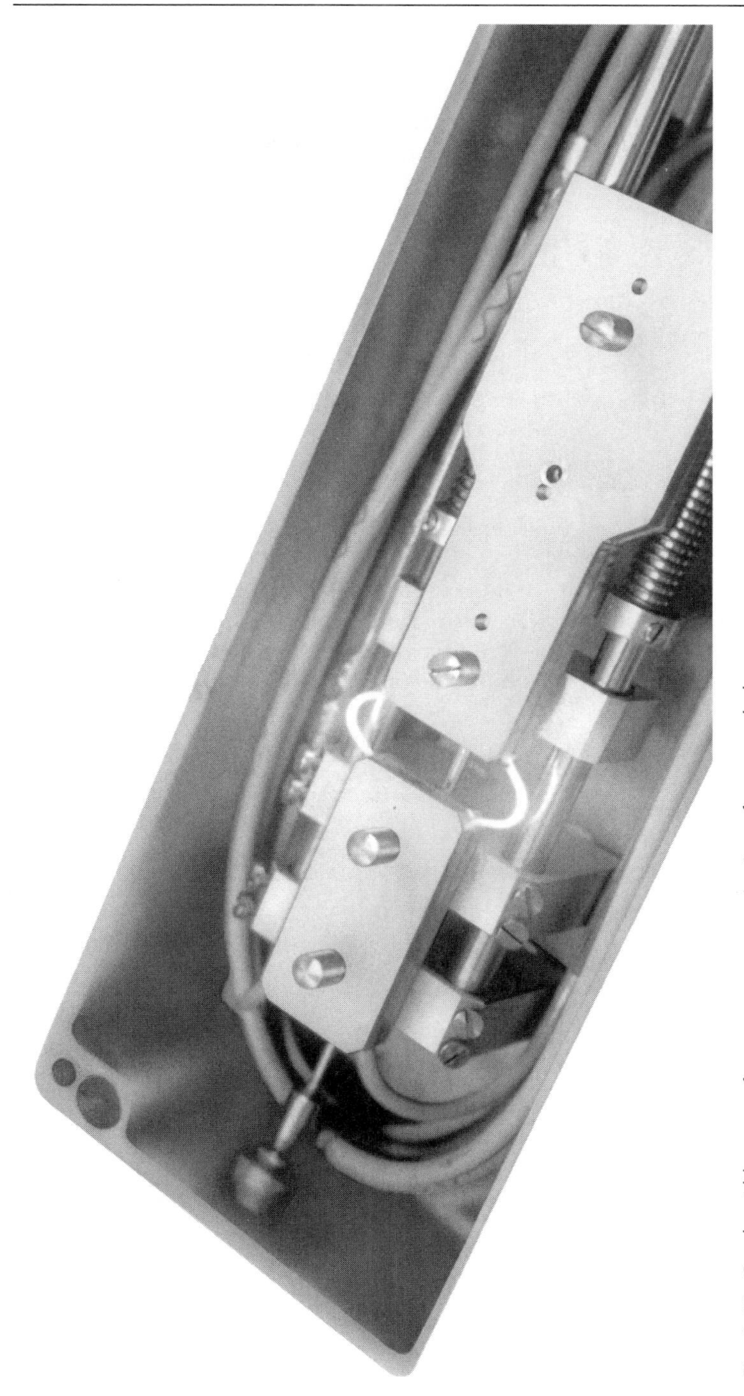

Fig. 3.62. Crack width sensor for permanent monitoring of structural elements

Fig. 3.63. Light emission due to bending in an optical fibre sensor

visible to the eye. For SICOM sensors the infrared light passes through a laser diode that transmits a single wavelength of light. Any changes in the path length calibrated when the optical fibre cable was installed

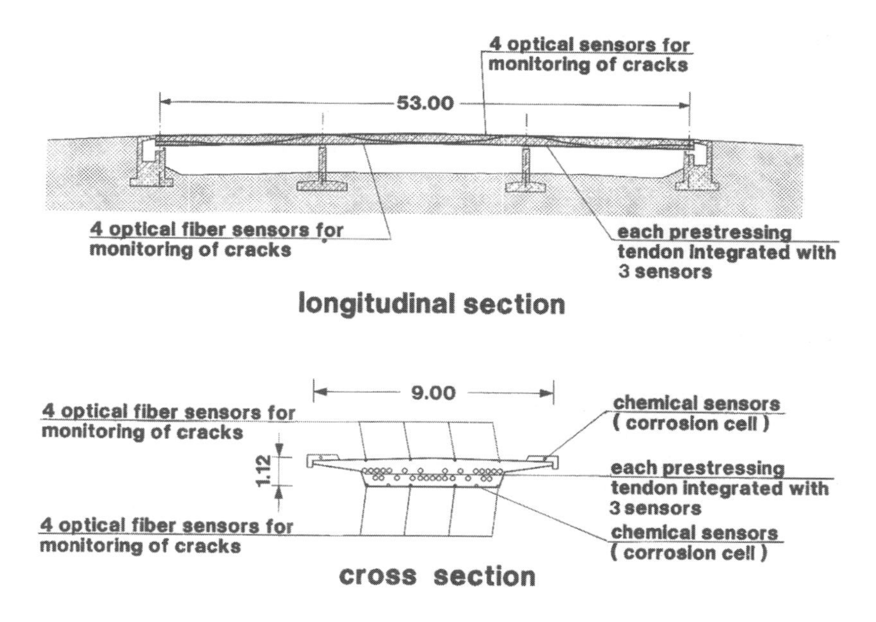

Fig. 3.64. Placing of sensors in a German bridge

will be picked up. Special receivers convert the light signal into electrical impulses. The sensors transport information over large distances at the speed of light and measure strains and deformations with 1/100 mm accuracy.

The advantages of optical fibre sensors over traditional measuring instruments are high durability, electromagnetic neutrality and the ability to function both as measurement transducers and information receptors. The result of this is that we can now monitor internal structural changes and anticipate the early onset of cracking due to electrolytic action (chlorides and corrosion) or material degradation such as alkali–silica reaction. The resulting structure can justly be referred to as intelligent (Fig. 3.64).

When taking into account the economic advantages of non-ferrous reinforcement and prestressing strand, the total cost of the structure including long-term maintenance or life cycle costing, as well as the primary cost of the alternative material, should be considered.

References and further reading

ANDRA, H-P, BAUR, H and STIGLAT, K (1984). Structural behaviour analysis and design of flat-slab floors. *Beton- und Stahlbetonbau*, October, 258–263, 303–310, 328–334.
BEKAERT LTD (1990a). *Industrial floors with Dramix steel fiber reinforced concrete.* Technical booklet.
BEKAERT LTD (1990b). *Steel fibre concrete under flexural load.* Dramix technical pamphlet.
BEKAERT LTD (1990c). *The properties of Dramix steel fiber concrete.* Dramix technical pamphlet.
BENNETT, D F H and MACDONALD, L A M (1992). *Economic assembly of reinforcement.* British Cement Association, Wexham Springs. Publication 97.321.
BRITISH CEMENT ASSOCIATION (1990). *Shearheads. Flat slab shear reinforcement manual.* Wexham Springs.
BRITISH CEMENT ASSOCIATION (2000). *Rationalisation of flat slab reinforcement.* Crowthorne. Publication 97.506.
CHANDLER, J W E (1982). *Design of floors on ground.* Cement and Concrete Association, Wexham Springs. Technical Report 550.
CLARKE, J L (1993). *Alternative materials for reinforcement and prestressing of concrete.* Blackie, London.
CLARKE, J L (1998). Reinforcement: where do we go from here? *8th Annual Conference on higher education and the concrete industry*, July.
CONCRETE SOCIETY (1994). *Concrete industrial ground floors: a guide to their design and construction.* Technical Report 34.
CORLEY, W G and HAWKINS, N M (1968). Shearhead reinforcement for slabs. *ACI Journal*, October. Title 65-59.
GEILINGER STRUCTURAL SYSTEMS. *Flat slabs with Geilinger steel shearheads.* Switzerland. Technical pamphlet.
GOODCHILD, C H (1993). *Cost model study.* British Cement Association, Crowthorne.
HUGHES BROTHERS LTD. *Glass fiber reinforced polymer rebar.* USA. Technical brochure.
MATTHEW, P W and BENNETT, D F H (1990). *Economic long-span concrete floors.* British Cement Association, Wexham Springs. Publication 97.311.

MODFORM SYSTEMS LTD (1984). *Design manual for use of Modform shear comb reinforcement in flat slab.*
PATERSON, W S and RAVENHILL, K R (1981). *Reinforcement connector and anchorage methods.* CIRIA Report 92.
STEVENSON, A M (1994). *Post-tensioned concrete floors in multi-storey buildings.* British Cement Association, Crowthorne. Publication 97.347.
WOLFF, R and MIESSELER, H-J (1990). Application and experience with intelligent prestressing systems based on fibre composite materials. *FIP 11th International Congress on prestressed concrete,* Hamburg, June.

4 Project profiles

A number of case study exercises on buildability, carried out with the help of consultants and contractors on 'live' building projects, are summarised in this chapter. They encapsulate many of the efficient construction techniques and concrete design options described in the previous chapters. All of the studies except 88 Wood Street were managed by the author during the building boom of the 1980s and were part of a promotional drive by the British Cement Association to improve the efficiency of concrete-framed structures.

Arrowhead project study

General

The Arrowhead development is situated in the Isle of Dogs, London, on the westerly corner of South Quay, overlooking West India Docks. The site covers an area of 2500 m^2, and forms phase 1 of the next stage of the Waterside Development plans for South Quay.

The Arrowhead development (Fig. 4.1) provides 13 levels of office space within the central block, and 11 levels of office space within each of the wing blocks, offering around 13 000 m^2 of total floor area above ground. A two-level basement structure provides parking for up to 160 cars. Architectural features include a granite facing to column and beam panels, large bay windows in tinted bronze, and a granite facade to the access core at the end of each wing block. There are two central lift and stair lobbies on each floor, in addition to a stairwell on the end of each wing block.

Floor-to-floor height has been given as 3.37 m, assuming a 325 mm slab and allowing 2.7 m for clear ceiling height. Each of the floors, up to level 11, has an area of 1200 m^2.

Suggestions on the original scheme layout

Suggestions put forward on alternatives to the original scheme layout attempt to improve the constructability of the frame without changing the intrinsic architectural character of the building shape and form.

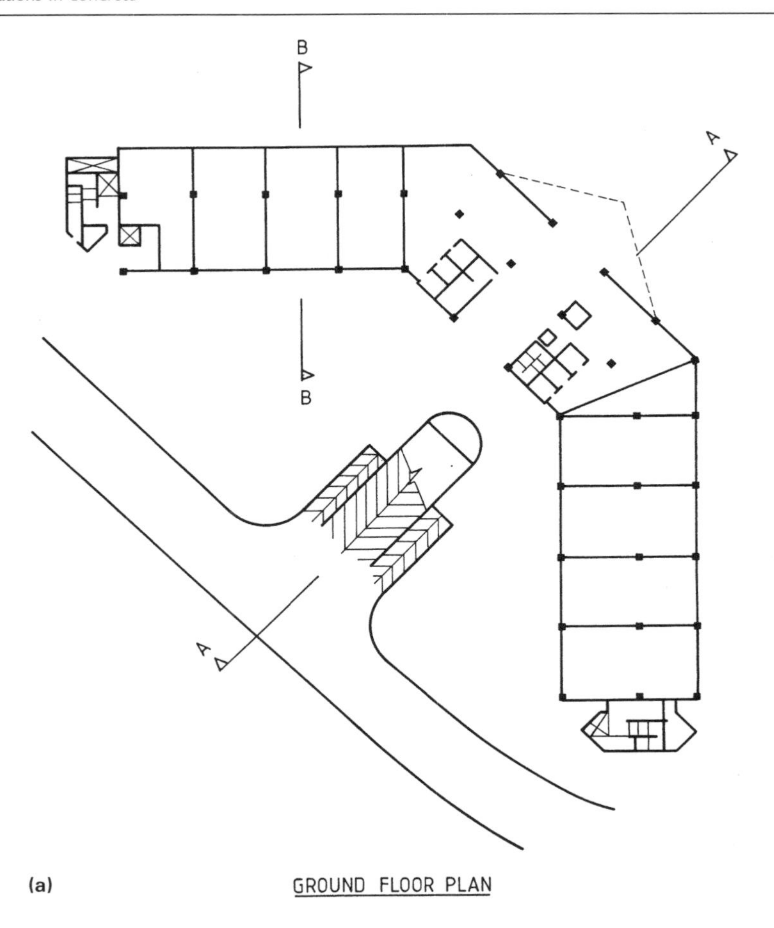

Fig. 4.1. Arrowhead project: floor plans and building section

The two central lift core areas appear to be of solid wall construction and are expected to provide the lateral stability. Built with in situ concrete, they represent a major item of work which will influence the floor cycle time. The two perimeter staircase cores also present the same type of construction difficulty and will dictate the critical path.

Central lift and stair core option

Design the structure as a moment frame with stability developed from the two central cores, adopting beam and column construction, and from closely spaced columns with an edge beam at the end of each

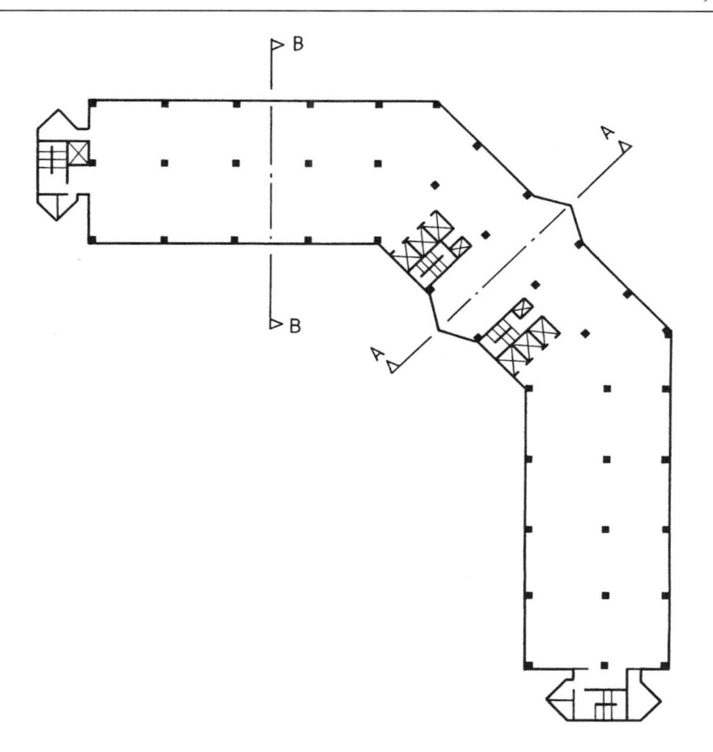

TYPICAL SUPERSTRUCTURE FLOOR PLAN
(1 - 11)

(b)

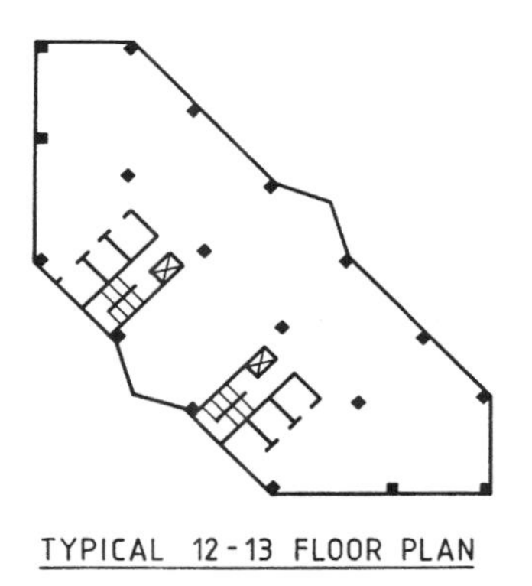

(c) TYPICAL 12 - 13 FLOOR PLAN

Fig. 4.1 (continued)

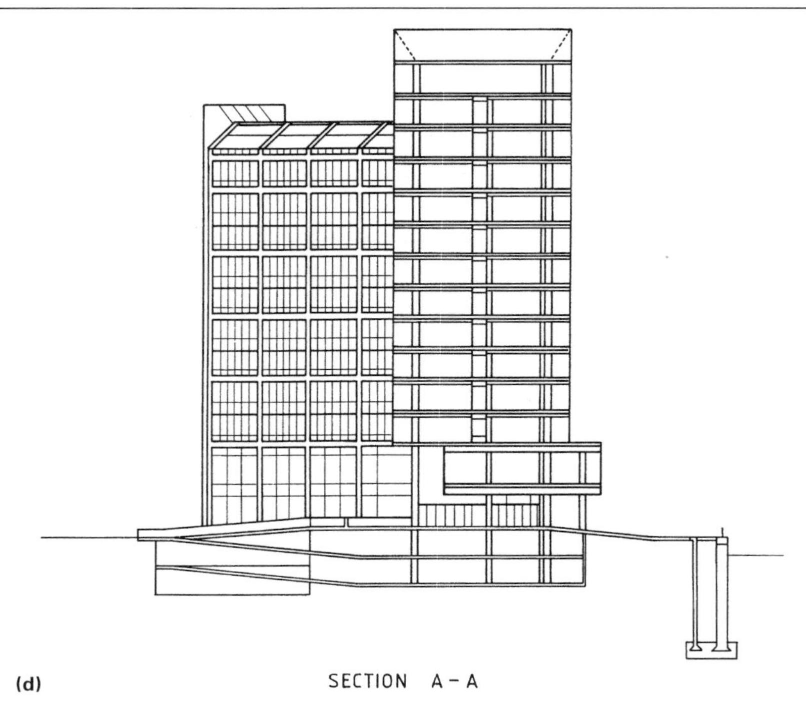

Fig. 4.1 (continued)

wing block (Fig. 4.2). By eliminating solid wall construction and keeping all vertical construction to columns only, the floor cycle time can be improved.

Wing core construction option

Precast the wing core structure and frame it with the cladding panels, independently from the construction activity for the main structure (Fig. 4.3). For practical reasons it would be difficult to frame the core structure with in situ concrete using flying or table forms. Adopting traditional construction techniques would become labour intensive and, for such a small volume of work, it offers no advantages over a precast solution. For the purpose of this study a typical floor layout as shown in Fig. 4.2 has been assumed.

Superstructure

Two schemes are considered, both well suited to the requirements of the site, the limitations on the use of cranage and the restricted access.

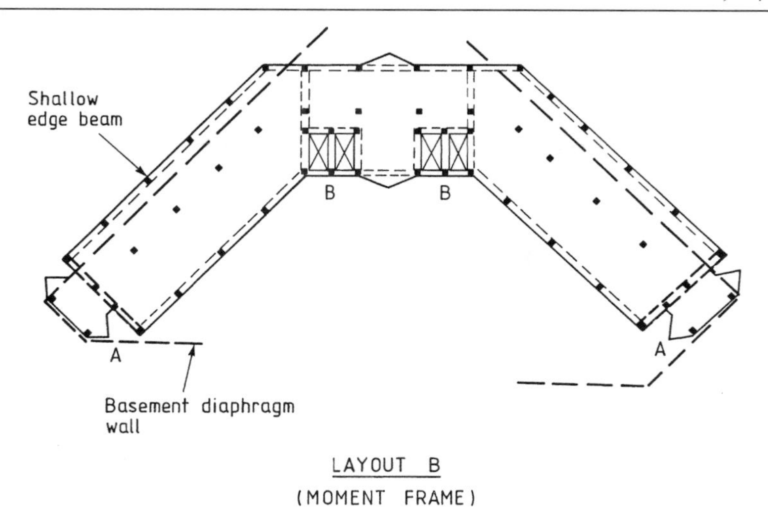

Fig. 4.2. Arrowhead project: moment frame option

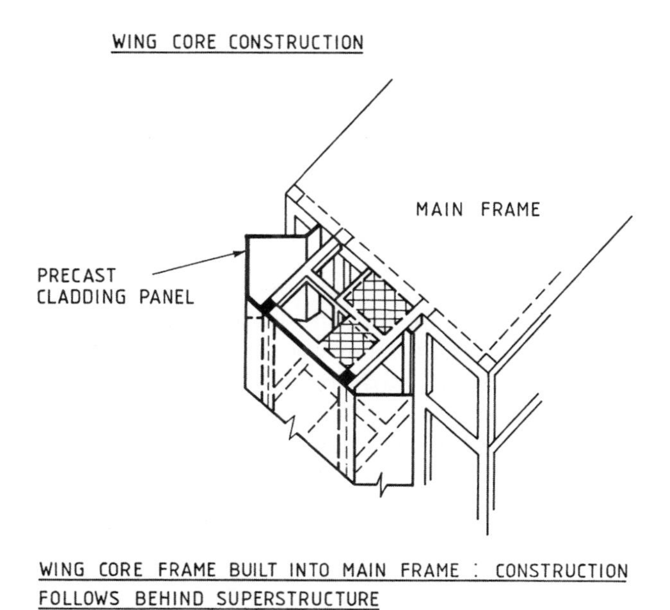

Fig. 4.3. Arrowhead project: wing core option

The overall plan has been to speed critical elements of the construction cycle, reduce the formwork requirement and minimise the labour content of operations where possible. Faster frame construction leads to earlier release of the crane for cladding work, which is on the critical path for longer than the frame; this will shorten the overall project time. The preliminary work of blinding and then casting a ground slab is common to all schemes and follows the piling work. Basement excavation begins after the upper floor is cast and the ground floor is free of props, formwork and the like. The two schemes are:

A — Post-tensioned construction;
B — Trough slab construction.

Other options such as composite construction are also worth evaluating, but time did not permit consideration in this study.

Scheme A: post-tensioned construction

Design

The structure has been designed as a moment frame, with internal stability provided by the two central core areas and by the close-centred columns and edge beam at each end of the wing blocks (Fig. 4.4). Floor loading is taken as kN/m^2 (live), with edge loading of 1.42 kN/m^2 from the cladding. Wind load has been calculated in accordance with the specification. For design purposes, the lightweight concrete density assumed is 19 kN/m^3, and that of normal concrete 24 kN/m^3.

Structural notes

Suspended floor. Designed as a flat slab, 200 mm deep using lightweight concrete grade C40 or 225 mm deep using C40 normal-weight concrete. Unbonded tendons have been specified because they are faster and cheaper to install, arriving on site factory assembled, colour coded and ready to lay. The full 100% stress will be applied to the slab once the concrete has achieved a minimum compressive strength of 20 MPa. This will be possible 36 h after casting of the floor slab.

Intermediate anchors will be positioned on alternate tendons at the first construction joint to allow separate stressing of the first and second pour areas for quick release of the flying form panels. Between the second and third pour areas an infill strip, about 1000 mm wide, will be left until after the second pour has been stressed and the third pour

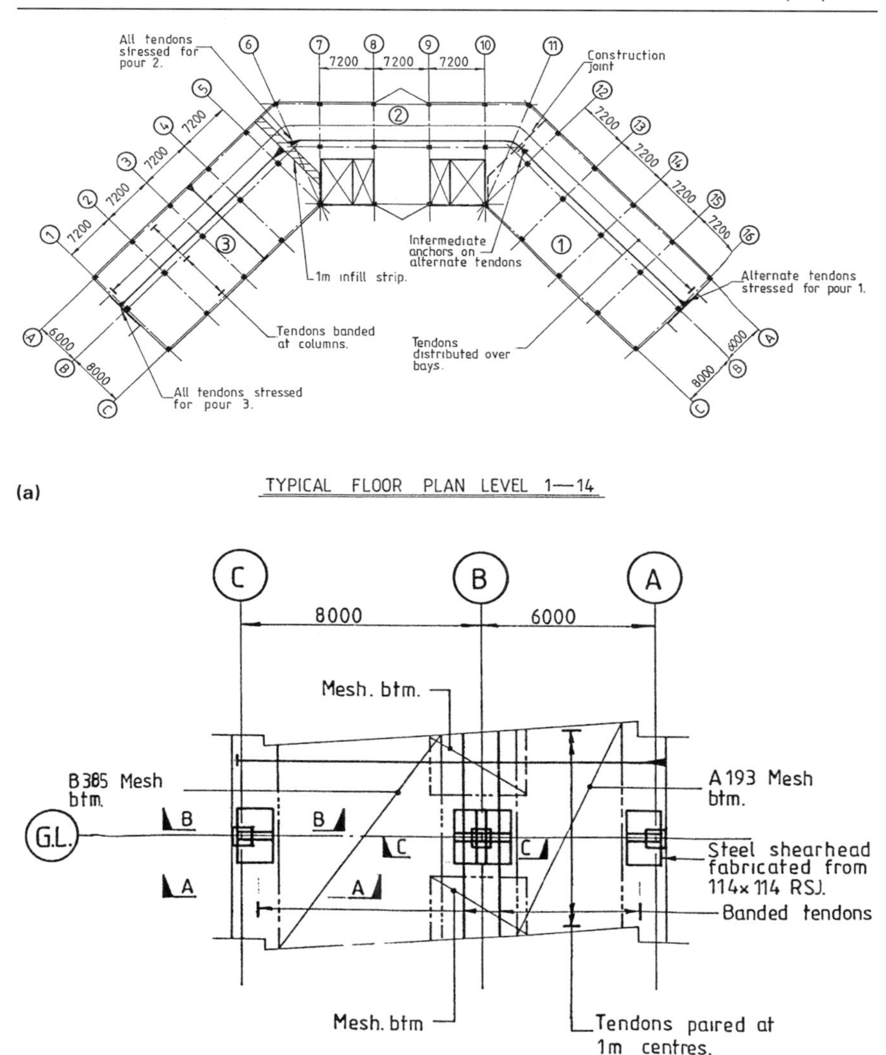

Fig. 4.4. Arrowhead project, Scheme A: (a) typical floor plan; (b) typical bay layout

area cast. Each pour area will have about 1800 m of strand which will require up to 50 man-hours to fix, in addition to fixing 3.5 t of rebar. A certain amount of prefabrication of rebar and workforce management of the post-tensioned operation will ensure an effective operation. In the initial stages, the floor will be completed on a four-day cycle, allowing

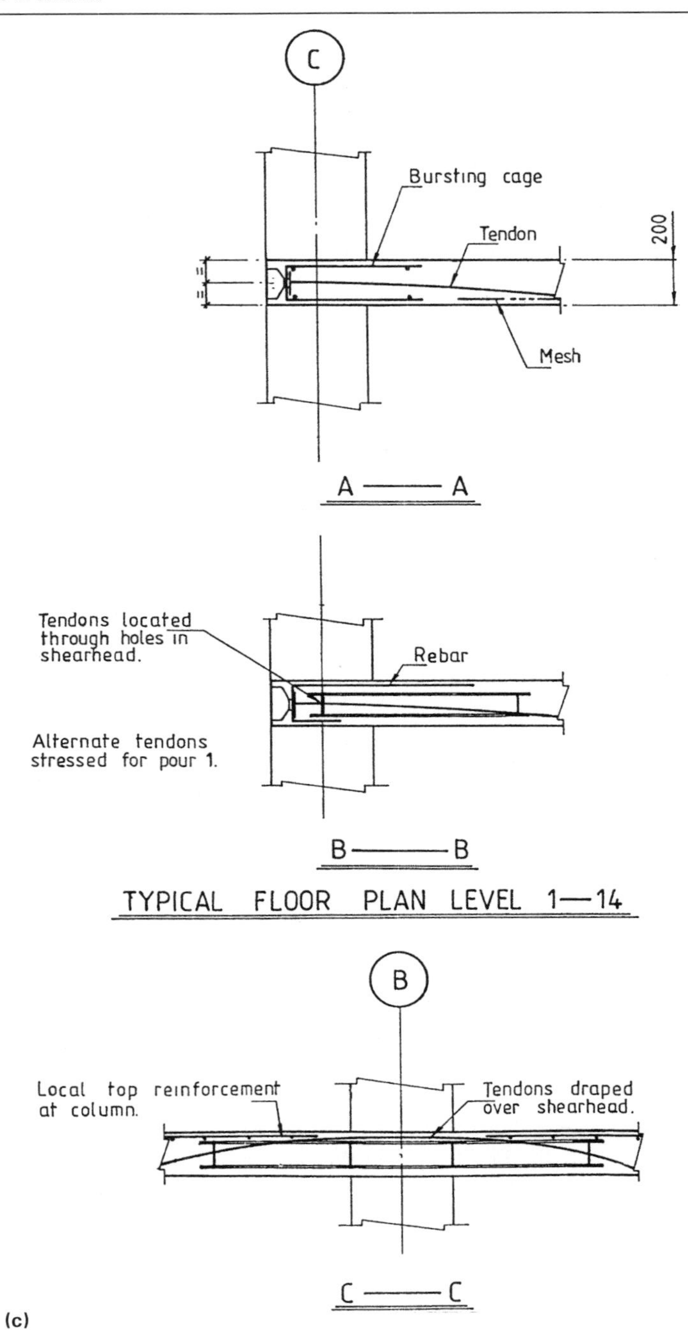

Fig. 4.4 (continued). (c) Slab–column connections

two working days to fix tendons and rebar in each pour area. Thereafter, as the routine becomes familiar, we would expect the tendon laying and rebar fixing to be complete within a normal working day, making it possible to hit the three-day cycle by the time the structure is up to, say, the fifth floor.

Columns. The column sizes have been standardised as 450 × 450 mm square throughout the structure, utilising the benefits of high-strength concrete. For simplicity of site operation and economic consideration, it is proposed that all columns are cast with 80 MPa concrete, and the rebar quantity adjusted accordingly.

Shearheads. To avoid drop panels in the slab over the column position, and to reconcile the high shear stress, a metal shearhead similar to a Geilinger type, successfully developed in Switzerland, will be incorporated (Fig. 4.4(b, c)).

Construction notes

Pour sizes. The distribution of the workload over the central area and the wing blocks suggests that each floor level is divided into three pour areas, the maximum pour being around 500 m^2. Adopting ground-floor construction techniques for the concreting work on the suspended slab — using a long vibrating screed rail with 'Permaban' type support rails, and allowing a tamping bay width of up to 9 m — each pour area is expected to be finished within a 3 h period. Concrete will be placed by lorry-mounted mobile concrete pump rated to 60 m^3/h, using a 38 m boom and static line extension to reach the 12th and 13th floors.

Formwork. Full bay-width flying forms plus a few short and medium length table forms should be specified to shorten the forming time to target the four-day floor cycle. With a crew of nine men, an area of about 600 m^2 can be formed in less than a working day, making it possible to start rebar and tendon laying on the same day.

Column construction. Kickerless construction should be adopted for column construction. For the four-day cycle, columns will be cast the day after concreting the floor slab. However, if there is a need to increase tendon fixing time from one to one and a half days, it is possible to cast the column on the same day as the floor slab, once the base concrete has reached an initial set. This technique has been used successfully in the USA. For the Arrowhead project, preliminary site trials would be advisable to acquaint the trade contractor with this method of construction.

Programme

Budget costs have been obtained from contractor's and manufacturer's (mid-1980s) prices. The guide price given below is a realistic costing for the Arrowhead project. Based on a net 13 000 m^2 of floor area, excluding crane hire charges and management contractor charges, the cost is:

(1) Lightweight concrete floor (200 mm)	£85/m^2
(2) Normal-weight concrete floor (225 mm)	£83.8/m^2
Estimated superstructure cost	£1 100 000

Scheme B: trough slab construction

Design

The structure has been designed as a moment frame, with the beam strips and column across the short width accepting the wind load in addition to taking gravity loads. On this basis there would appear to be no special requirement to stiffen the ends of the wing blocks as for Scheme A.

Floor loadings are taken as 5 kN/m^2, and 1.42 kN/m^2 for the cladding acting on the slab perimeter. Wind loading is taken as per specification. Either lightweight concrete C40 grade or normal-weight concrete C40 could be used. For structural purposes a lower 28-day compressive strength would prove adequate, but it would not achieve 20 MPa within 36 h of casting, to permit early release of formwork.

Structural notes

Suspended floor. This is designed as a 320 mm deep trough slab, with ribs of thickness 250 mm spaced at 1200 mm centres and having a 95 mm topping (Fig. 4.5). The ribs span 6.2 m onto 1000 mm × 320 mm deep beam strips. The beam strips span across the short width of the frame, supported on columns at 8 m and 6 m spacing. The structure is normally reinforced using 460 MPa high-yield bars. Soffit formwork could be struck to release flying forms once the concrete has achieved 20 MPa compressive strength. It is anticipated that the creep factor will be higher for this scheme; therefore, careful planning of the de-propping and re-propping sequence will ensure deflections within the specified limits. A further detailed design check on this point is advisable. Refer to the notes on structure in Scheme A, as the other structural elements are the same, except for shearheads, which are not required for this scheme.

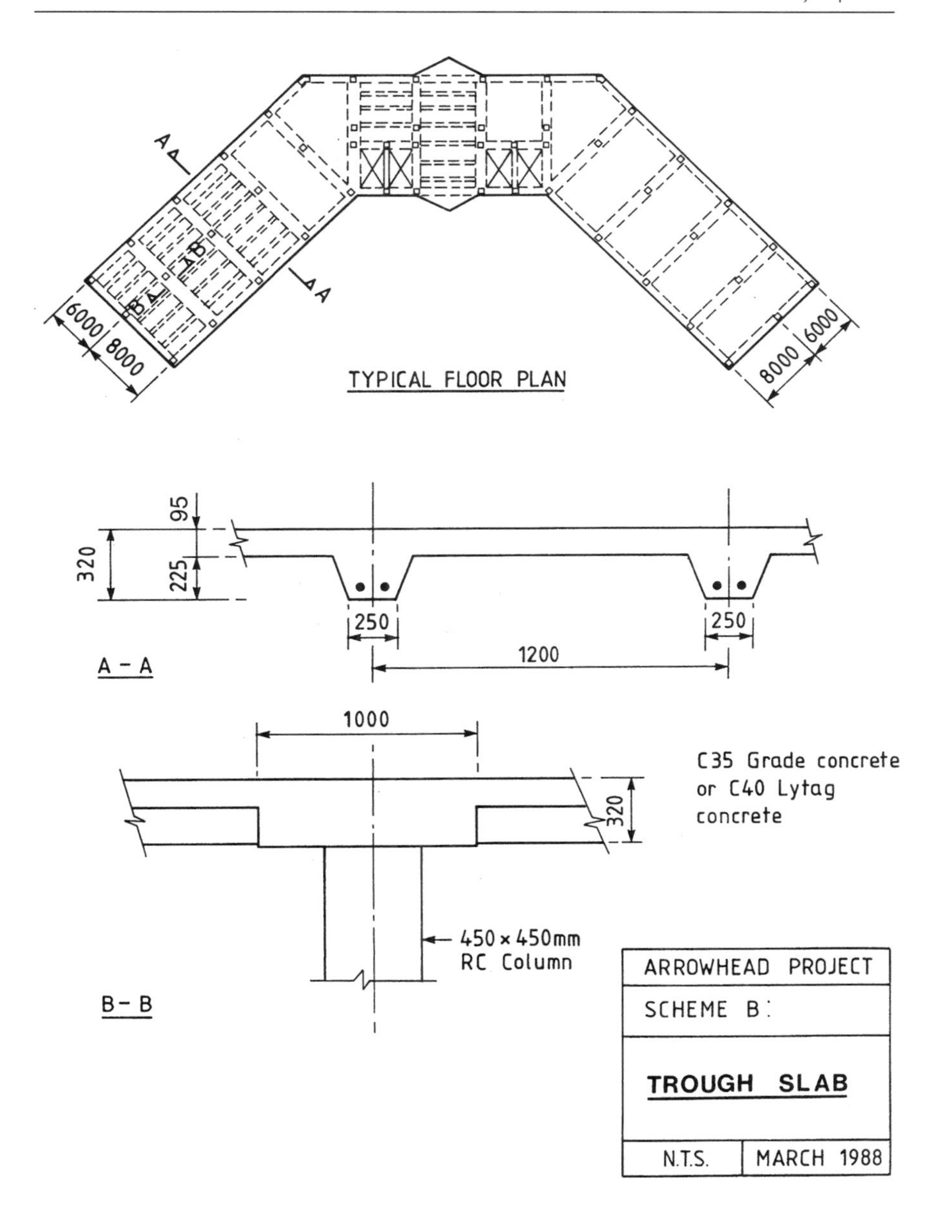

Fig. 4.5. Arrowhead project, Scheme B: trough slab

Construction

Formwork. The trough slab profile will be formed using glass-fibre-reinforced plastic (GRP) moulds fixed integrally to the flying forms and table forms. A compressed air line is advised for releasing the GRP moulds prior to striking the formwork assembly. Generally, the removal and re-positioning of the flying forms will take longer than the flat slab option in Scheme A. But it is expected that each pour area can be formed within a working day.

Pour size and column construction. As in Scheme A.

Reinforcement. Extensive prefabrication of reinforcement cages for beam strips and the 95 mm topping over the trough floors will improve productivity and minimise labour content. However, allocation of crane work time for this critical activity would need careful planning. It will delay the start of rebar laying until after flying forms are installed. Reinforcement fixing may require 1.5 days per pour area, improving to one working day as the routine becomes familiar. A four-day cycle would appear to be achievable, but a three-day cycle may prove unrealistic.

Cost

Detailed budget costings for this scheme have not been completed. However, it may be reasonable to assume the cost figures given in Scheme A, as the two schemes overlap on construction activities. Any changes or extras are nullified by the omission of shearheads and marginally less concrete volumes for the floor.

Assume: cost/m^2 as in Scheme A, i.e. £85/m^2.

Estimated superstructure cost. £1 100 000 (excluding crane charges and management contractor's fee).

The four-day floor cycle

On the basis of previous BCA case studies on projects of similar scope, and after detailed discussion with specialist manufacturers and contractors, it appears that the four-day floor cycle is achievable, as indicated on the bar chart (Fig. 4.6).

It has been assumed that two cranes are available exclusively for the frame construction and are sited to give the best sweep over the floor plan. Allowing for an initial five-week period to complete the structure up to first-floor level, the four-day floor cycle could then begin.

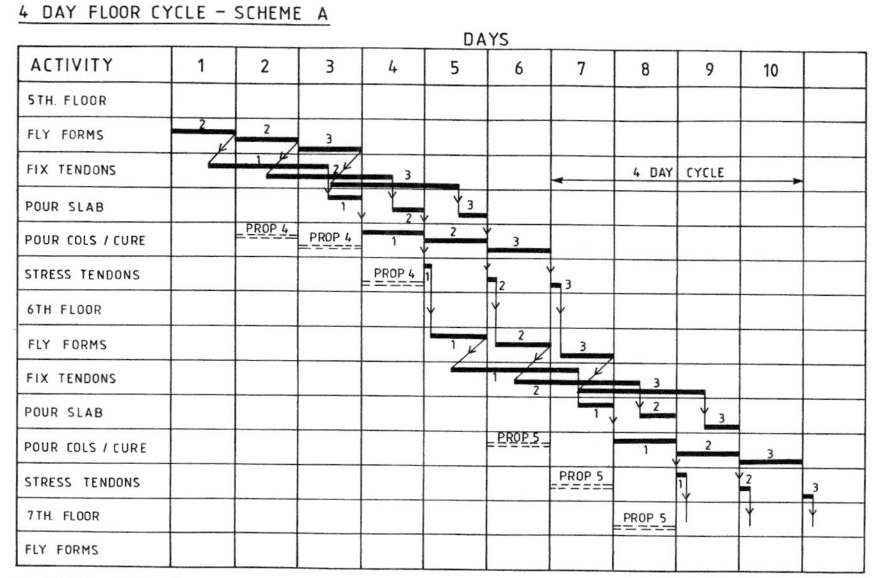

1 = POUR AREA 1
2 = POUR AREA 2
3 = POUR AREA 3

Fig. 4.6. Arrowhead project: four-day floor cycle (Scheme A)

A day-by-day resource scheduling and typical activity plan for work on the fifth floor is illustrated in Fig. 4.6. The floor area is divided into three bays. This work schedule was prepared by Tom Scott, Chief Engineer of CECO Construction of Chicago. The programme suggests it is feasible to complete the superstructure in 15 weeks.

There are seven gangs, each gang being a three- or four-man team, but they double up to form up to 10-man teams on certain activities. They work an average 8 h day, but occasionally they may stay later to finish an activity.

Typical resourcing for the four-day cycle is shown in Table 4.1 and Fig. 4.7.

Towards the three-day floor cycle

If it proves feasible to complete the tendon and rebar fixing in one working day, then a three-day floor cycle is possible. The three-day cycle should best be attempted once the site team have successfully 'hit' the four-day cycle on three or more successive floors. It is only at this point the team will have gained sufficient confidence, know-how and experience of all the construction activities to schedule and plan a three-day cycle.

Table 4.1. Arrowhead project: typical resourcing for the four-day cycle

Work activity on day 3, 5th floor (Fig. 4.7(a))				
Bay 1	Man-h	Men	Hours	Gang
Fix tendons (finish)	9	3	3	1
Miscellaneous formwork	12	4	3	3
Permaban rail	16	4	4	4
Pour slab	24	8	3	4
Bay 2				
Fix tendons	42	6	7	2
Miscellaneous formwork	21	3	7	5
Prop level 4	20	4	5	6[a]
Bay 3				
Fix tendons (start)	12	3	4	1
Miscellaneous formwork	16	4	4	3
Fly form 7 @ 3 man-h	21	4	5	6[a]
	193			

Work activity on day 4, 5th floor (Fig. 4.7(b))				
Bay 1	Man-h	Men	Hours	Gang
Set and pour 18 columns @ 4 man-h each	72	9	8	7
Bay 2				
Fix tendons (finish)	9	3	3	1
Miscellaneous formwork	12	4	3	1
Permaban rail	16	4	4	4
Pour slab	24	8	3	4
Bay 3				
Prop level 4	20	4	5	6[a]
Fix tendons	42	6	7	2
Miscellaneous formwork	21	3	7	5
	216			

Work activity, day 5, 5th floor	(Fig. 4.7(c))			
Bay 1	Man-h	Men	Hours	Gang
Fly form 7 @ 3 man-h	21	4	5	6
Fix tendons (start)	12	3	4	1
Miscellaneous formwork	16	4	4	3
Bay 2				
Set 20 columns @ 4 man-h	80	10	8	7
Bay 3				
Fix tendons (finish)	9	3	3	1
Miscellaneous formwork	12	4	3	3
Permaban rail	16	4	4	4
Pour slab	24	8	3	4
	190			

[a] Note: Gang 6 have to work a ten-hour day on day 3, but reduce work hours on day 4

It is probable that the Arrowhead project team's combined expertise, backed with specialist knowledge from trade contractors and manufacturers, could achieve the three-day floor cycle after level 7.

Acknowledgements
CECO Corporation, USA
David Ramsay, Trevor Crocker & Partners
PSC Freysinnet
Pioneer Concrete
Alumna Systems

Corsham Street development, London

Scope

This preliminary report is based on information provided on the scheme drawings and on discussions held with the client and architect on the building specification.

As far as possible, realistic programme times have been stated for the options proposed, together with costs based on preliminary design of the structure and past experience. Where appropriate, advice has been sought from specialist manufacturers and contractors for pricing and programming requirements.

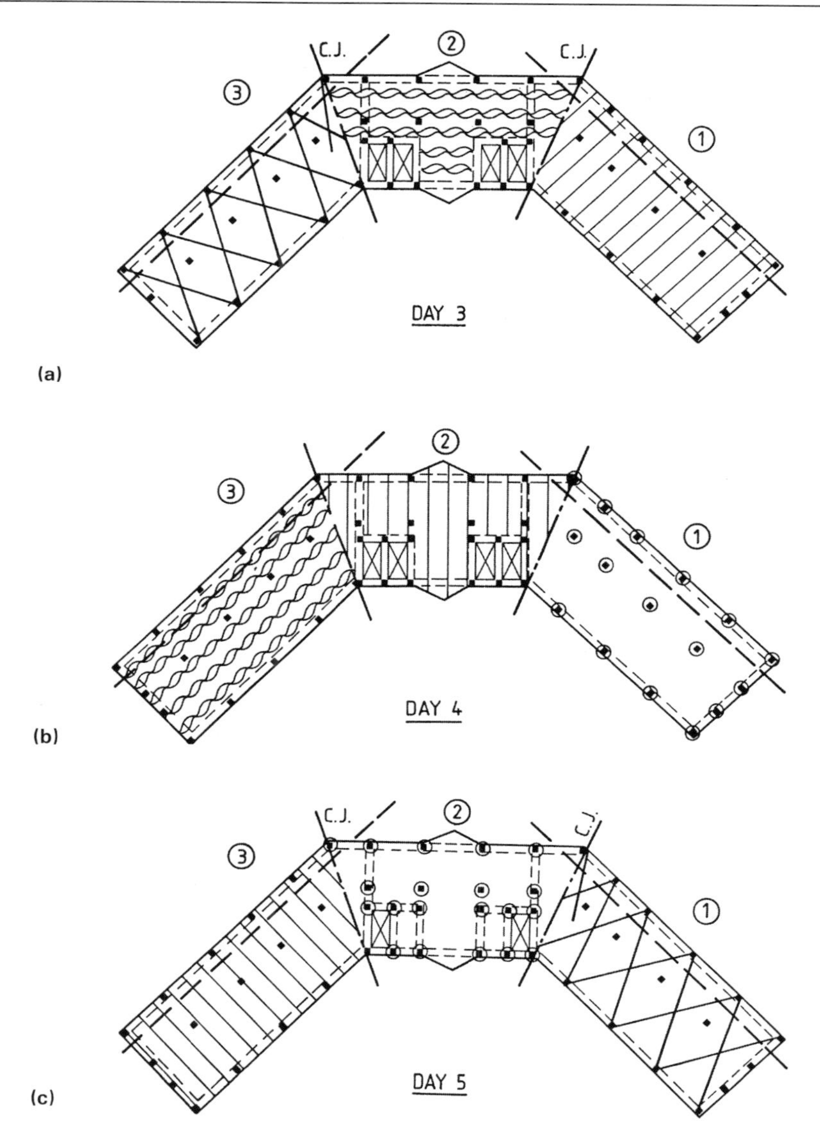

Fig. 4.7. Typical resourcing for the four-day cycle: (a) day 3; (b) day 4; (c) day 5

The substructure and superstructure design does not take full account of any planning restrictions or of the soil conditions on the site. Such information was not available, but certain assumptions have been made, based on knowledge of similar sites in the locality.

Detailed consideration of the cladding and service installations falls outside the scope of this report, but the cost of installation has been outlined.

Basic buildability principles

In the appraisal of Corsham Street (Fig. 4.8), basic buildability principles have been pursued in the concept design and followed in all the construction processes, to arrive at a more efficient and cost-effective way of building. The following guidelines have been used.

Design for building repetition

Keep the floor layout of the building repeating from bay to bay and from floor to floor. If work is repetitive and simple, labour and supervisory staff are likely to become familiar with it and the learning curve will be short and steep; therefore productivity will be higher.

Standardisation of member sizes

Keeping to standard member sizes and maintaining a constant depth of construction, constant storey height and constant column dimensions makes construction fabrication easier, simpler and interchangeable. It will usually be more cost effective to increase the strength or the amount of reinforcement to accommodate differing loads and spans than to vary the size of structural members.

Simplicity of construction

Simplicity reduces the chances of errors in the design or on the site and encourages economic construction. For example, solid core wall construction for lift and stairs may require as much labour resources as the rest of the floor. Economy of construction is achieved by rationalising the building process, sequencing critical activities to reduce downtime.

Minimisation of sequential working

Reducing site operations by adopting efficient construction techniques helps to bring down cost and speed up the construction process. Prefabrication of reinforcement and formwork panels should be considered, kickerless construction adopted and precast prefabrication introduced to minimise traditional formwork procedures.

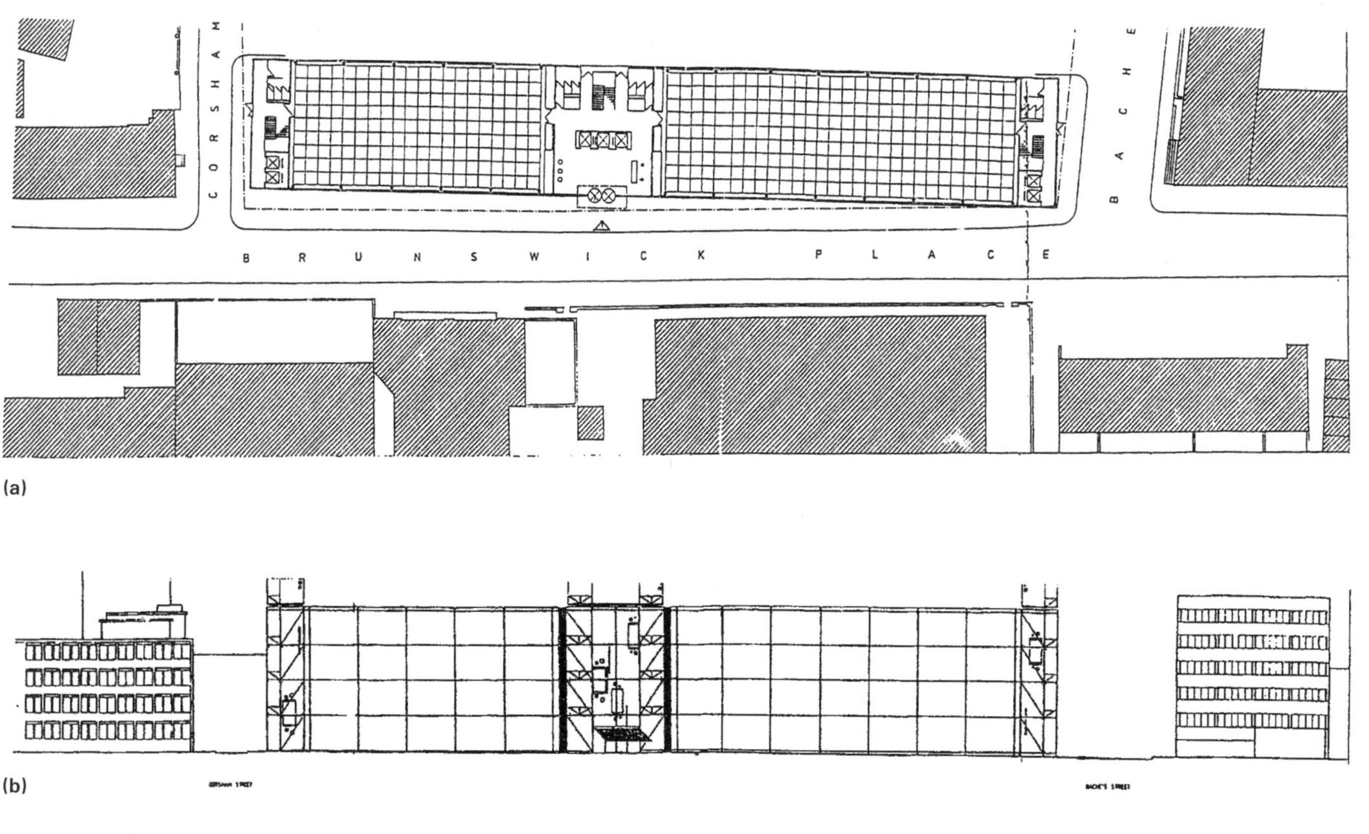

Fig. 4.8. Corsham Street development: (a) ground floor plan; (b) typical section

Modifications to the original scheme layout

In order to meet the objectives of buildability and minimum cost, the scheme layout has been modified in that bay sizes are maintained at constant geometry and core areas are centralised, where possible.

Points on frame design

Methods of the construction approach will consider:

(a) incorporating precast concrete or composite precast elements, thus providing better visual quality and off-site prefabrication;
(b) adopting lift slab construction, to eliminate the cost of formwork and falsework as all construction work is carried out at ground level;
(c) using proprietary structural elements as 'off-the-shelf' products and not specials, to keep the cost of units to a minimum.

Points on services

The four-storey office building will have exposed floor soffits, with all main services runs—both primary and secondary ductwork—and sprinklers exposed, to eliminate the cost of a false ceiling. Heating of the building is provided by a low-pressure hot water system with low-level radiator units placed by windows on the long sides of the building. In summer natural cross-ventilation may be used, assisted by ceiling fan units.

A 'computer floor' with a clear depth of 100 mm has been provided, allowing for the distribution of telecommunication cabling and computer wiring. Electrical sockets are floor mounted.

High-level natural lighting through the window openings is supplemented by fluorescent strips, with diffusers at 2 m centres.

Points on cladding

It is proposed that any exterior cladding could be either a precast double-T beam or loadbearing or non-loadbearing brickwork A tilt-up concept was considered but has not been reported.

Points on foundation

The structure is to be built on a pad foundation; there is no basement level.

Structural options

General

A number of options were considered for the construction of the building. Designs have been developed and costed for three schemes which are likely to come within the building budget that has been set. The schemes are:

(1) lift slab: flat slab in situ construction;
(2) lift slab: composite precast construction;
(3) tilt-up precast walls and floor construction.

Generally, the design loads are taken as 4 + 1 kN/m^2 with an edge loading allowance on the perimeter from the cladding panels. Deflection checks have been made for the temporary conditions during construction and, over the long term, under service load. The estimated deflections for all three options satisfy BS 8110.

Construction notes

In all three schemes the central lobby area has been built separately from the main floors. In order to keep the costs down and to simplify construction, the floors on each side of the central lobby are built in advance. The central and wing lobby areas are then constructed with infill slabs and masonry walls to enclose the stairwells and lift lobbies. The details of construction and design are covered under each option.

Option 1: lift slab (flat slab in situ construction)

General

The structure consists of a 275 mm in situ lightweight concrete, post-tensioned flat slab spanning 10 m between precast concrete columns. There is a cantilever overhang of 2.0 m, giving an overall width of 14 m to the building. In the longitudinal direction the building measures 96 m. The columns are precast, 450 mm square, and spaced at 6 m intervals in the longitudinal direction. For the lift slab construction the floors are cast on the ground and then hydraulically lifted into position and supported off the precast columns (Figs. 4.9 and 4.10). The central lobby area and end staircases are constructed after the main floors have been lifted into position.

Design

The floor is designed for a superimposed dead load of 1 kN/m^2 and a live load of 4 kN/m^2 with an additional allowance of 1 kN/m^2 for parti-

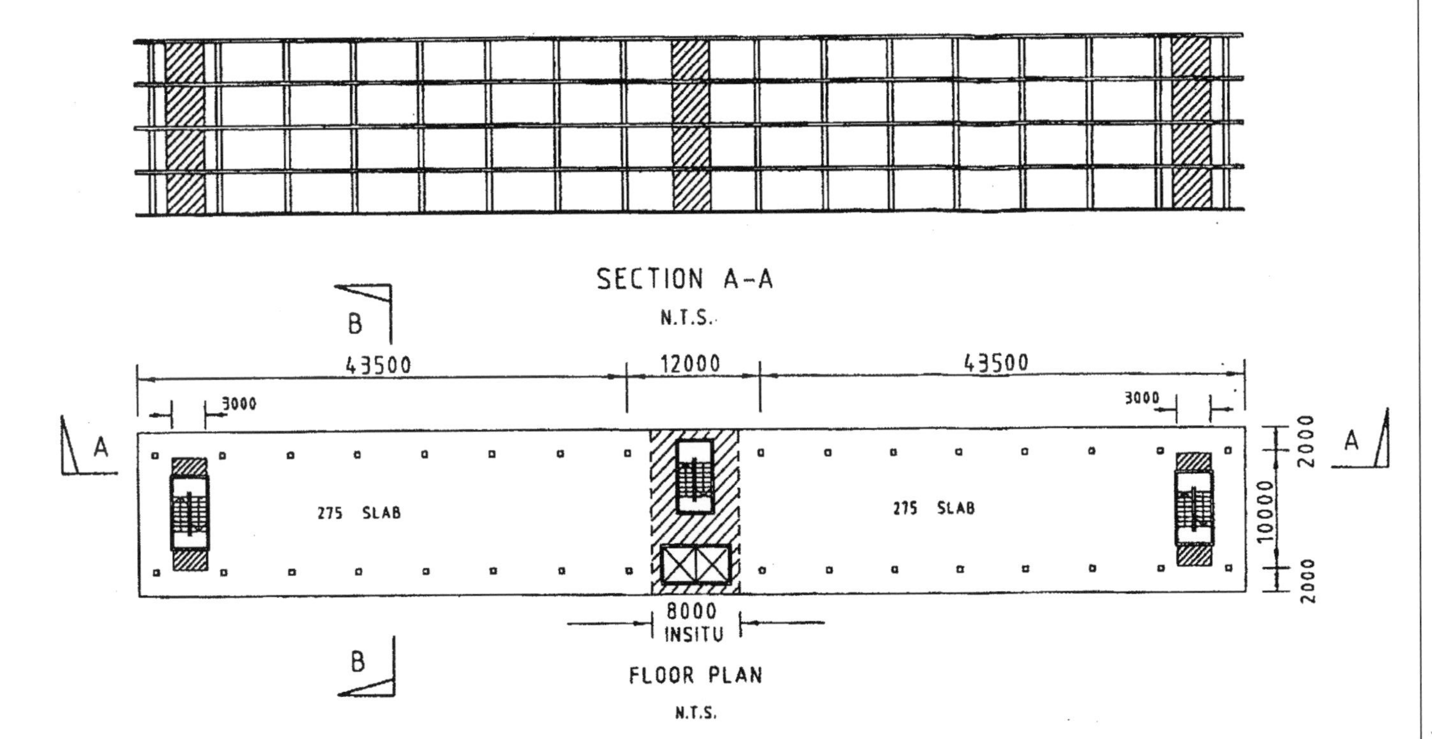

Fig. 4.9. Corsham Street development, option 1, lift slab: floor plan and longitudinal section

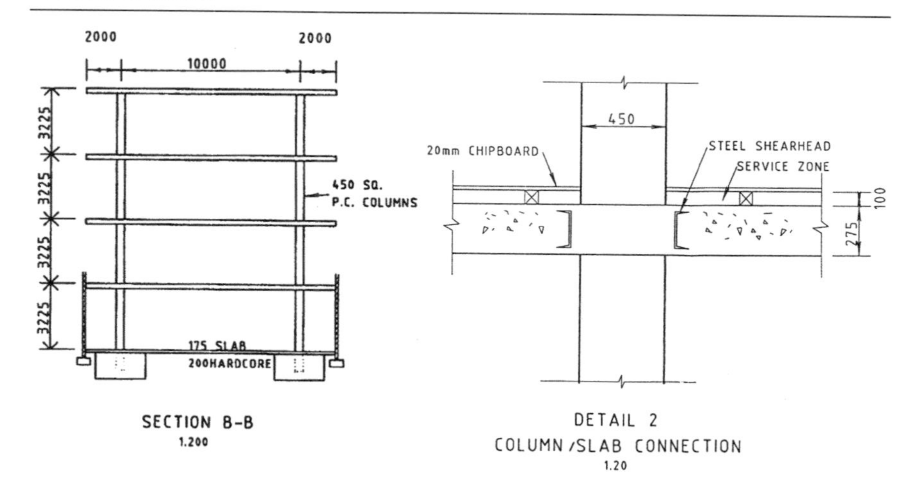

Fig. 4.10. Option 1, lift slab: transverse section and column–slab connection detail

tions. The floor has been divided into two equal halves for ease of lifting. The floors over the central lobby and wing lobby areas are of in situ construction with bar reinforcement, which is then structurally tied to the post-tensioned slab.

The liftwell and stairwell enclosures are formed by loadbearing masonry walls, with precast lintel beams to frame over doorways and access ways. Foundations for core walls are strip footings. For the precast concrete columns, mass foundations have been designed, using a 20 MPa concrete.

Precast column construction uses a 60 MPa concrete mix, and slabs are a 40 MPa lightweight concrete mix. The stability of the building in the longitudinal direction is provided by frame action between the precast columns and the floor slab. In the transverse direction, stability is provided by the masonry wall panels in the central and wing lobby areas.

Building services

Heating is provided by a low-pressure hot water system, with low-level radiator units placed by the windows on the long side of the building. In summer a natural cross-ventilation system could be used, assisted by ceiling-mounted fan units. Horizontal distribution of services is immediately below slab level. Penetration of the slab for pipes is allowed, but away from column positions. Vertical distribution takes place adjacent to the stair or lift cores. A computer floor with a clear depth of 100 mm allows for distribution of telecom wiring.

Electrical socket outlets are floor mounted. The computer floor consists of lightweight concrete blocks supporting 38 mm chipboard. High-level natural lighting is supplemented by fluorescent strips, with diffusers at 2 m centres.

Construction sequence (Figs 4.11 and 4.12)

Mass concrete bases with pockets to receive precast concrete columns are cast first. Then the precast columns are erected and plumbed into position. After grouting and completion of infill work, the ground-floor slab is then cast and the surface floated to receive the lift-slab floors. The lift slabs are poured, one on top of the other, with a separating membrane between them. Post-tensioning is applied in two stages: the first applied with 50% of the final stress when the concrete achieves a strength of 15 MPa, in order to control cracking from early-age thermal contraction of the concrete. Final-stage stressing is applied once the concrete has achieved the specified 28-day strength of 40 MPa.

Once all the slabs are cast, they are lifted hydraulically into position off the columns and parked at the designated level before the shear-heads are grouted. Stability in this temporary condition is provided by the cantilever action of the precast columns. The masonry walls to the lift shaft, stairwells and the central and wing areas are then built. The infills to the floors are then cast and tied to the lift-slab floors to complete the superstructure.

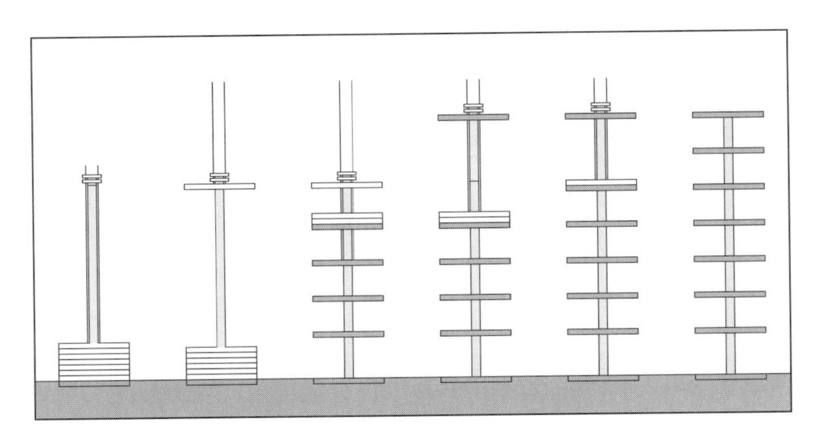

Fig. 4.11. Lift slab construction sequence

(a)

(b)

Fig. 4.12. (a) Typical lift slab column layout showing shearheads. (b) Hydraulically lifting a floor slab to the upper floor level

Programme

The overall time for completion of the foundations, ground slab, frame and core areas, including site set-up, is 12 weeks. Generally this will require 20 men on site, with one JCB 3C for excavations, two mobile cranes, a Bomag 90 and a mobile concrete pump.

Cost

Total cost for the structure is £706 000; this includes preliminaries, insurance, contingency and profit.

Gross usable floor area = 5540 m^2; price = £127.4/m^2.

Option 2: lift slab (precast composite construction)

General

The structure consists of 450 mm square precast columns at 6 m centres, with in situ/precast composite floors built at ground level and then hydraulically lifted into position, precast rib beams spaced at 3 m centres, spanning 11 m between the supporting edge beams (Fig. 4.13). There is a cantilever of 1.5 m, giving an overall width of 14 m to the building. In the longitudinal direction the building measures 96 m.

Structural design

The floor is designed for a superimposed dead load of 1 kN/m^2 and a live load of 4 kN/m^2, with an additional allowance of 1 kN/m^2 for partition loads. The floor has been divided into two equal halves for lifting. The floors on the lobby areas and central section are of in situ construction with bar reinforcement. The lift shaft and stairwell enclosures are formed by loadbearing masonry walls, with precast concrete lintel beams to frame over access ways. The building requires strip foundation bases.

The floor slab consists of precast prestressed beams 535 mm deep which act compositely with a precast and in situ floor slab with an overall depth of 115 mm (Fig. 4.14). The prestressed beams are supported by a 2.0 m wide edge beam which runs along the building perimeter. The precast columns require a 60 MPa concrete, whilst the precast beam and slab elements use 50 MPa concrete. Concrete strength for the 2 m edge beam and the in situ topping is 25 MPa, while the bases and strip foundations require 20 MPa concrete. The stability of the building in the longitudinal direction is provided by frame action between the precast columns and the floor slab. In the transverse direction, stability is provided by the masonry walls of the wing and core areas.

Building services

Heating is to be provided by a low-pressure hot water system with low-level radiator units placed by the windows on the long sides of the building. In summer a natural cross-ventilation system may be used, assisted by ceiling-mounted fan units. Horizontal distribution of services is immediately below slab level. Penetration of the slab for pipes is allowed away from column positions. Vertical distribution takes place adjacent to the stairs or lift cores. A computer floor with a clear depth of 100 mm allows for distribution of telecom wiring. Electrical socket outlets are floor mounted. The computer floor consists of lightweight concrete blocks supporting 38 mm chipboard. High-level natural lighting is supplemented by fluorescent strips, with diffusers at 2 m centres.

Construction sequence

Mass concrete bases with pockets to receive precast columns are poured first. Then follow the strip foundations for the lobby and stairwell cores. Precast columns are then erected, plumbed and grouted in. After the necessary preparatory work for the infill areas and joints, the ground-floor slab is cast with a power float finish to receive the precast floor planks. A polythene membrane for the in situ perimeter beam strips is laid before the reinforcement cage is positioned, and precast rib beams are placed at 3 m centres. Formwork is then placed on the external face and the inside faces before the precast soffit slabs are positioned. Once the reinforcement is finally tied, the 100 mm topping of concrete is cast. The remaining slabs are constructed in similar fashion, separated by a polythene membrane.

On achieving the required minimum strength, the lift slabs are hoisted slowly into position using hydraulic jacks and special Diwidag rods (Fig. 4.15). Stability in this condition is provided by the cantilever action of the precast columns. The slabs are parked in position at the correct level, and then grouted in at the shearheads which are rigidly connected to the columns. During this period the masonry walls for the cores are started and the in situ infill is stitched to the floor slabs.

Programme

The overall time for completion of the foundations, ground slab, lift slabs and core areas, including site set-up, is 12 weeks. Generally, twenty operatives will be required with one JCB for excavations, two 40 t crawler cranes, a Bomag 90 and a mobile concrete pump.

Cost

Total cost for the structure, inclusive of preliminaries, is £718 000.
Gross usable floor area = 5540 m^2; price/m^2 = £129.50.

Option 3: tilt up (precast double-T wall and floor units)

General

The structure uses precast concrete loadbearing wall units, which support 2.4 m wide precast double-T floor units spanning 14 m across the width of the building (Fig. 4.16). The double-T ribs of the vertical wall units face inwards, providing a flush external face that can receive an architectural finish. It is envisaged that 800 mm wide glass panels will run the height of the building between the vertical precast units.

Structural design

The floor is designed for a superimposed dead load of 1 kN/m^2 and a live load of 4 kN/m^2 with an additional allowance of 1 kN/m^2 for partitions. The floor and wall units are standard prestressed units with a nominal width of 2.4 m (Fig. 4.17). Stability of the building in the temporary condition is developed by the rigid connections provided between the floor and wall units and by the masonry panels around the lift and stair core in the permanent condition. All the precast units use a 60 MPa concrete mix, whilst a 25 MPa mix is used for the 100 mm topping. The lift and stair core areas are enclosed by masonry loadbearing walls with precast lintel beams spanning doorways and access openings. The core walls require strip foundation bases, similar to the strip foundation for the external precast units.

Building services

Heating is to be provided by a low-pressure hot water system with low-level radiator units placed by the windows on the long sides of the building. In summer a natural cross-ventilation system may be used, assisted by ceiling-mounted fan units. Horizontal distribution of services is immediately below slab level. Penetration of the slab for pipes is allowed away from column positions. Vertical distribution takes place adjacent to stair or lift cores. A computer floor with a clear depth of 100 mm allows for distribution of telecom wiring. Electrical socket outlets are floor mounted. The computer floor consists of lightweight concrete blocks supported on 38 mm chipboard. High-level natural lighting is supplemented by fluorescent strips with diffusers at 2 m centres.

Construction sequence

Strip foundations are poured first, followed by the ground-floor slab. Each wall unit, weighing 12 t, is lifted into position and then propped. After plumbing of the wall, the double-T floor units are erected, level by

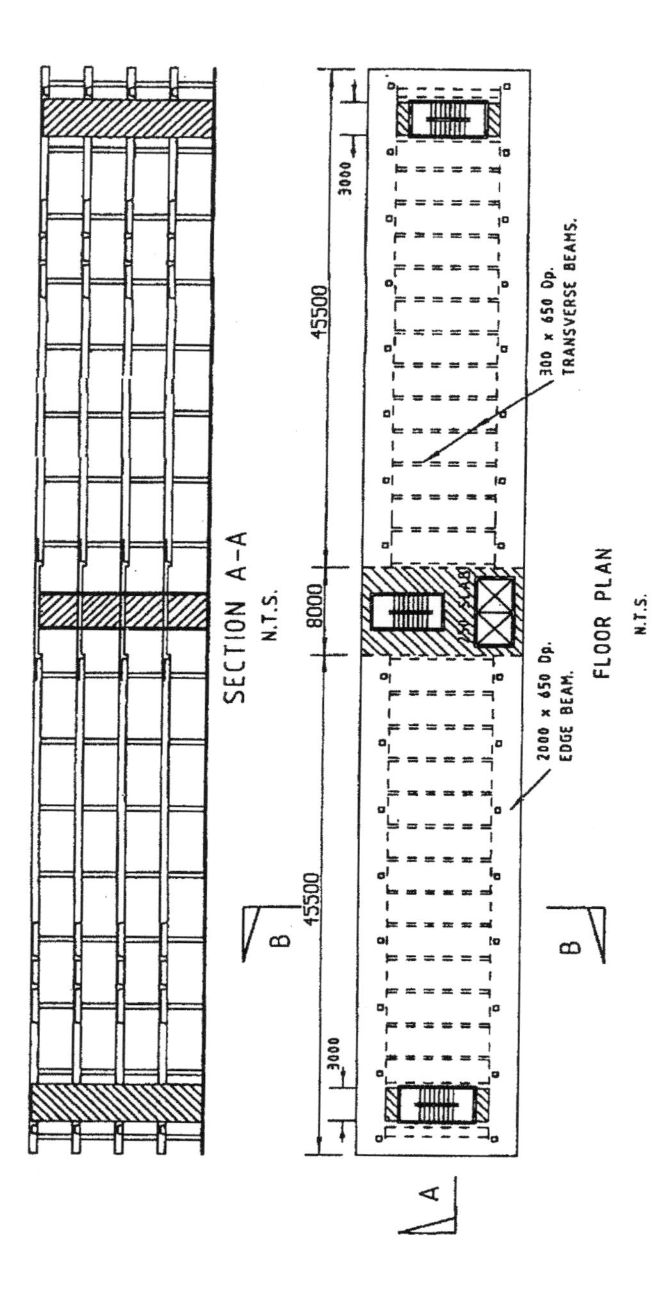

Fig. 4.13. Corsham Street development, option 2, precast composite lift slab: floor plan and longitudinal section

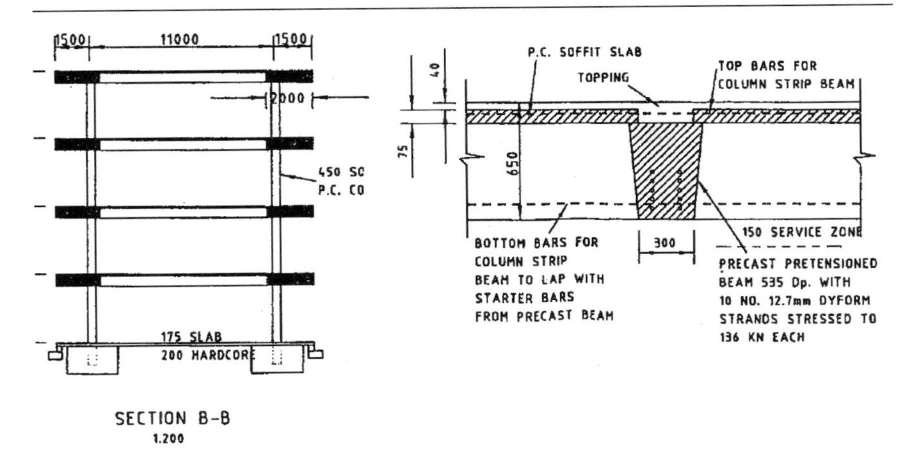

Fig. 4.14. Option 2, precast composite lift slab: transverse section and beam–slab connection detail

level. Stability of floor and wall is provided by site-welding the ends of the floor units to the wall bearings. The topping of 100 mm is then poured to complete the construction.

Programme

For the site erection of the wall units and double-T beams, including foundation and ground slab, construction will be completed in 9 weeks. However, the total time, including design time and precast production time, requires a lead time of about 15 – 20 weeks prior to the start of site work. Two crawler cranes capable of 40 t lifts will be used for precast erection, with two erection crews.

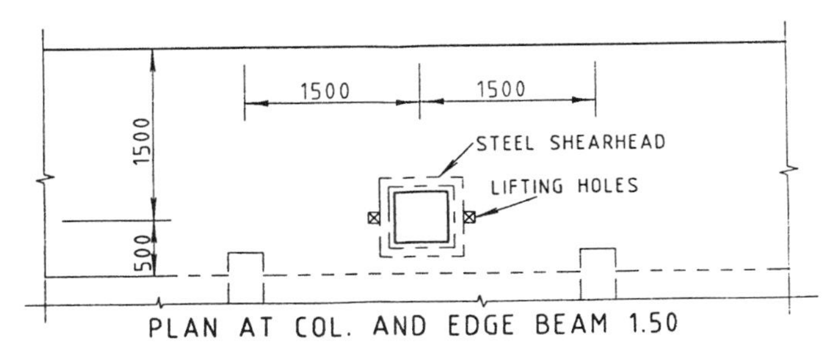

Fig. 4.15. Option 2: plan at column and edge beam

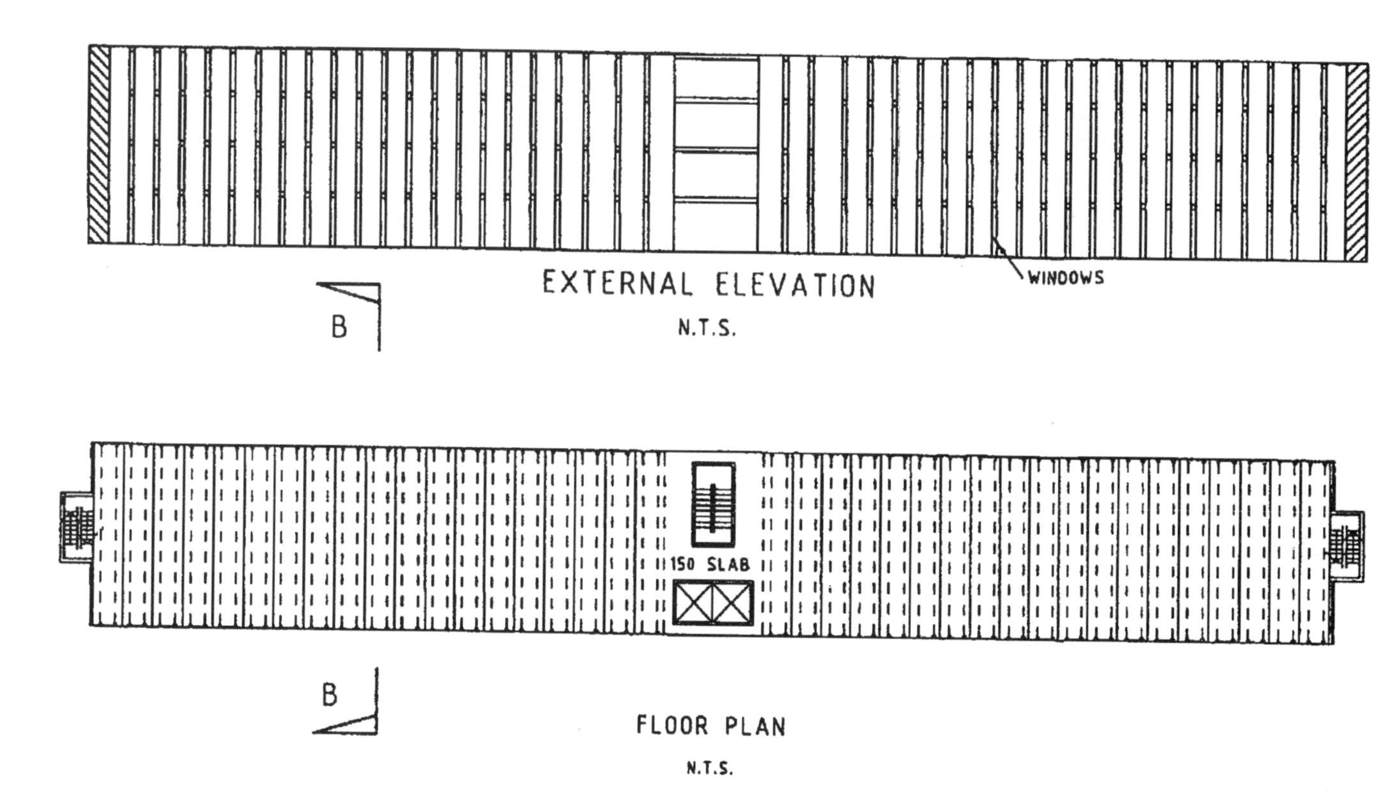

Fig. 4.16. Corsham Street development, option 3, tilt up (with double-T wall and floor units): floor plan and longitudinal section

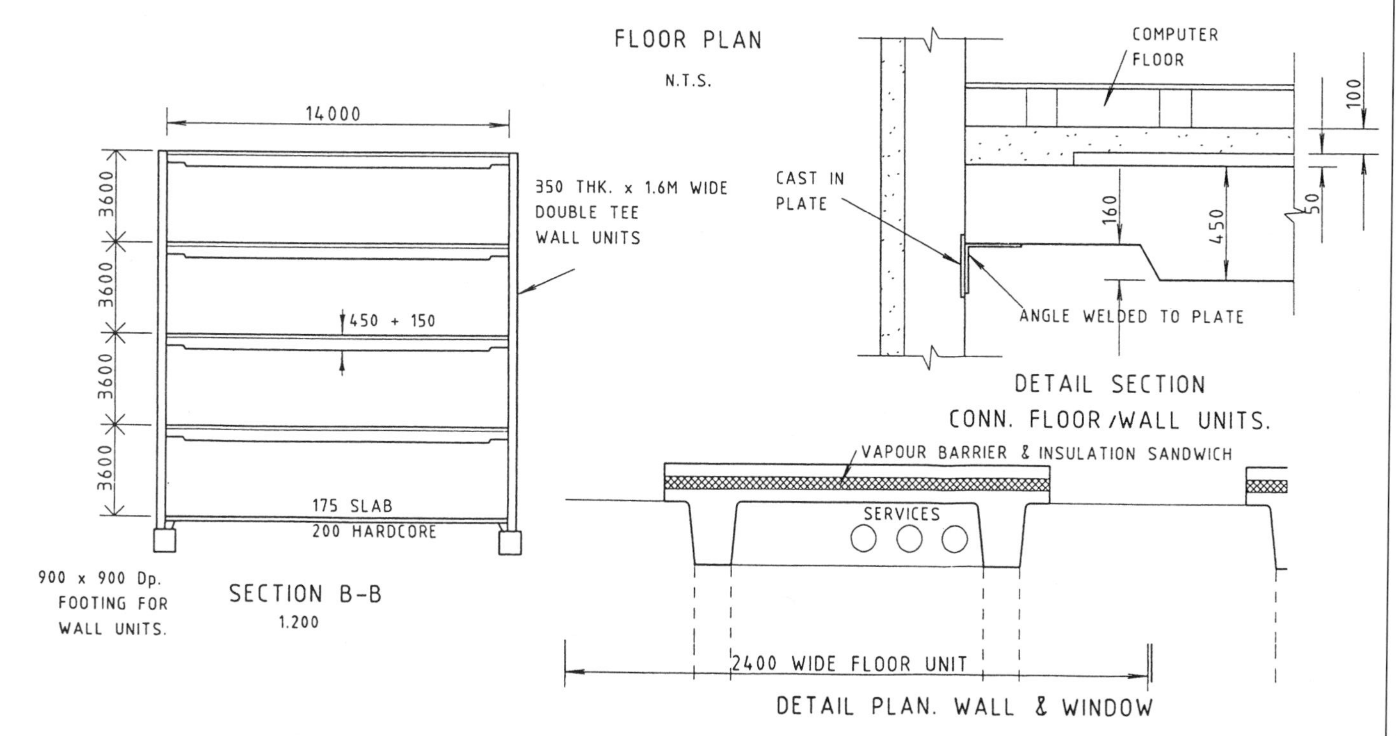

Fig. 4.17. Option 3, tilt up: transverse section and connection details

Cost

Inclusive of precast wall and floor beams, in situ topping, infill areas and bases, but not the masonry wall core areas, the price is £1 185 000.

Gross floor area = 5540 m^2; price/m^2 = £213.90.

Summary

The schemes outlined in this appraisal each have particular advantages which meet the preliminary objectives of least cost. However, it is recognised that at this preliminary stage further design development is necessary to reconcile the building specification and the end use.

As regards the cladding for building options 1 and 2, costs have been estimated for an exterior brickwork skin 100 mm thick having a strength of 15 MPa, and an internal lightweight blockwork skin 150 mm thick with a minimum strength of 3.5 MPa. The budget cost of £140/m^2 (elevation) includes support scaffold and ancillary work. Glazing units will either be full-storey-height panels for option 3, or window sizes for options 1 and 2, both covering about 30% of the elevation area. Double-glazed units in aluminium frames to W/20 specification have been priced. The budget cost is £220/m^2 (elevation).

The rates for glazing and the brickwork/blockwork wall have been factored to show costs/m^2 per unit of floor area. As the total elevation of the building is 2000 m^2 and the gross usable floor area 5540 m^2, the factored cost for glazing units is £27/m^2, and for brickwork/blockwork it is £36/m^2.

Scheme costs

Costs for each of the three options are summarised in Table 4.2.

Conclusions

It is feasible to build a four-storey office building to the specification described in this report for less than £55/ft^2. However, the least-cost approach on the services and raised floor specification may require review, as the building may not match the high quality required for letting space in some commercial markets.

Nevertheless, with the traditional cost of building construction lying between £75 and £90/ft^2 — including professional fees, service installation etc. — the three prototype building designs, exploiting lift slab and precast composite construction, show considerable potential for cost saving. It would be reasonable to expect the built cost of these prototype schemes, inclusive of professional fees, services, etc., to be between £50 and £75/ft^2.

Table 4.2. Corsham Street development: scheme costs

Option 1: lift slab (flat slab in situ construction)	£/m^2
(a) Structure cost	127.4
Allowance for	
(b) Brickwork and cladding	36.0
(c) Glazing	27.0
(d) M & E services	100.0
(e) Raised floor	25.0
Contingency 5%	16.6
Estimated budget price	332.0 (£/ft^2 = 30.6)

Option 2: lift slab (precast composite construction)	£/m^2
(a) Structure cost (inclusive)	129.5
Allowance for	
(b) Brickwork and cladding	36.0
(c) Glazing	27.0
(d) M & E services	100.0
(e) Raised floor	25.0
Contingency 5%	16.5
Estimated budget price	334.0 (£/ft^2 = 30.8)

Option 3: Tilt-up (precast wall and floor construction)	£/m^2
(a) Structure and cladding cost	214.0
Allowance for	
(b) Glazing	27.0
(c) M & E services	100.0
(d) Raised floor	25.0
Contingency 8%	30.0
Estimated budget price	396.0 (£/ft^2 = 36.6)

88 Wood Street, London

Richard Rogers' three interlinked glass towers (Figs. 4.18 and 4.19), rising 10, 14 and 18 floors and set back from the main street, were designed as a speculative office building for Daiwa Securities. The

Fig. 4.18. 88 Wood Street (courtesy: Richard Rogers Partnership)

building provides 22 300 m^2 of net usable floor space in a comfortable, spacious and efficient working environment for a staff of 2000 people. According to the City of London's Planning Department, it is easily the most exciting and imaginative building design seen in London for some

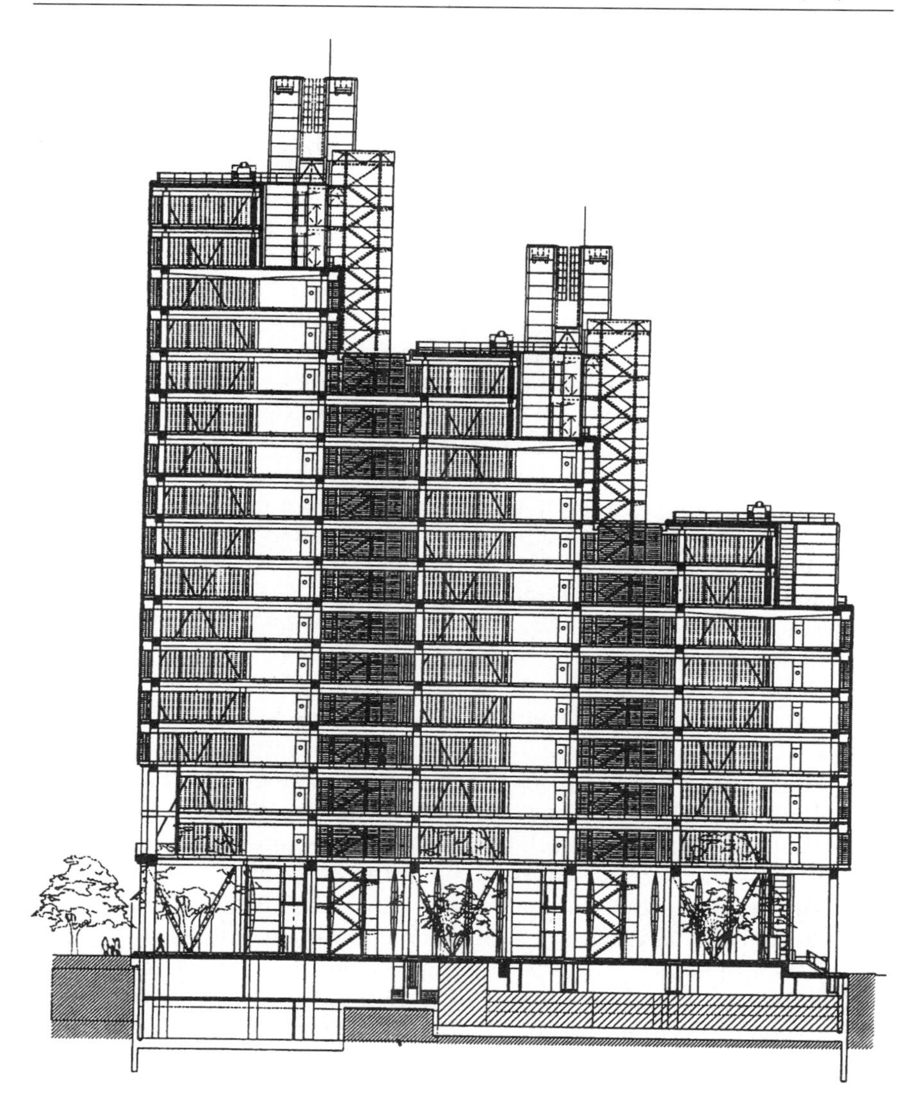

Fig. 4.19. 88 Wood Street: front elevation

while. Jonathan Glancy, architecture critic of the *Guardian*, thinks the building might be the finest speculative office block yet built in Britain.

It is full of high-strength concrete in the columns, prestressed post-tensioned concrete in the slender wide-span floors, and exposed concrete in the core walls and the floor soffit over the entrance lobby. Yet such an innovative structure came within a hair's breath of being cancelled because of foundation problems. Ironically, the substrata soil was stable and well defined and presented no problems; it was the warren of subterranean tunnels and a large chamber below the ground that forced the foundations to perform structural gymnastics. A Royal Mail tunnel, two BT tunnels and a large BT bunker obstructed the location of piles from their optimum support positions (Fig. 4.20). It was a matter of sinking a pile where one could find the space. The budget only allowed for a conventional under-reamed piling scheme within the London clay, not for deep straight piles into the Thanet sands, nor for super-enlarged under-reamed piles which would now be necessary. In one corner of the site where an under-reamed pile could be augured, it had to be 5.4 m in diameter — twice the diameter of a conventional under-ream — and the biggest ever formed in the City of London. The sheer size of the BT bunker below ground level, measuring 20 m wide and 40 m long, forced the building superstructure to be supported above this 'live exchange' by six huge steel transfer beams, one of which was 19.5 m long and weighed 82 t.

Each of the tower structures is transversely stiffened by externally cross-braced, concrete-encased, steel H frames. A robust concrete building frame was chosen for added stability, after the client had asked the designers to squeeze a further 30% out of the site by increasing the floor area. Post-tensioned ribbed floors were preferred, enabling an additional floor to be added within the planning height by saving as much as 300 mm per floor over competing alternatives (Fig. 4.21). Post-tensioning also minimises the dead load carried to the foundations as well as increasing the floor span. In situ construction was adopted as this provides inherent stability from frame action in the short-span direction of the 6 m × 15 m column grid. A span of 15 m was now possible, reducing the number of columns and allowing the floor to cantilever 1.5 m beyond the perimeter column line to capture additional floor space (Fig. 4.22). Although a composite steel floor would bridge the same span length and fit within the required storey height, the problem was the constraint of distributing the services through notches in the steel beams. The minimum 600 mm floor depth of the post-tensioned rib floor allows services to pass cleanly underneath with the maximum of flexibility.

As the edge columns were now within the floor plate because of the cantilever, it was impractical to cross-brace between them. An external stiffening structure was introduced on the façade and the original tubular steel section replaced with slender Macalloy bar above the ground-floor lobby, to enhance the visual appearance. The cores and stairwells are stabilised by the building frame. The two cores of the 18- and 14-storey blocks are entirely glazed and carefully detailed as the elevations were considered to be of high architectural importance. The stair flights

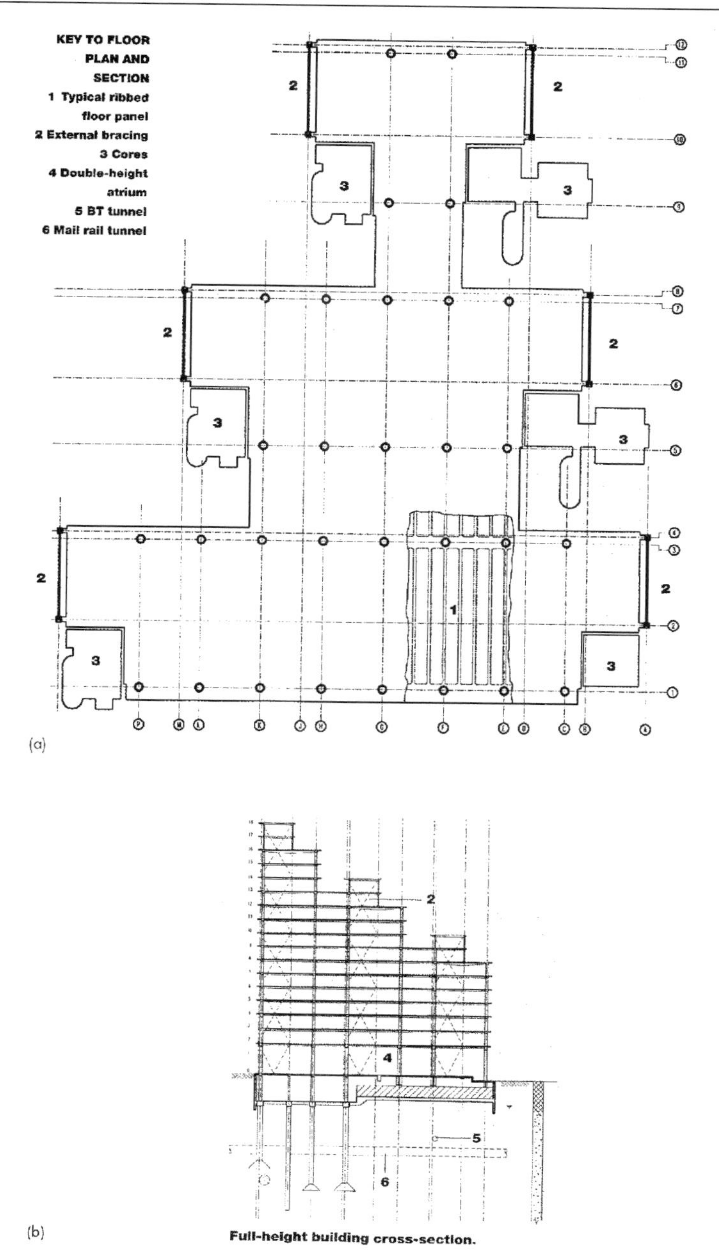

Fig. 4.20. 88 Wood Street: (a) typical floor plan; (b) building cross-section, showing piling—figures 5 and 6 on the diagram show the locations of the BT and Royal Mail tunnels respectively

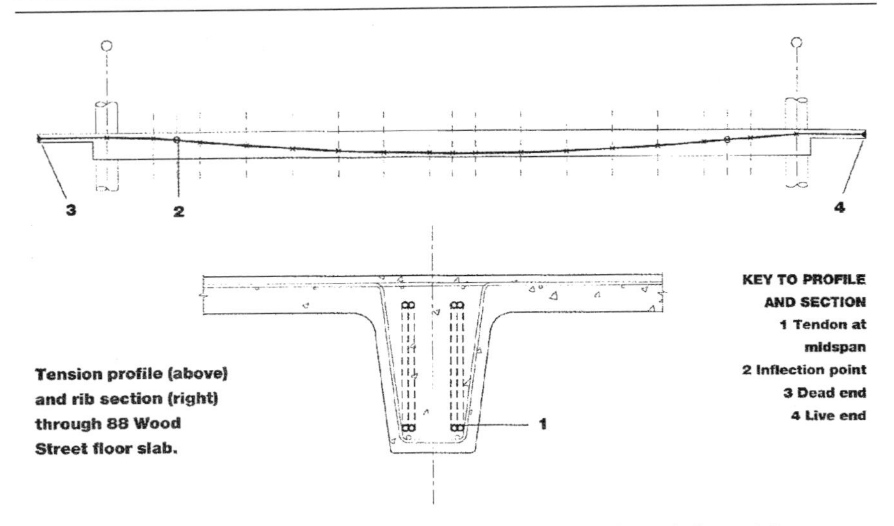

Fig. 4.21. 88 Wood Street: tension profile and rib section through floor slab

were pre-tensioned to stabilise the supporting columns, and pre-tensioned rods were used as glazing mullions.

Two free-standing tower cranes, one 95 m under the hook and the tallest in the city, and a 31 m high Schwing concrete pump with hydraulic boom, serviced the construction of the floor slabs, which were up to 2100 m^2 in plan area. The floor pours were restricted to 240 m^2 because of working restrictions imposed on the contractor due to the proximity of the Barbican residential buildings. It is probable that the prestressing subcontractor CCL could only lay that area of tendons — some of which were 70 m long — in a working period. In all, fifty-two 18 m-long crane-handled table forms manufactured by SGB were used to construct the floor slabs. Tendons were initially stressed, 36 h after casting, to 50% of their full design force when the concrete strength had reached 20 MPa. This is sufficient to carry the dead load and allow table forms to be moved after 48 h, completing the larger floor plates on a 2-week cycle. The final stressing took place some weeks later, after the concrete had reached 40 MPa and before the cladding and services were installed.

There were doubts about the suitability of pumping C80 microsilica concrete to the top of the 18th floor for the columns. In the end a C60 concrete without microsilica was substituted and the columns were additionally reinforced to retain their orginal diameter. Pioneer Willment produced a pumpable concrete with a Portland cement content of 500 kg/m^3, using 20 – 5 mm graded limestone coarse aggregate from Shepton Mallet and marine sand from Greenwich (Fig. 4.23). A

Fig. 4.22. Thin cantilevered floors are a feature of 88 Wood Street (courtesy: New Civil Engineer*)*

Fosroc water-reducing admixture was added to keep the water/cement ratio below 0.40 and to provide a workability of 125 mm slump. However, it was difficult to compact such a 'sticky mix' to a high standard of finish with zero making good, especially for the 9.8 m double-storey-high columns in the entrance lobby. After trying conventional pneumatic internal poker vibrators, an Ishoko high-frequency poker vibrator manufactured in Japan was used to produce a blemish-free surface finish. The high-frequency vibration of the Ishoko remains constant when the poker is immersed in concrete. Many pneumatically driven internal vibrators lose power and frequency of vibration when they are immersed in cohesive high-

Fig. 4.23. 88 Wood Street: concreting an upper-storey floor slab (courtesy: New Civil Engineer*)*

performance concrete. From 14 000 cycles per minute they can slow down to 2000 cycles per minute, which is very ineffective and will not compact the concrete.

The circular metal column formwork, manufactured by Form Fab, was air blasted with soft grit to remove the shiny surface of the steel shutters and give a uniform matt concrete finish on striking. All seam welds were ground smooth so that no unwanted joint lines would show. Site trials were conducted to produce the same finish, using GRP moulds supplied by Zeta Form, for the ribbed floor slab. The moulds were shot blasted with tiny pieces of sponge and, with careful selection of the release agent, a matt concrete surface finish was achieved. Columns were poured carefully, using a skip and tremie, to minimise entrapment of air pockets and to allow sufficient time for air pockets to escape during internal compaction (Fig. 4.24). Alignment of the columns proved a bit tricky with the floor being post-tensioned. Stressing the tendons over such long spans produces elastic shortening of the floor slab. The perimeter columns had to lean 10 mm out of plumb at the top of each pour to compensate for this.

Fig. 4.24. 88 Wood Street: atrium columns (courtesy: Richard Rogers Partnership)

Despite the care taken by the design team to bring the cost of construction down to £48 million, the project eventually went over budget due to the late decision by the client to install the power plant in the basement rather than on the upper floor as was originally intended. The contractor had to completely retrofit the basement, which was already packed with service plant, and break through the completed floor slab in the basement to build two new service tunnels under the existing BT exchange. The tunnels were required for running the feed pipes to and from the cooling tower on the roof to the generator plant now located in the basement. It is quite feasible to cut openings and make alterations to post-tensioned concrete floors without great difficulty and no more onerous or time consuming than with reinforced concrete floors. Procedures for de-tensioning, propping and restressing cables have to be followed and the work carried out by prestressing contractors.

The facades of the main office floors are glazed from floor to ceiling, to maximise on the daylight and the views. The long east–west façades facing the main road are clad in the largest double-glazed units in the world. Expensive diamond glass was specified to give maximum transparency. Panoramic lift towers link the office floors at levels 8, 12 and 16, leading directly onto the roof terraces with spectacular views of the city skyline.

Visitors arrive at the main entrance in Wood Street and enter the spacious 8 m high reception area with its stone-covered floor and 54 m long concrete feature wall (Fig. 4.25). The reception area runs the full length of the building and gives an east–west view of St Olave's Gardens. The glazed north wall overlooks a landscaped garden that extends into the building, creating an impressive meeting place.

88 Wood Street: project details

Original client: Daiwa Europe Property
Project manager: Kajima UK Engineering
Architect: Richard Rogers Partnership
Structural and services engineer: Ove Arup & Partners
Management contractor: Laing Management
Concrete frame contractor: John Doyle Construction
Piling: Bachy-Soletanche
Services contractor: Rosser & Russell
Start on site: Autumn 1994
Practical completion: August 1999

Fig. 4.25. 88 Wood Street: reception area (courtesy: Richard Rogers Partnership)

A new concept in air rights building construction

(Author collaboration with Andrew Carruthers (Maunsell McIntyre, Australia) and David Ramsay (White Young Green))

Why air rights?

Air rights developments are concerned with the air space above an existing structure, which is usually owned by a third party. They are particularly attractive to modern developers, as they can unlock development value from what is perceived as dead unusable space. Classic examples are buildings erected over railways and roads.

The unique characteristic of an air rights development is that it is a construction above, around or below an existing structure or undertaking (Fig. 4.26). Moreover, the use of the existing structure must be maintained during the construction period, with often a third-party landowner with whom an agreement has also to be reached.

Building air rights in city centres is no easy task because of site confinement, the proximity of nearby buildings and, not least, the added complexity of long-span construction. However, air rights is not a recent innovation. It goes back at least to the Middle Ages with an obvious example, London Bridge.

An air rights building is primarily a building supported by a bridge. The bridge closes the gap over the road, rail, water or other obstacle, to allow a modern office block to sit in the air space above it. Naturally, air rights designs will draw inspiration from bridge engineering.

Engineers have been building bridges since Roman times, using a wide variety of materials and construction techniques. Bridge technology has advanced from timber and masonry to cast iron, wrought iron, and to steel and concrete, both reinforced and prestressed construction (Fig. 4.27). The first known bridge to be built using the 'balanced cantilever' approach was in Japan in the fourth century. Considerable expertise exists in launching structures to bridge short and long spans

Fig. 4.26. Ponte Vecchio, Florence — An ancient 'air rights' bridge

Fig. 4.27. The Bosphorus bridge under construction

without disruption to the flow of traffic below. Logically, any air rights designer will take his lead from the wealth of bridge building technology at his disposal.

Bridge building and air rights

From a bridge engineer's perspective it seems odd that the majority of 'air rights' buildings put up in the past 50 years have largely ignored concrete. In fact, most of them, with one exception, have used structural steel. Here are a few eye-catching examples. Note the obvious adoption of bridge technology in all of them.

Federal Reserve Bank, Minneapolis

Although there are several office blocks built with floors suspended from the top of the building, the Federal Bank must be unique by being suspended 83 m. The ten floors of the bank are supported by a post-tensioned catenary system, stretched between concrete cores at the ends (Figs. 4.28 and 4.29). The horizontal thrust of the catenary is resisted by deep steel trusses positioned at roof level. The end cores are designed as vertical cantilevers, stiffened by the catenary suspension. Much the same principles are used in the construction of suspension bridges — for example, the Humber Bridge.

Fig. 4.28. Federal Reserve Bank, Minneapolis, under construction

Charing Cross, London

Anyone in London attending an evening concert on the South Bank will see the beautifully lit façade and arched canopy of Charing Cross Station, rising out of the Thames like a giant 'art deco' juke box (Fig. 4.30). The roof-level steel arches span 36 m and carry ten floors of office

Fig. 4.29. Federal Reserve Bank, Minneapolis — the completed building

(a)

(b)

Fig. 4.30. Charing Cross Station: (a) the steel arch canopies; (b) steel lattice of the glass curtain walling (courtesy: Ove Arup & Partners)

accommodation below them. The arch transfers the floor loads to the caisson foundation located outside the station footprint and away from underground tunnels. Although the arch feature is visible, the design intent is obscured by the curtain wall detailing. This form of construction has been adapted from conventional bridge engineering and has been used for a number of air rights buildings, e.g. Broadgate Phase 11.

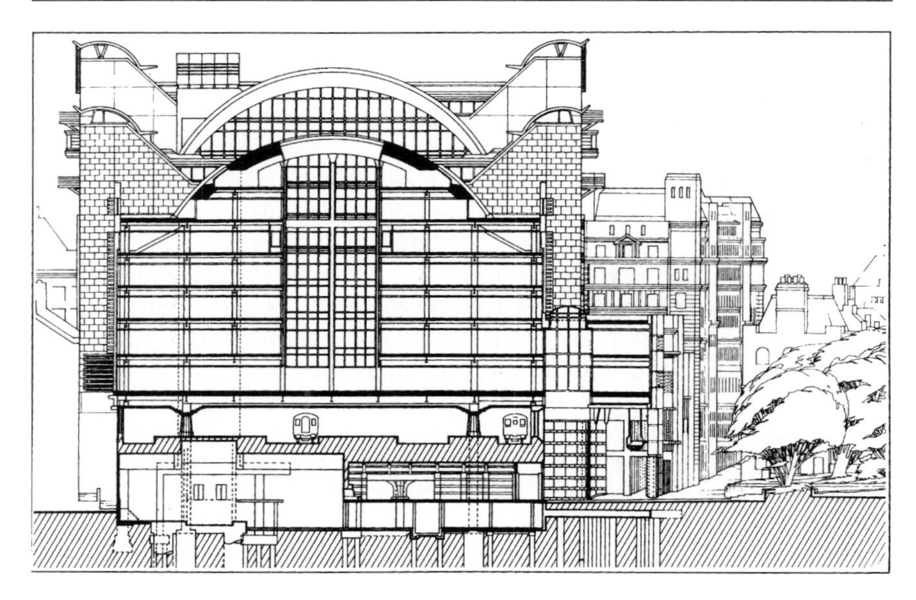

Fig. 4.30 (continued). (c) Typical section (courtesy: Ove Arup & Partners)

The Galleria, Hatfield, London

For those motorists travelling north on the A1, the monotony of carriageway and kerb is pleasantly disrupted when approaching the Hatfield Tunnel. The gleaming 'Galleria Shopping Centre' offers momentary respite, with hints of Hollywood and Universal Studios in Hertfordshire, as the foot eases off the accelerator pedal. The curved roof canopy, spanning the full width of the tunnel roadway and more, is a cable-stayed structure. The roof canopy is hung from cables attached to the lattice framework of the roof and stayed by the steel masts at the end of the structure. The steel masts transfer the cable forces to the foundation.

The principles are exactly the same as in cable-stayed bridge construction, a recent example of which is the Dartford Crossing.

La Grande Arche, Paris

It is notable that none of the buildings so far highlighted have exploited the versatility of concrete in bridge engineering. The exception is La Grande Arche, but then this is an exceptional building. Whilst the concept could not possibly work economically for air rights development in any city centres, the principles nevertheless are taken from basic bridge building technology.

La Grande Arche, nicknamed 'The Cube', is a vast hollow cube 110 m long on each of its faces, with a central void measuring 70 m in each direction (Fig. 4.31). It is a post-tensioned concrete structure with not one structural or expansion joint to be seen.

The primary elements of the Cube comprise four vast square frames, each a complete ring of prestressed concrete rising vertically and running horizontally through the structure. To stiffen these frames, or 'megas' as they are called, there are four thickened structural floors spaced equally on each of the vertical legs, and four girder beams to bridge the top and bottom of the horizontal 'deck' sections spanning the 70 m void. The Cube is also a twin 36-storey office block linked by concrete bridges, top and bottom, to allow the structure to straddle the metro, roadway and mainline rail tunnels that pass beneath it.

Why are there so few examples of air rights developments using concrete? Why have air rights developers ignored the vast wealth of expertise in concrete bridge technology? Here is a solution taken from bridge construction technology.

Concrete — a logical choice for air rights

An air rights concept in concrete is a logical extension of bridge construction engineering. For this conceptual design exercise, Broadgate

Fig. 4.31. La Grande Arche, Paris

Phase Eleven was chosen as the reference. This was done in order to highlight any real advantages and economic benefits to be gained from a concrete alternative. The air rights building spans 78 m over rail lines and accommodates a ten-storey office block with a gross floor area of 43 000 m^2.

The designers wish to make it clear that there is no criticism of the original steel design (Fig. 4.32), which at the time was appropriate for that particular building.

Conventional steel arch construction requires considerable scaffolding and temporary support during erection of the arch. The steel arches were fabricated in 6 m sections and thus were required to be fully supported during erection. In addition, the whole area below the construction had to be protected with a crash deck to eliminate risk of injury or damage to the workings of the main line station below.

The development of the arch construction in concrete has led to obvious differences in approach, with potential for major cost and time savings. The special feature of the concrete arch solution is the total elimination of support scaffold during construction, and positioning of the arch sections over the rails (Fig. 4.33). By 'tilting in' the arch segments, it means that no cranage or heavy lifting devices are required.

The arches have been designed with a parabolic taper, in 80 MPa concrete, and constructed using climbing forms. Cable stays are fixed to the arches at the two-thirds point and tied back to tension piles via the suspension arms. The tension pile is required to counter the out-of-balance forces on the pier support during the tilting operation. The stays also provide lateral stability to the structure and control its movement during tilting.

Arch tilting will be carefully synchronised by the use of hydraulic jacks located at the pier points. The jacks initially displace the arch out of equilibrium; the self-weight will then take over, to tilt the arch into position. The cable stays, via the suspension arm, will stabilise and control progress of the tilting operations. Arches will be tilted in pairs, to phase the construction, allowing an early start to the building element. The stays remain in position until the arch boom is constructed. They are removed before floor slab construction begins. The suspension arm can be left in place as a feature or removed, to suit client and third-party wishes.

The floors within the arch are suspended from Macalloy rods in tension (Fig. 4.34). The Macalloy rods or hangers have shearheads fabricated at floor level to carry the post-tensioned primary floor beams. Above the arch, the building construction is a conventional column and floor design (Fig. 4.35). The primary floor beams carry 17.5 m and 14 m long precast beams and a soffit slab floor. An example of arch construction in bridges (Fig. 4.36) demonstrates how the tilting arches of

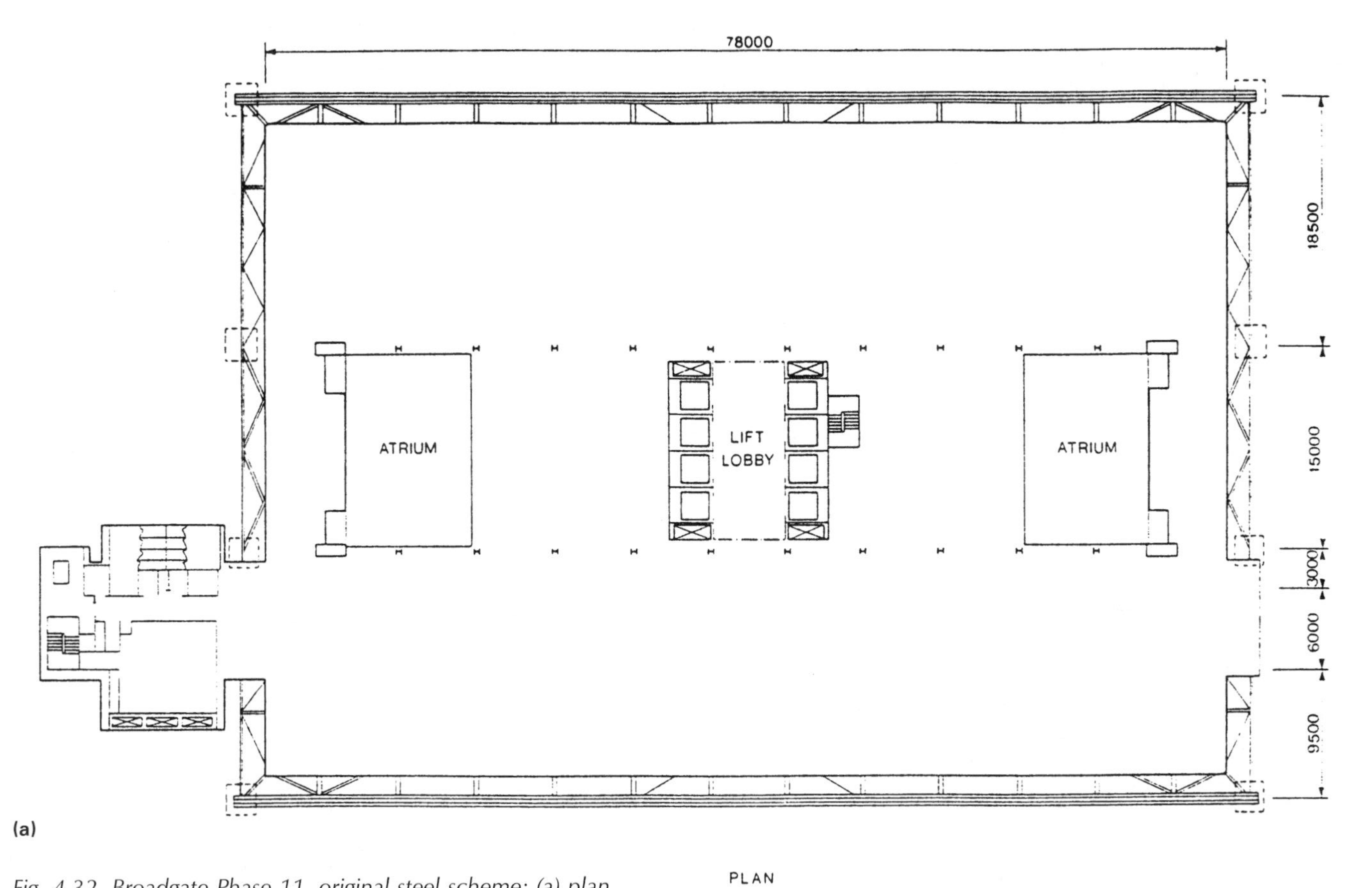

Fig. 4.32. Broadgate Phase 11, original steel scheme: (a) plan

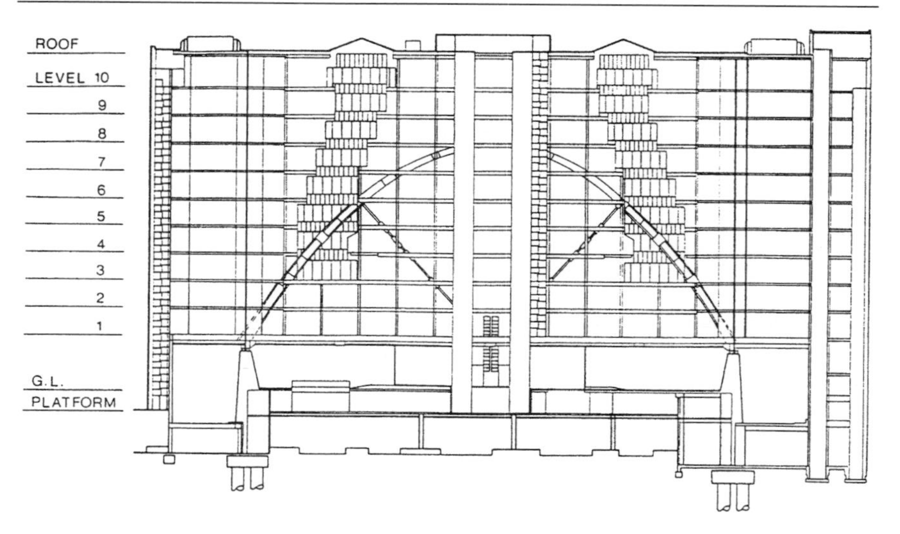

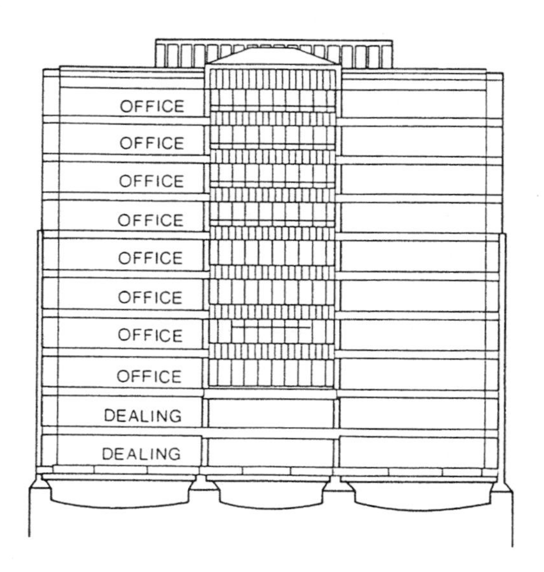

NORTH-SOUTH SECTION

(b)

Fig. 4.32 (continued). (b) Sections

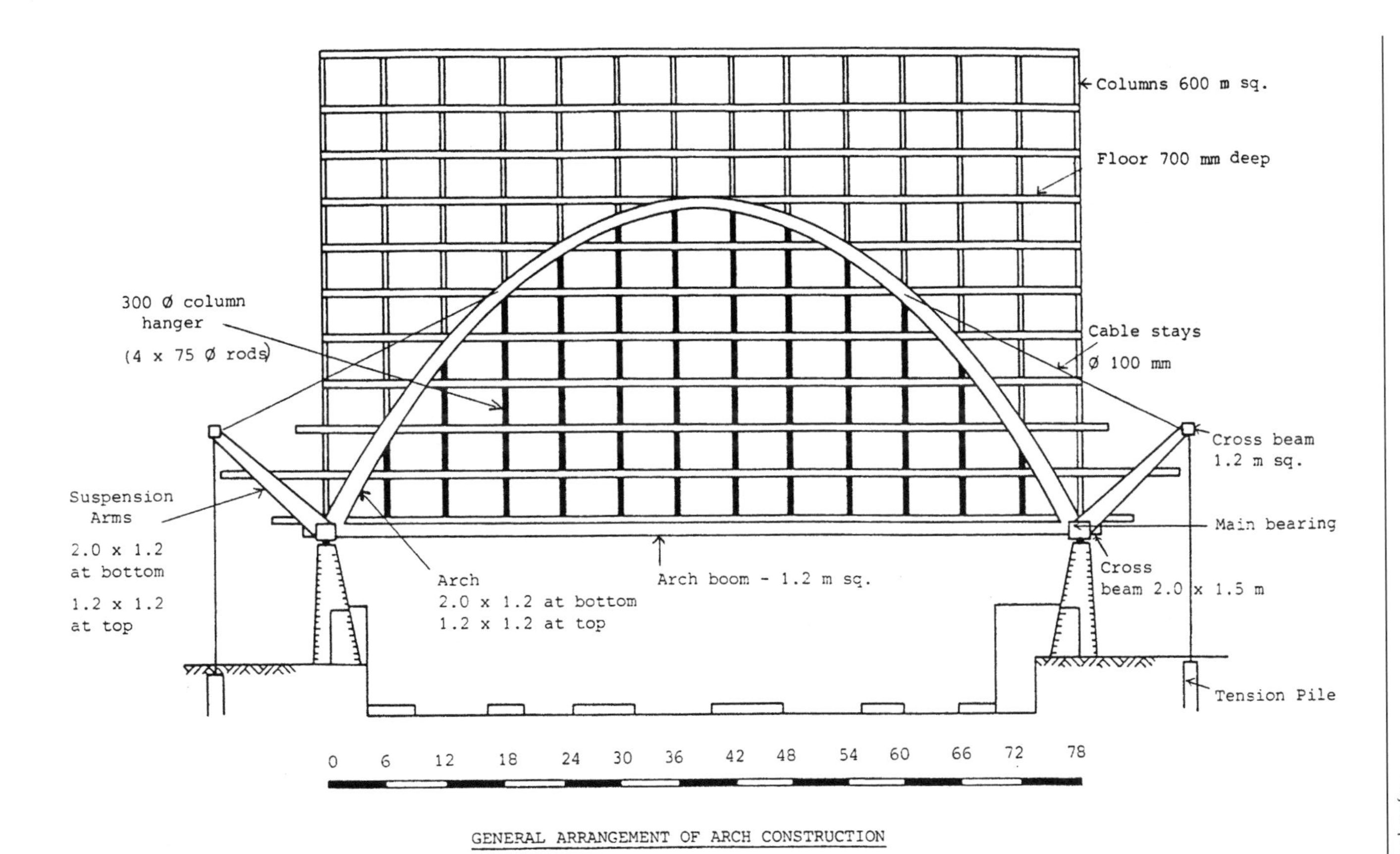

Fig. 4.33. Broadgate Phase 11: concrete arch solution

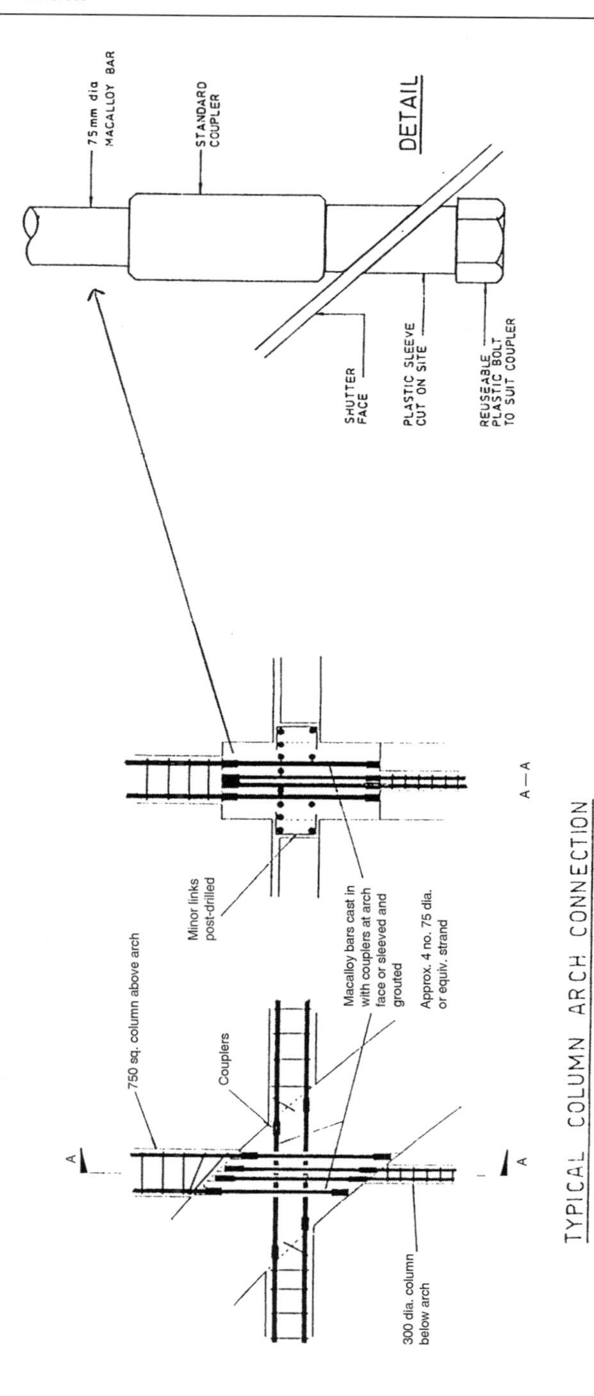

Fig. 4.34. Broadgate Phase 11: typical column–arch connection using Macalloy bars

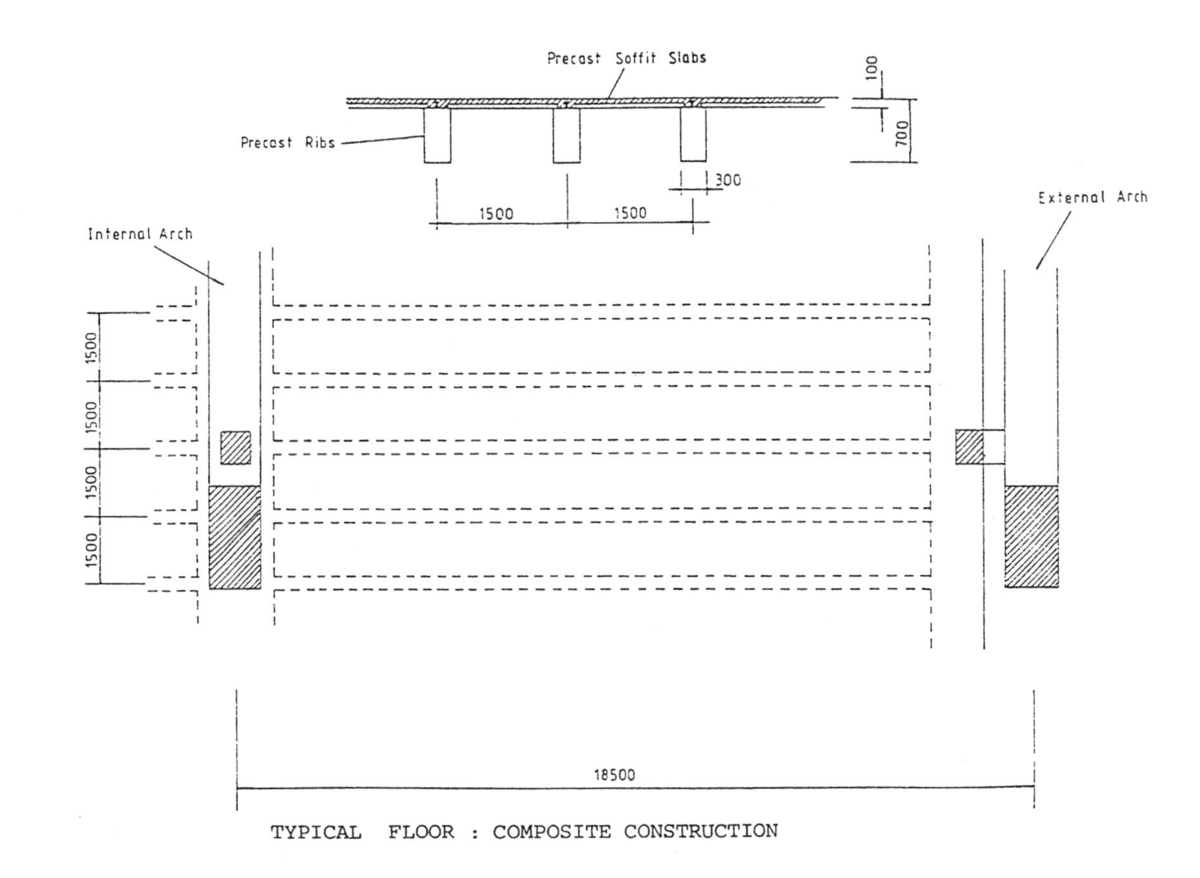

Fig. 4.35. Broadgate Phase 11: (a) typical floor (composite construction)

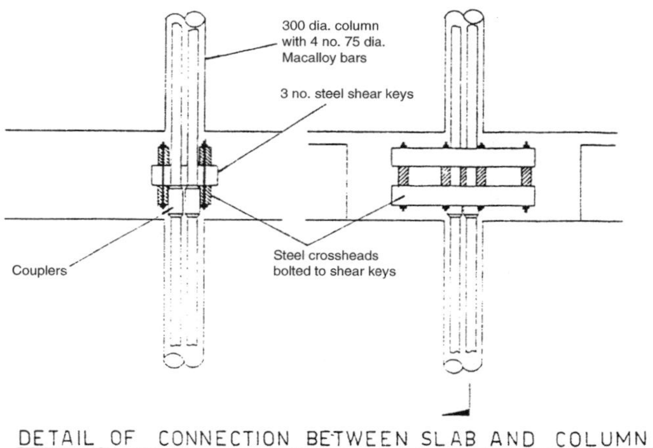

Fig. 4.35 (continued). (b) Slab–column connection detail

Broadgate Phase 11 would be formed. The construction sequence of the air rights building is shown in Fig. 4.37. Figure 4.38 shows computer-generated images of the proposed scheme.

Notes on construction detail

Main bearings. Eight massive bearings will take the entire load of the 10-storey structure. They will be pinned to allow thermal movement of the structure without severe strain on the bearing mountings. The bearings will be mounted on elastomeric anti-vibration pads to minimise noise and vibration transmitted from the railway workings below.

Suspension arm. Work will proceed with the construction of the suspension arm and main beam spanning between the arch bearings.

Arch construction. The arches will be formed by casting a pair of half-sections and then tilting them down into position (Fig. 4.36). The arches will be constructed using a Peri-type SKS climbing form system. The fair-faced concrete will be cast in ten lifts, each 4 m in length. The formwork dimension will be varied by adjusting the width of a tapered internal panel. All reinforcement will be prefabricated and assembled from the platform on the climbing frame.

Macalloy hangers. Hangers will be fixed to the arch at 6 m intervals and suspended down to the boom in 20 m lengths. Each hanger group consists of four 75 mm diameter rods, with shearhead connec-

Fig. 4.36. An example of how climbing formwork is used in the arch construction of bridges

(a)

Fig. 4.37. Broadgate Phase 11 construction sequence: (a) forming the suspension arms and the arches

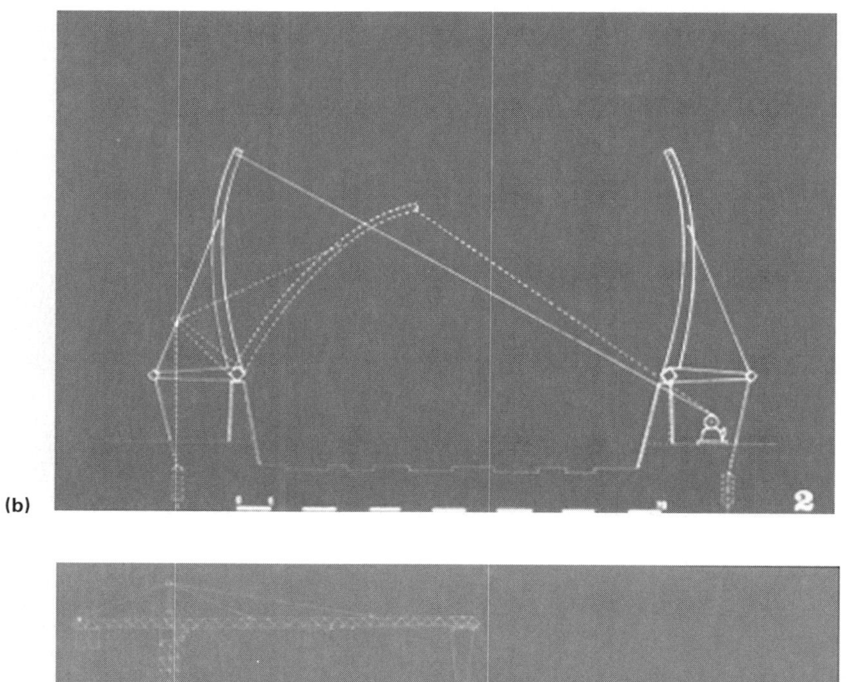

(b)

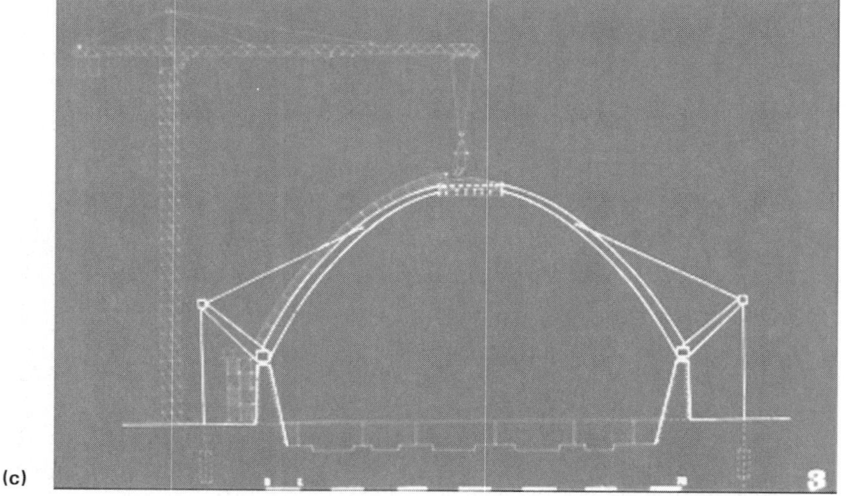

(c)

Fig. 4.37 (continued). (b) Tilting the arches once the stays are anchored in the ground; (c) stitching the arches together—stays remain anchored to ground

tors welded on, to form a structural tie for the primary beam at floor level.

Boom construction. The boom will be cast in 6 m lengths, using a cantilevered formwork jacket. The unbonded post-tensioned strand

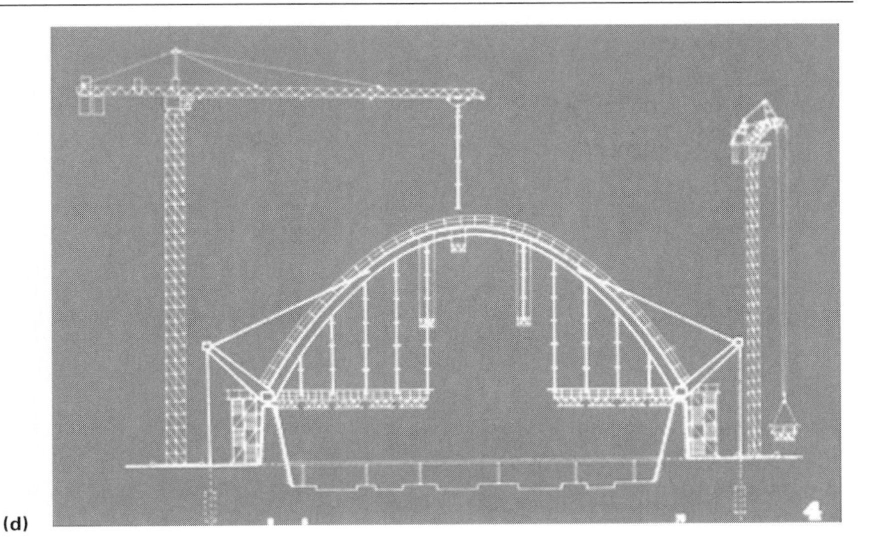

(d)

(e)

Fig. 4.37 (continued). (d) Constructing the boom, which is suspended from Macalloy rods; (e) casting the floors, which are supported by Macalloy rods within the arch and by concrete columns above the arch

will be carried through each beam section, initially without stressing. There is sufficient beam reinforcement to maintain structural integrity under dead load. Once the final 6 m section is cast and the concrete has achieved 40 MPa strength, stress is applied to the strands to half

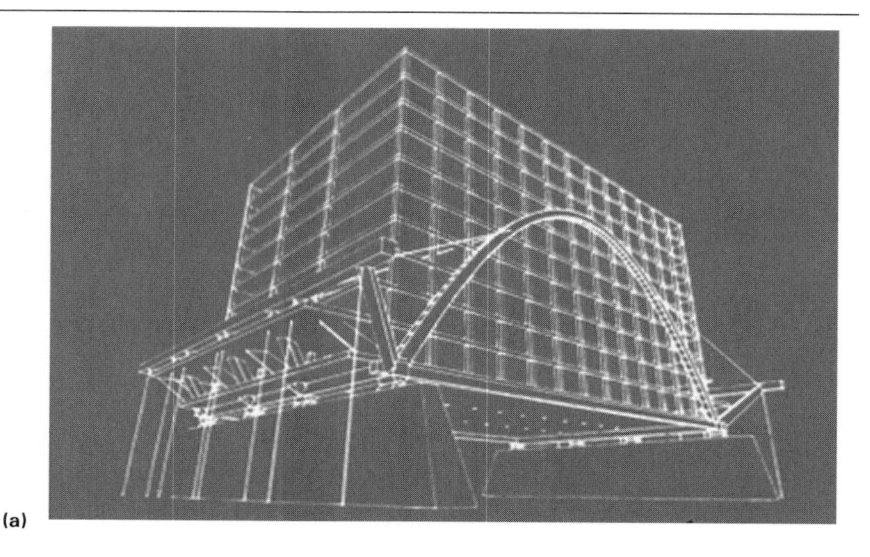
(a)

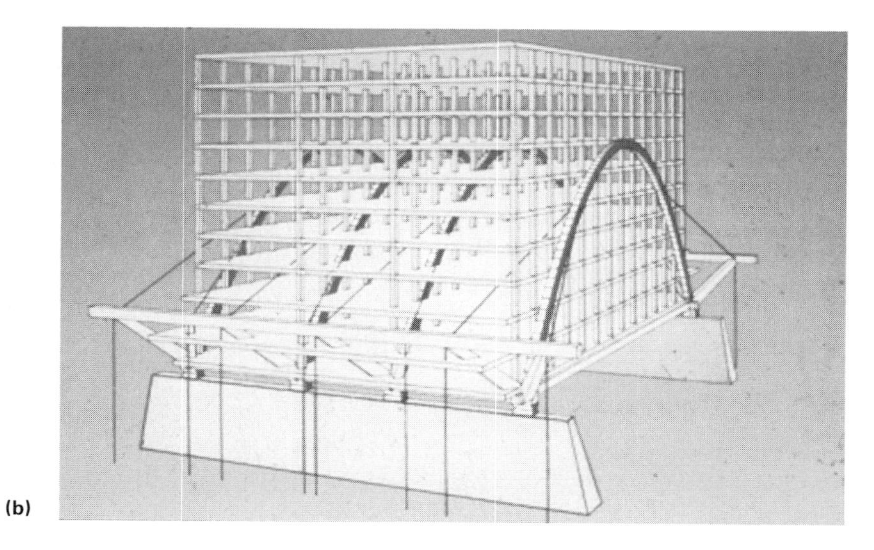
(b)

Fig. 4.38. Computer-generated images of the Broadgate Phase 11 proposal

working load. Floor construction will then begin, with the prestress in the boom increased in stages to match the increase in support load as the building gains in height.

Quantities. In terms of material quantities, the four arches and booms will require 2040 m^3 of 80 MPa concrete, 240 t of rebar, 42 t of prestressing strand, 180 t of Macalloy rods and 220 t of bearing plate, before allowance can be made for the superstructure of the building.

Concrete air rights structures

Whatever the task, bridge engineers have always sought to appraise all the economic options from the material choices available. Concrete is extensively used and is the dominant option in bridge construction. It is a logical choice for air rights developments, for the following reasons:

- ☐ *appearance*—easy to shape and mould, texture and profile for aesthetic qualities;
- ☐ *structural efficiency*—utilising the compressive strength of concrete and high-tensile properties of prestressing;
- ☐ *economy*—properly detailed and specified, concrete will produce an economically competitive alternative to steel. It is likely that the savings in temporary works, lead times and material cost for the tilt-in arch structure will result in a most cost-effective air rights building.

Concrete also offers real advantages in air rights construction by reducing building congestion and minimising construction complexity. There are obvious difficulties in transporting large prefabricated steel sections across the country and subsequently erecting them in a confined space.

An air rights building based on the 'tilt in' arch principle is currently under consideration by consultants in Sydney, Australia, for a major new development for the Road Transport Authority of New South Wales.

References and further reading

ORTON, A (1988). *The way we build now.* Van Nostrand Reinhold.
RICHARDSON, V (1999). Stressful. *RIBA Journal,* May. (Article on 88 Wood Street.)

Index